Progress in Molecular and Subcellular Biology

Series Editors: W.E.G. Müller (Managing Editor), Ph. Jeanteur, I. Kostovic, Y. Kuchino, A. Macieira-Coelho, R. E. Rhoads

28

Springer-Verlag Berlin Heidelberg GmbH

A.-P. Arrigo · W.E.G. Müller (Eds.)

Small Stress Proteins

With 47 Figures

 Springer

Professor Dr. ANDRÉ-PATRICK ARRIGO
Laboratoire Stress Oxydant
Chaperons et Apoptose
Centre de Génétique Moléculaire et Cellulaire
CNRS UMR-5534
Université Claude Bernard Lyon-I
43, Bd. du 11 Novembre
69622 Villeurbanne, France

W.E.G. MÜLLER
Institut für Physiologische Chemie
Abteilung für Angewandte Molekularbiologie
Johannes Gutenberg-Universität
Duesbergweg 6
55099 Mainz, Germany

ISSN 0079-6484
ISBN 978-3-642-62708-8

Library of Congress Cataloging-in-Publication Data
Small stress proteins / André-Patrick Arrigo, Werner E.G. Müller (eds.).
    p.   cm. – (Progress in molecular and subcellular biology ; 28)
    Includes bibliographical references and index.
    ISBN 978-3-642-62708-8   ISBN 978-3-642-56348-5 (eBook)
    DOI 10.1007/978-3-642-56348-5
    1. Heat shock proteins. I. Arrigo, André-Patrick, 1947–   II. Müller, Werner E.G.   III. Series.
    QP552.H43 S63 2002 572'.6 – dc21

http.//www.springer.de

© Springer-Verlag Berlin Heidelberg 2002
Originally published by Springer-Verlag Berlin Heidelberg New York in 2002
Softcover reprint of the hardcover 1st edition 2002
The use of general descriptive names, registered names, trademarks, etc. in this publication does not imply, even in the absence of a specific statement, that such names are exempt from the relevant protective laws and regulations and therefore free for general use.

Cover design: Meta Design, Berlin
Typesetting: Best-set Typesetter Ltd., Hong Kong
SPIN 10766941 39/3130 – 5 4 3 2 1 0 – Printed on acid-free paper

Preface

Studies have identified important families of proteins (denoted: heat shock or stress proteins, Hsps) which display an enhanced expression in response to heat shock or other physiological stresses. Besides the characterization of the genes encoding Hsp and the mechanisms of their induction, recent studies have concentrated on the function of these proteins. It was shown that the expression of Hsp protects the cell against different types of aggressions. In addition, Hsp can regulate essential biochemical processes in unstressed cells. For example, members of the Hsp60 and Hsp70 families act as ATP-binding proteins allowing the folding of nascent or denatured proteins as well as the assembly or disassembly of protein complexes. These observations have led to the discovery of the molecular chaperone concept (Ellis and Hemmingsen 1989).

Amongst the proteins whose expression is up-regulated by heat shock or other types of stresses are the small stress proteins also denoted (sHsps, sHsp or sHSP). Small stress proteins encompass a large numbers of related proteins which are represented in virtually all organisms, including prokaryotes. These polypeptides share a structural domain, often referred to as the α-crystallin domain, common to the lens protein alpha-crystallin (Ingolia and Craig 1982; Wistow 1985). In addition to being increased in response to several types of stresses, the Hsp level is also upregulated during development and correlates with the differentiation and oncogenic status of the cell. In spite of the fact that sHsp can confer cellular protection against stresses, their molecular function has remained enigmatic for years. Recently, major findings have clarified this point. First, it has been shown that an ATP-independent chaperone activity is associated to sHsp (Jacob and Buchner 1993). A second important observation concerns the ability of sHsp to form large oligomers which interact with misfolded polypeptides. These structures probably act as reservoirs of folding intermediates that are subsequently presented to the ATP-dependent chaperones (Hsp70 and other co-chaperones) (Ehrnsperger et al. 1997; Lee et al. 1997). sHsp have therefore been called the "forgotten chaperones".

In addition to their protein chaperone property, several other observations have contributed to the renewed interest in sHsp. For example, human Hsp27 is expressed in numerous cancer cells and induces protection against a variety of toxic chemicals used in chemotherapy. Hsp27 and αB-crystallin are also expressed in several pathological situations, such as neurodegenerative diseases. Moreover, a mutation in αB-crystallin has been shown to be responsible for a special type of myopathy. Recent data also suggest homeostatic functions of sHsp at the level of signal transduction, growth, differentiation and transformation processes. Another important clue was the recent finding that sHsp can protect against cell suicide or "apoptosis" (Arrigo 1998, 2000).

This book is the first one to be devoted only to small stress proteins. It surveys the current knowledge concerning the expression and function of sHsp in different organisms ranging form prokaryotes to human, except plant sHsp which are not described here. Each chapter gives an overview of a specific subject area and includes current results from each author's laboratory as well as a overview of future research directions. This book will be of value to researchers and graduate students interested in the field of cell stress biology, ranging from the molecular to clinical level.

The book is organized in several parts describing different biological systems that are currently used to analyze the expression and function of sHsp. The first section of the book, which contains two chapters, deals with sHsp genetic diversity. The first chapter provides an overview of the complex evolutionary diversity of sHsps, particularly prokaryotic sHsp which are now considered as ancestors of eukaryotic sHsps. The second chapter deals with sHsp diversity in a desert fish model. The second section of the book is related to the function and expression of sHsp in different eukaryote models. It begins with a chapter describing the chaperone activity of sHsp. Then, sHsp from *C. elegans*, *Drosophila* and different mammalian systems are presented. The third section of the book contains three chapters which describe pathological states associated with sHsp expression and gene therapy approaches aimed at modulating sHsp level of expression.

We would like to express our gratitude to all authors who contributed to this book for their cooperation and patience and to Springer for its support and strong interest in publishing this book.

In this book, small stress proteins (cognate or stress induced and abbreviated sHsp, sHsps or sHSP depending the authors) are defined as those proteins possessing the so-called α-crystallin protein domain. The name used for individual sHsp is HspXX where XX corresponds to the two first digits of the apparent molecular weight, excluding αA and αB-crystallins which are called by their own name.

Pers-Jussy, France   A.-P. ARRIGO, W.E.G MÜLLER
August 2001

References

Arrigo AP (1998) Small stress proteins: chaperones that act as regulators of intracellular redox state and programmed cell death. Biol Chem 379:19–26

Arrigo AP (2000) sHsp as novel regulators of programmed cell death and tumorigenicity. Pathol Biol (Paris) 48:280–288

Ehrnsperger M, Graber S, Gaestel M, Buchner J (1997) Binding of non-native protein to Hsp25 during heat shock creates a reservoir of folding intermediates for reactivation. EMBO J 16: 221–229

Ellis RJ, Van der Vies SM, Hemmingsen (1989) The molecular chaperone concept. Biochem Soc Symp 55:145–153

Ingolia TD, Craig E (1982) Four small *Drosophila* heat shock proteins are related to each other and to mammalian alpha-crystallin. Proc Natl Acad Sci USA 79:2360–2364

Jakob U, Gaestel M, Engel K, Buchner J (1993) Small heat shock proteins are molecular chaperones. J Biol Chem 268:1517–1520

Lee GJ, Roseman AM, Saibil HR, Vierling E (1997) A small heat shock protein stably binds heat-denatured model substrates and can maintain a substrate in a folding-competent state. EMBO J 16:659–671

Wistow G (1985) Domain structure and evolution in alpha-crystallins and small heat shock proteins. FEBS Lett 181:1–6

Contents

Chaperone Function of sHsps
M. Haslbeck and J. Buchner

The Small Heat Shock Proteins of the Nematode *Caenorhabditis elegans*: Structure, Regulation and Biology
E. Peter M. Candido

Drosophila Small Heat Shock Proteins: Cell and Organelle-Specific Chaperones?
S. Michaud, G. Morrow, J. Marchand, and R.M. Tanguay

The Developmental Expression of Small HSP
S.M. Davidson, M.-T. Loones, O. Duverger, and M. Morange

**Expression and Phosphorylation of Mammalian Small
Heat Shock Proteins**
K. Kato, H. Ito, and Y. Inaguma

**sHsp-Phosphorylation: Enzymes, Signaling Pathways
and Functional Implications**
M. Gaestel

Small Stress Proteins: Modulation of Intracellular Redox State and Protection Against Oxidative Stress
A.-P. Arrigo, C. Paul, C. Ducasse, O. Sauvageot, and C. Kretz-Remy

Small Stress Proteins: Novel Negative Modulators of Apoptosis Induced Independently of Reactive Oxygen Species
P. Arrigo, C. Paul, C. Ducasse, F. Manero, C. Kretz-Remy, S. Virot, E. Javouhey, N. Mounier, and C. Diaz-Latoud

Hsp27 as a Prognostic and Predictive Factor in Cancer
D.R. Ciocca and L.M. Vargas-Roig

Cytoskeletal Competence Requires Protein Chaperones
R. Quinlan

Hsp27 in the Nervous System: Expression in Pathophysiology and in the Aging Brain
A.M.R. Krueger-Naug, J.-C.L. Plumier, D.A. Hopkins, and R.W. Currie

Protection of Neuronal and Cardiac Cells by HSP27
D.S. Latchman

Evolution and Diversity of Prokaryotic Small Heat Shock Proteins

Guido Kappé[1], Jack A.M. Leunissen[2], and Wilfried W. de Jong[1]

1
Introduction

To understand the evolutionary mechanisms that led to the diversification of the various types of heat shock proteins(Hsps) and their functioning in multichaperone networks is a great challenge (Feder and Hofmann 1999). Considerable information is already available on the evolution of the Hsp60 and Hsp70 families (e.g., Gupta 1995; Budin and Philippe 1998; Karlin and Brocchieri 1998; Macario et al. 1999; Archibald et al. 2000; Brocchieri and Karlin 2000). Relatively less is known about the early evolution of the small heat shock proteins (sHsps), which are considerably more divergent in structure and function than the Hsp60s and Hsp70s.

The sHsps are found in bacteria, archaea and eukaryotes. They range in monomer size between 12 and 43kDa, and are characterized by a conserved "α-crystallin" domain of about 80 residues (reviewed in Plesofsky-Vig et al. 1992; Caspers et al. 1995; Waters and Vierling 1999). The N-terminal domains and C-terminal extensions, if present, are not conserved in sequence and length. The sHsps generally form high molecular weight complexes, between 150 and 800kDa in size, although some occur as dimers or tetramers. The only published crystal structure of any sHsp is as yet that of *Methanococcus jannaschii* Hsp16.5 (Kim et al. 1998). This archaeal sHsp forms a hollow sphere, ~120 Å in outer diameter, and composed of 24 subunits with an Ig-like β-sandwich folding. The characteristic property of most sHsps is their ability to suppress the in vitro aggregation of denaturing proteins, while in vivo their expression protects cells during stress (Ehrnsperger et al. 1997b). For detailed information about the broad variety of structural and functional properties of the sHsps we refer to the other chapters in this Volume.

It is the purpose of this chapter to describe the evolutionary diversity of prokaryotic sHsps. In terms of sequence divergence, this diversity is broader than that amongst plant or animal sHsps (de Jong et al. 1998; Waters and

[1] Department of Biochemistry, University of Nijmegen, P.O. Box 9101, 6500 HB Nijmegen, The Netherlands
[2] Centre for Molecular and Biomolecular Informatics, University of Nijmegen, P.O. Box 9101, 6500 HB Nijmegen, The Netherlands

Progress in Molecular and Subcellular Biology, Vol. 28
A.-P. Arrigo and W.E.G. Müller (Eds.)
© Springer-Verlag Berlin Heidelberg 2002

Vierling 1999). In fact, both plant and animal sHsps are suggested to be mono-phyletic groupings, which may have originated separately from different ancestral prokaryotic sHsps. Knowing the evolutionary history of the prokaryotic sHsps, and comparing their properties, may also help to understand the structure-function relationships of eukaryotic sHsps. This will provide an insight in the pathways along which they diversified, and how they acquired their characteristic structural and functional properties.

2
sHsps in Prokaryotes

Most prokaryotic sHsps function as chaperone-like proteins in the cytoplasm, but some have become structural components of the spore coat, as for *Bacillus subtilis* COTM (Henriques et al. 1997), while others associate with membranes, like *Mycobacterium tuberculosis* Hsp16.3 (Cunningham and Spreadbury 1998) and *Oenococcus oenii* Hsp (Jobin et al. 1997). In addition to rigidifying and sta-bilizing membranes under heat stress, such associations may constitute a feed-back mechanism for the regulation of heat shock genes, as has been proposed for *Synechocystis sp.* SP21 (Horváth et al. 1998). As in eukaryotes, many prokaryotic sHsps are developmentally regulated and inducible by heat and other stress. Their expression can reach high levels, e.g., up to 22% of total protein for *Streptococcus thermophilus* Hsp16.4 (González-Márques et al. 1997). This Hsp16.4 and other sHsps are plasmid-encoded, which may help to achieve such high expression levels.

There is good evidence that sHsps facilitate the refolding of denatured pro-teins in conjunction with other chaperones (Ehrnsperger et al. 1997b; Lee et al. 1997). Indeed, *Escherichia coli* IBPB interacts in a multichaperone network with Hsp60 and Hsp70 (Veigner et al. 1998). Such a functional relation is also suggested by a heat-shock cluster comprising genes for sHsps and GroES in *Bradyrhizobium japonicum* (Narberhaus et al. 1996), by an operon encoding an sHsp as well as a DnaK in *Thermotoga maritima* (Michelini and Flynn 1999), and by an operon encoding DnaK and two sHsps in *Porphyromonas gin-givalis* (Yoshida et al. 1999). If such multichaperone networks are a universal requirement for cellular functioning, one might expect to find the essential components in all organisms. Yet, the gene for Hsp70 is lacking in several Archaea (Macario et al. 1999). Although it is generally taken for granted that sHsps occur in all organisms, this remains to be established. The availability of a rapidly increasing number of complete prokaryote genome sequences now makes it possible to determine whether this is indeed the case. This also enables us to assess the number and variety of sHsp genes in different bacte-ria and archaea, as described in the following sections.

3
Are sHsps Dispensable in Some Pathogenic Bacteria?

We searched the complete, known genomes of 15 bacteria and 4 archaea for sequence coding for sHsps. These genomes were selected to represent a broad variety of prokaryotic lineages. For comparison, we also included the genome of the eukaryote *Saccharomyces cerevisiae*. To minimize the chance of over-looking sHsp-related sequences, five different search profiles were used, based on the α-crystallin domain of known sHsps. Four of these profiles were calcu-lated from a previously published alignment of the α-crystallin domains of a broad array of sHsps (de Jong et al. 1998). The alignment was divided into two subalignments, one comprising the animal sHsps and the other all remaining sHsps, because considerable differences were observed between these two groups of sequences. To calculate the profiles, two different programs were used on each subalignment, PROFILEMAKE v4.4 (Wisconsin Package Version 8.1) and PROFILEWEIGHT v2.1 (Thompson et al. 1994a). These profiles are available from the CMBI FTP server (ftp://ftp.cmbi.kun.nl/pub/molbio/-kappe/). The fifth profile was downloaded from the PROSITE database (acces-sion no. PS01031). This profile has been calculated using 140 sHsps from a wide range of organisms, including animals, plants, fungi and prokaryotes.

The completely sequenced genomes that were used in the search for sHsps are listed in bold in Table 1. All open reading frames (ORFs) that code for at least 50 amino acids were extracted from these genomes. These ORFs were taken as a database that was searched with the five sHsp profiles as a query, using Compugen's implementation of the PROFILESEARCH program (GenCore users manual, Compugen). Only those ORFs with a similarity over at least 60 positions, with fewer than 11 gaps in the region of similarity, and with more than 25% identity to a profile were considered as possible sHsps. These ORFs were compared with the SWISSPROT database, using Compugen's implementation of the FASTA program (GenCore users manual, Compugen), and were only considered as genuine sHsps when the best scoring proteins turned out to be known sHsps.

Interestingly, no sHsp-related sequences could be detected in eight of the bacterial genomes (indicated by n.d. in Table 1). In four bacterial genomes a single sHsp gene was found, *E. coli* and *M. tuberculosis* have two copies, and *B. subtilis* three. Three of the archaeal genomes had one sHsp gene, the fourth (*Archaeoglobus fulgidus*) had two copies. In yeast, as expected, only the genes for the known Hsp26 and Hsp42 were found. All 16 detected prokaryotic sHsp-related sequences are present in the protein databases, although *Pyrococcus horikoshii* 172aa had not been recognized as an sHsp; this demonstrates the sensitivity of our search procedure.

Since sHsp genes have been reported to also reside on bacterial plasmids (see Table 1), we similarly searched the plasmids of *Borrelia burgdorferi*, the extra-chromosomal element 1 of *Aquifex aeolicus*, and the small and large

Table 1. Prokaryotic sHsps found in completely sequenced genomes and in protein databases

Species[a]	sHsp[b]	Accession number	Subunit size[c]	Complex size[d]	Functional properties[e]	References[f]
Bacteria						
Aquifex aeolicus (1.55)	HspcI	O67316	144/17.2			Deckert et al. (1998)
Azotobacter vinelandii	IBPB	P96193	147/16.3			
Bacillus subtilis (4.21)	COTM	Q45058	130/15.2			Kunst et al. (1997) ; Henriques et al. (1997)
	YDFT	P96698	143/17.0			
	YOCM	O34321	158/18.4			
Borrelia burgdorferi (1.23)	n.d.					Fraser et al. (1997)
Bradyrhizobium japonicum	HspA	P70917	152/17.2		ind+	Narberhaus et al. (1996)
	HspB	P70918	153/17.2		th-, ind+	Narberhaus et al. (1996)
	HspC	P70919	166/18.6		th-, ind+	Narberhaus et al. (1996)
	HspD	O69241	151/17.3		ind+	Münchbach et al. (1999)
	HspE	O69242	150/17.2		ind+	Münchbach et al. (1999)
	HspF	O69243	163/18.6		ind+	Münchbach et al. (1999)
	HspH	O86110	151/17.1		ind+	Münchbach et al. (1999)
Buchnera aphidicola	HspA*	O31288	153/18.0			van Ham et al. (1997)
Chlamydia pneumoniae (1.23)	n.d.					Kalman et al. (1999)
Chlamydia trachomatis (1.04)	n.d.					Kalman et al. (1999)
Clostridium acetobutylicum	Hsp18	Q03928	151/17.7		ind+	Bahl et al. (1995)
Escherichia coli (4.64)	IBPA	P29209	137/15.8		th-, ind+	Blattner et al. (1997); Thomas and Baneyx (1998)
	IBPB	P29210	142/16.1	600*	ch+, th-, ind+	Veigner et al. (1998); Thomas and Baneyx (1998); Shearstone and Baneyx (1999)
Haemophilus influenzae **Rd** (1.83)	n.d.					Fleischmann et al. (1995)
Helicobacter pylori 26695 (1.67), J99 (1.64)[g]	n.d.					Alm et al. (1999)
Lactobacillus delbrueckii	Hsp	P94867	123/13.6			

Organism	Protein	Accession	Residues/MW	Oligomer	Expression	References
Legionella pneumophila	GSPA	S49042	166/19.0		ind+	Abu Kwaik et al. (1997)
Mycobacterium avium	18 kDa1	P46729	147/16.5			
	18 kDa2	P46731	138/15.9			
Mycobacterium intracellulare	18 kDa1	P46730	149/16.5			
	18 kDa2	P46732	140/15.7			
Mycobacterium leprae	Hsp167	P12809	148/16.7		ind+	Nerland et al. (1988); Booth et al. (1993)
	18 kDa[h]	E980239	147/16.5			
Mycobacterium tuberculosis	Hsp163	P30223	143/16.1	145 (9)*	ch+, ind+	Cole et al. (1998); Chang et al. (1996); Yang et al. (1999)
(4.41)	Hsp20p	O53673	159/17.8			
Mycoplasma genitalium (0.58)	n.d.					Fraser et al. (1995)
Mycoplasma pneumoniae (0.82)	n.d.					Himmelreich et al. (1996)
Oenococcus oenii	Hsp	P94898	148/16.9		ind+	Guzzo et al. (1997)
Rickettsia prowazekii (1.11)	Hsp22	E71682	163/18.7			Andersson et al. (1998)
Stigmatella aurantiaca	SP21	Q06823	188/21.3	(2)	ind+	Heidelbach et al. (1993)
Streptococcus thermophilus	ASP	P80485	142/16.4		ind+	González-Márquez et al. (1997)
	Hsp164*	O30851	142/16.4		th-, ind+	Somkuti et al. (1998)
	1SThsp*	O52190	150/17.4		ind+	O'Sullivan et al. (1999)
	SEChsp*[h]	O52192	142/16.4		ind+	O'Sullivan et al. (1999)
Streptomyces albus G	Hsp18	Q53595	143/16.1	Granules	th+, ind+	Servant and Mazodier (1996)
Synechococcus vulcanus	HspA	O82825	145/16.7	400 (24)	ch+, ind+	Roy et al. (1999)
Synechocystis sp. (3.57)	SP21	P72977	146/16.6		th+, ind+	Kaneko et al. (1996); Horváth et al. (1998)
Thermotoga maritima (1.86)[i]	Hsp	Q9ZFD1	142/17.0	400–450 (23–26)*	ch+	Nelson et al. (1999); Michelini and Flynn (1999)
Treponema pallidum (1.14)[i]	n.d.					Fraser et al. (1998)
Archaea						
Archaeoglobus fulgidus (2.18)	Hsp201	O28973	174/20.4			Klenk et al. (1997)
	Hsp202	O28308	140/16.5			Klenk et al. (1997)
Methanobacterium thermoautotrophicum (1.75)	Hspcl	O26947	145/17.1			Smith et al. (1997)
Methanococcus jannaschii (1.66)	Hsp165	Q57733	147/16.5	400 (24)*	ch+	Bult et al. (1996); Kim et al (1998)
Pyrococcus horikoshii (1.74)	172aa	O59514	172/20.1			Karawabayasi et al. (1998)

Table 1. *Continued*

Species[a]	sHsp[b]	Accession number	Subunit size[c]	Complex size[d]	Functional properties[e]	References[f]
Representative eukaryotic sHsps included in the phylogenetic analyses						
Caenorhabditis elegans	Hsp16	P06581	143/16.3			
Chlamydomonas reinhardtii	Hsp22	P12811	157/16.7			
Homo sapiens	AACr	P02489	173/19.9			
Neurospora crassa	Hsp30	P19752	228/25.3			
Pisum sativum	Hsp181	P19243	158/18.1			
	Hsp22	P46254	202/22.9			
Saccharomyces cerevisiae **(12.93)**	Hsp26	P15992	213/23.7			
	Hsp42[h]	Q12329	375/42.8			
Toxoplasma gondii	Hsp30	Q27354	229/25.0			

[a] In bold: species with completely sequenced genomes (genome size in parentheses; Mb). These genomes were downloaded from the NCBI (ftp://ftp.ncbi.nlm.nih.gov/genbank/genomes/bacteria/). For three of these species, extrachromosomal elements (*A. aeolicus*, AE000667; *B. burgdorferi*, accession no. AE000784–AE000794 and AE001575–AE001584; *M. jannaschii*, L77118 and L77119) were also searched.
[b] sHsps are designated as indicated in the databases; n.d., no detectable sHsps in genome; *, sHsp encoded on plasmid.
[c] Number of residues/subunit mass (kDa).
[d] Complex size (kDa) and number of subunits (in brackets); *, obtained from recombinant proteins.
[e] Presence (+) or absence (-) of in vitro chaperone-like activity (ch), cellular thermoprotection (th), and stress inducibility (ind).
[f] These references, if not included in the reference list in this paper, can be found at ftp://ftp.cmbi.kun.nl/pub/molbio/kappe/
[g] The complete genomes of two *H. pylori* strains are known and both lack sHsps.
[h] Not included in the phylogenetic analyses; *M. leprae* 18 kDa and *S. thermophilus* SEChsp because they are almost identical to – and likely to be the same genes as – *M. leprae* Hsp167 and *S. thermophilus* ASP, respectively, and *S. cerevisiae* Hsp42 because *S. cerevisiae* Hsp26 and *N. crassa* Hsp30 were included as most divergent fungal representatives.
[i] The sHsp obtained from the complete genome of *T. maritima* as used in this study (accession no. Q9WYK7) has an insertion of five residues and two amino acid replacements as compared with a previously submitted sequence (accession no. Q9ZFD1).

extrachromosomal elements of *M. jannaschii*, all species with completely known genomes. In addition, we screened the huge plasmid NGR234a (536 kb) of *Rhizobium* sp. (accession no. AE000065 - AE000109). No sHsps were found in these extrachromosomal elements.

It thus appears that certain bacteria can do without sHsps, just as some archaea can do without Hsp70 (Macario et al. 1999). We obviously cannot exclude that some very deviating sHsp-related genes have been missed in the eight species without detectable sHsp genes in their chromosome; however, in that case, the encoded proteins, not having the characteristic sHsp sequence features, can hardly be considered as authentic sHsps. Also, we cannot exclude that sHsp genes are present on plasmids in some of these species. Since sHsps are represented in all kingdoms, their apparent absence in some bacteria must reflect a loss of the corresponding genes, rather than the ancestral state. Bacteria without genomic sHsps are found in different clades (e.g., *Mycoplasma* in the *Bacillus/Clostridium* group, and *Haemophilus influenzae* in the γ-proteobacteria) which also include species that possess sHsps (see below, Fig. 2). Loss of sHsps must thus have occurred repeatedly.

The absence of sHsp genes is not simply correlated with reduction in genome size, although sHsps are lacking in all but one of the seven genomes that are smaller than 1.3 Mb (Table 1). Bacteria without genomic sHsps are all highly host-adapted and often intracellular pathogens in vertebrates. This would suggest that only under those conditions sHsps may become dispensable; it may even be advantageous to get rid of sHsps as major surface antigens. On the other hand, various pathogenic bacteria do have sHsps, often even in multiple copies, as in *M. avium* and *M. intracellulare*.

4
Phylogenetic Analysis of Prokaryotic sHsps

To have a broad representation of prokaryotic sHsps for phylogenetic analyses, the SWISSPROT, TREMBL and PIR databases were also searched for additional prokaryotic sHsps, using the PROFILESEARCH program and the same five sHsp profiles. We selected a further 26 prokaryotic sHsps, as listed in Table 1. Striking are the seven sHsps in *B. japonicum*, with at least five more to be characterized (Münchbach et al. 1999). An initial alignment of the retrieved prokaryotic sHsps, complemented with eight eukaryotic sHsps to represent the animal, plant, fungal and protist kingdoms (Table 1), was made using the CLUSTALW v1.5 program (Thompson et al. 1994b). After manual editing, we divided the alignment into two parts, representing the variable N-terminal and the more conserved α-crystallin domains. In the α-crystallin domains (Fig. 1), the two regions with the highest homology (positions 27–61 and 94–127) and the typical IxI motif at positions 141–143 contain the characteristically conserved sequences as reported earlier for various sets of mostly eukaryotic sHsps (Plesofsky-Vig et al. 1992; de Jong et al. 1998; Waters and Vierling 1999).

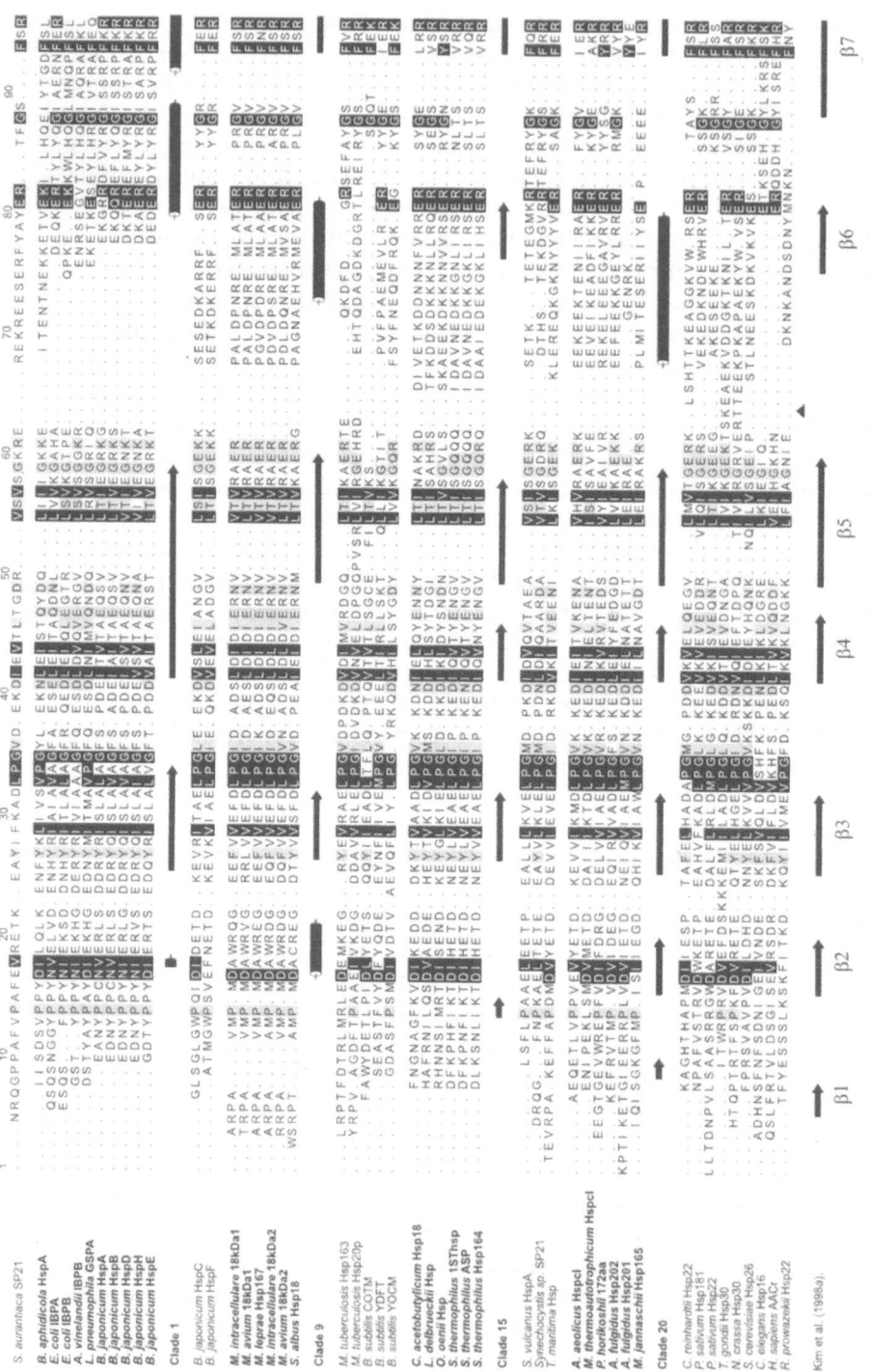

Fig. 1. Alignment and secondary structure predictions of the C-terminal α-crystallin domains of prokaryotic and representative eukaryotic sHsps. To demarcate the N- and C-terminal domains we took the structure of *M. jannaschii* Hsp16.5 as a reference (Kim et al. 1998), considering the β₁-strand that follows the structurally undetermined N-terminal domain as the beginning of the α-crystallin domain. For full names, accession numbers, and some structural and functional characteristics (see Table 1). Species are clustered in accordance with their grouping in the phylogenetic tree as shown in Fig. 2. A stretch of 64 residues has been omitted from the *N. crassa* Hsp30 sequence (*upright arrowhead* between alignment positions 64 and 65). *Black and gray* show positions where >75% and 50–75% of the residues are similar, respectively;

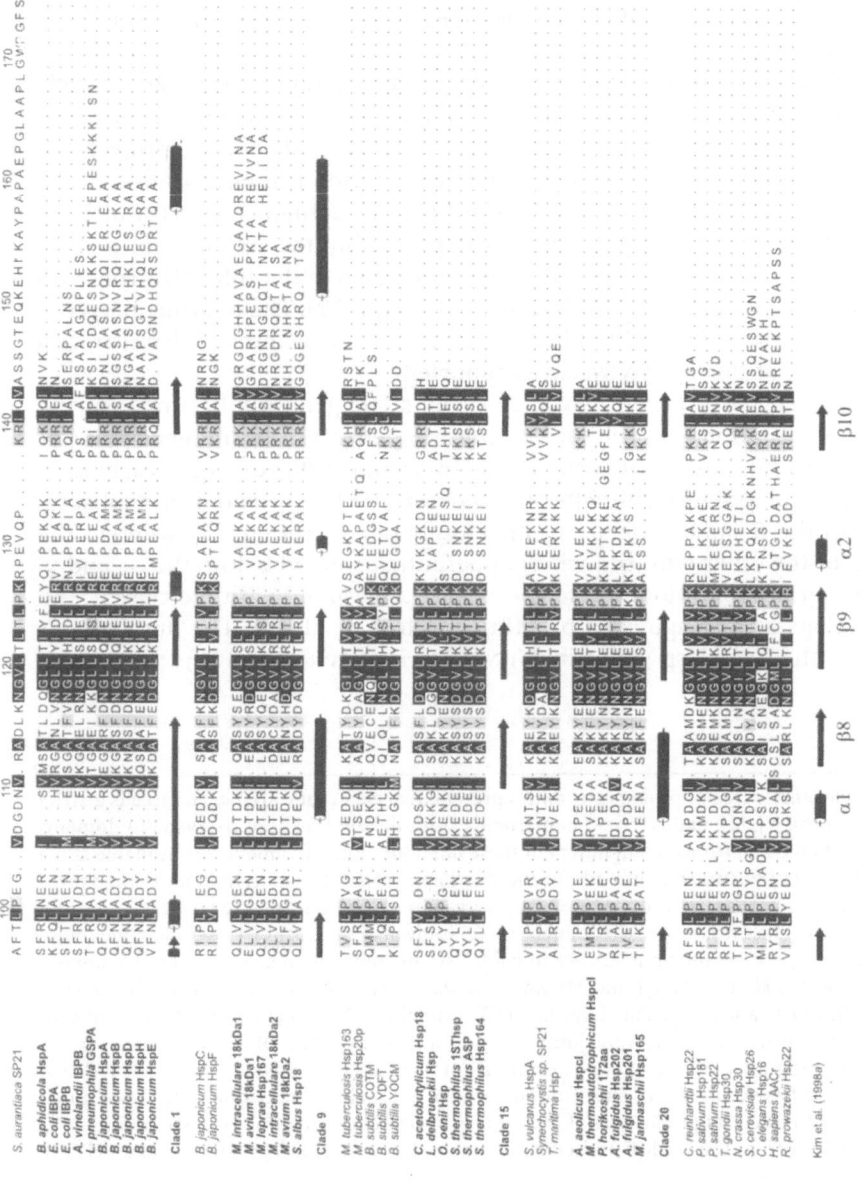

Fig. 1. *(Continued)* EQ, DN, KR, ST, ILMV and FWY are considered similar. Secondary structure predictions for the sHsps in the four major clades (1, 9, 15 and 20), as observed in Fig. 2, are depicted below the respective species. These secondary structure predictions were performed using PHD (Rost and Sander 1994) at the PredictProtein server of the University of Columbia (http://dodo.cpmc.columbia.edu/predictprotein/). *Arrows and cylinders represent β-strands and α-helices, respectively.* The *bottom line* presents the secondary structure of *M. jannaschii* Hsp16.5

Optimal alignment is essential for reliable phylogeny reconstruction. To reduce the effects of the inherent subjectivity of manual editing, the alignment as shown in Fig. 1 was realigned by CLUSTALW v1.7 after removing all gaps. The most conspicuous changes after the realignment concerned the altered positioning of gaps, indicating that aligning is ambiguous in those regions. Phylogenetic analyses were performed on both alignments with three methods: neighbor joining (NJ, Saitou and Nei 1987), maximum parsimony (MP Felsenstein 1993) and quartet puzzling (QP, Strimmer and von Haeseler 1996). The six resulting phylogenetic trees were combined in a consensus topology (Fig. 2).

The better supported clades in the sHsp consensus tree are largely compatible with the standard reference for prokaryote phylogeny as based on 16S rRNA sequences (see the ribosomal database project II, http://rdp.cme.msu.edu/). The most notable exception is the separation of the *M. tuberculosis* sHsps (branch 14) from the other actinobacteria (branch 9). The nesting of the bacterial *A. aeolicus* HspcI within the archaeal sHsp clade (branch 20) seems remarkable, but is also supported in the 16S rRNA tree, and in another recent sHsp tree (Waters and Vierling 1999). The α-proteobacterial *Rickettsia prowazekii* Hsp22 tends to group with the two animal sHsps (branch 24 in Fig. 2), but this is only weakly supported.

The tree suggests a complex history of sHsp gene duplications during prokaryote evolution (but see the complicating effects of lateral gene transfer below in Sect.°5). The two *E. coli* sHsps (IBPA and IBPB) diverged early in γ-proteobacterial radiation, but after the separation of α-, γ- and δ-proteobacteria. Five of the *B. japonicum* sHsps are closely related (branch 4), whereas the ancestor of the other two (branch 8) apparently diverged before the α- and γ-proteobacterial separation. These two clades correspond with the class A and class B sHsps as distinguished by Münchbach et al. (1999). The duplications leading to the two sHsp genes in *Mycobacterium avium* and *Mycobacterium*

──▶

Fig. 2. Phylogenetic tree of prokaryotic and representative eukaryotic sHsps. The tree is a strict consensus of the NJ, MP and QP trees constructed from the alignment of α-crystallin domains as shown in Fig. 1, and from a realignment of these domains. Both alignments are available at the CMBI FTP server, ftp://ftp.cmbi.kun.nl/pub/molbio/kappe/. Nodes supported in fewer than three of the six trees are collapsed; the averaged support values of the branches are *boxed*. A table with the six support values for each of the branches 1–25 is given at the CMBI FTP server. When more than 50% are supported, the branches are in **bold**. * sHsps from completely sequenced genomes; # plasmid-located sHsps. The NJ and MP analyses were carried out using the PHYLIP package v3.752c with 500 data sets (calculated by SEQBOOT) and the PAM250 matrix as the model of substitution. When using PROTPARS the jumble option was set to 4. The QP analyses were carried out using the program PUZZLE v4.0.2. The number of puzzling steps was set to 1000, the PAM250 matrix was used as the model of substitution, and the approximate maximum likelihood function was used to calculate parameters of the models of sequence evolution. The strict consensus tree was derived from the NJ, MP and QP trees using the CONSENSE program of the PHYLIP package

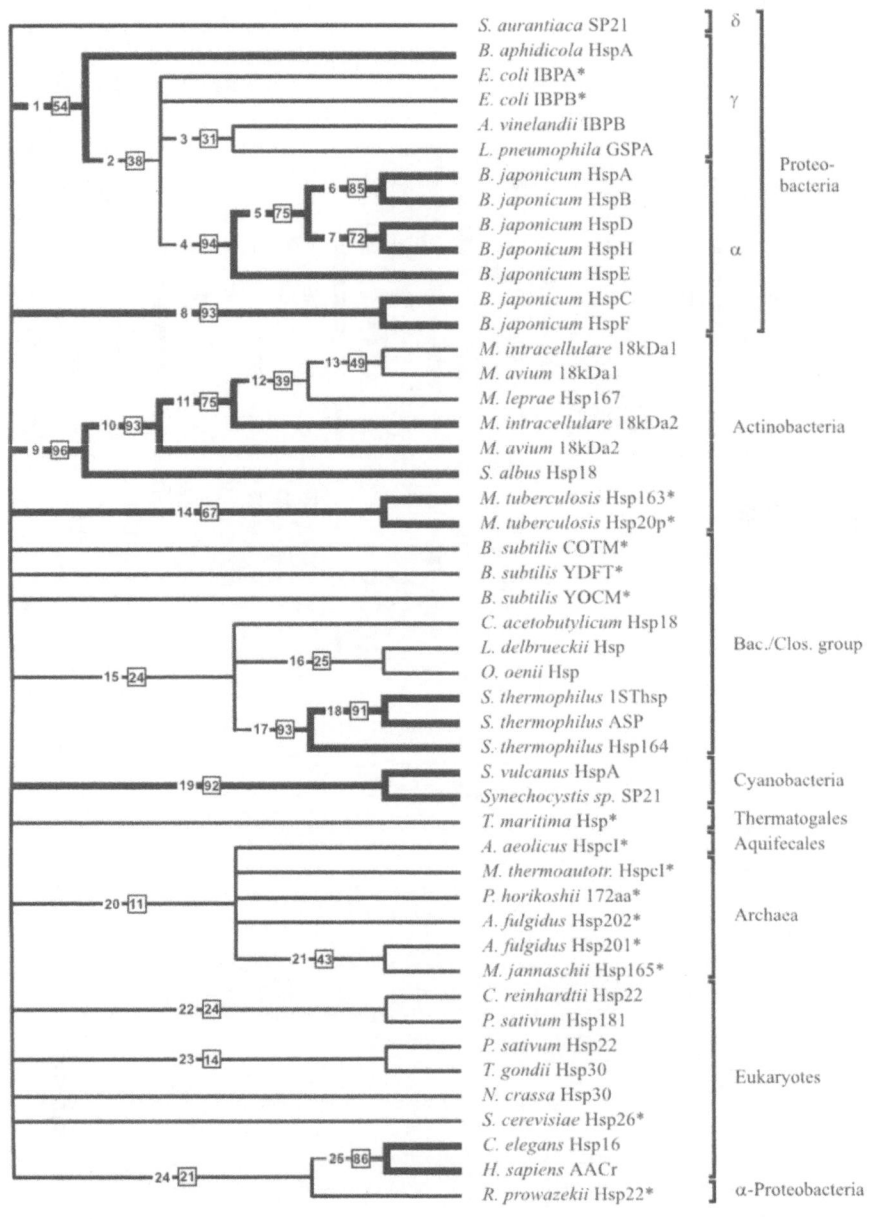

S. aurantiaca SP21

B. aphidicola HspA
E. coli IBPA
E. coli IBPB
A. vinelandii IBPB
L. pneumophila GSPA
B. japonicum HspA
B. japonicum HspB
B. japonicum HspD
B. japonicum HspH
B. japonicum HspE

Clade 1
B. japonicum HspC
B. japonicum HspF

M. intracellulare 18kDa1
M. avium 18kDa1
M. leprae Hsp167
M. intracellulare 18kDa2
M. avium 18kDa2
S. albus Hsp18

Clade 9
M. tuberculosis Hsp163
M. tuberculosis Hsp20p
B. subtilis COTM
B. subtilis YDFT
B. subtilis YOCM

C. acetobutylicum Hsp18
L. delbrueckii Hsp
O. oenii Hsp
S. thermophilus 1SThsp
S. thermophilus ASP
S. thermophilus Hsp164

Clade 15
S. vulcanus HspA
Synechocystis sp. SP21
T. maritima Hsp

A. aeolicus Hspcl
M. thermoautotrophicum Hspcl
P. horikoshii 172aa
A. fulgidus Hsp202
A. fulgidus Hsp201
M. jannaschii Hsp165

Clade 20
C. reinhardtii Hsp22
P. sativum Hsp181
T. sativum Hsp22
N. gondii Hsp30
N. crassa Hsp30
S. cerevisiae Hsp26
C. elegans Hsp16
H. sapiens AaCr
R. prowazekii Hsp22

intracellulare occurred before these species diverged (branches 10–13). The absence of clustering of the three *B. subtilis* sHsps is of interest, although traces of shared residues can be detected in the alignments of YDFT and YOCM in Figs. 1 and 3. This diversity of sHsps in *B. subtilis* becomes even more intriguing by the recent finding of a plasmid-borne sHsp in *B. subtilis* (accession no. Q9X3Z5), which does not cluster with any of the three highly divergent genomic sHsps in this bacterium, nor with any other sHsp in Fig. 2 (data not shown). The three *S. thermophilus* sHsps are very similar (branch 17). Four new *S. thermophilus* sHsps, all located on plasmids, have recently appeared in the database (Q9X9N3, Q9X9C3, AAF04359, AAF04361), and are again very similar to the three sHsps of this species in Fig. 2 (data not shown). It is remarkable that at least six of the sHsps found in *S. thermophilus* are plasmid-borne.

Also, the two cyanobacterial sHsps in Fig. 2 are very similar (branch 19). Within archaea, an early duplication has given rise to the two sHsps in *A. fulgidus*. Finally, it would appear that the duplication leading to the two included *Pisum sativum* sHsps may have preceded the origin of the plant lineage, and that the two fungal sHsps display no special relationship.

5
Lateral Transfer or Convergent Evolution of Prokaryotic sHsp Genes?

When highly divergent sHsps occur in the same or closely related species, it cannot readily be decided whether gene duplication or lateral transfer is involved, the latter being a common feature in prokaryotes (Lawrence 1999). Three of the groupings as depicted in Fig. 2 deviate from prevalent opinions about prokaryote relationships. The fact that the two *M. tuberculosis* sHsps have no detectable tendency to group with the sHsps of the other actinobacteria probably means that the *M. tuberculosis* sHsps are paralogs of the other actinobacterial sHsps. There are no compelling reasons to invoke lateral transfer in this case, since the *M. tuberculosis* sHsps do not resemble any other prokaryote sHsp. Similarly, the tendency of *A. aeolicus* HspcI to cluster with archaea might suggest lateral gene transfer, but could also reflect the common adaptation to extreme environmental conditions or even a genuine relation-

◄ _____

Fig. 3. Alignment and secondary structure predictions of the N-terminal sequences of prokaryotic and representative eukaryotic sHsps. This alignment is made using CLUSTALW and manually compressed since there is no overall sequence similarity. The first 30 residues of the *Toxoplasma gondii* Hsp30 sequence and 34 residues from the *A. fulgidus* Hsp201 sequence (located between the presented N-terminal and α-crystallin domains) were omitted to avoid undue lengthening of the alignment. Within the four clades 1, 9, 15 and 20, positions with over 65% identical residues are shaded in gray. *Black* shows residues corresponding to or reminiscent of (parts of) the F/WDPF motif as proposed by Ehrnsperger et al. (1997a, p. 538). Secondary structures predicted for the sequences in the four clades are represented by *cylinders* (α-helices) and *arrows* (β-strands)

ship. Finally, the clustering of the typhus bacterium *R. prowazekii* Hsp22 with human AACr and *Caenorhabditis elegans* Hsp16 (branch 26) is only marginally supported. Rather then assuming lateral transfer from its human host, a few scattered shared replacements may cause this clustering, possibly reflecting some parallel changes in *R. prowazekii* Hsp22 to elude the human immune system. However, just as for the archaeal Hsp60 and Hsp70 systems (Macario et al. 1999), lateral transfer remains a potential source of innovations for the prokaryote sHsps.

6
Secondary Structure Prediction of the α-Crystallin Domain

To estimate to what extent the secondary structure as observed in the crystal structure of *M. jannaschii* Hsp16.5 is conserved in other prokaryotic sHsps, we performed predictive studies with the PHD program (Rost and Sander 1994). The results are incorporated in Fig. 1. Because the outcome of PHD is much more accurate when using multiple alignments rather than single sequences, we performed the prediction only for the larger clades in Fig. 2 (branches 1, 9, 15 and 20). When we compare these predicted secondary structures with the published structure of *M. jannaschii* Hsp16.5 (bottom line), it appears that strands β_3, β_5, β_9 and β_{10} are highly conserved, and strands β_4 and β_7 somewhat less so. The other β-strands and the short α-helices of *M. jannaschii* Hsp16.5 are likely to be poorly conserved in the other prokaryotic sHsps. It is noteworthy, though, that a tendency for α-helicity is observed in particular regions (positions 68–89, 108–116 and 151–163).

Our earlier secondary structure prediction of sHsp α-crystallin domains (Caspers et al. 1995) turned out to be surprisingly correct when the experimentally determined structure of *M. jannaschii* Hsp16.5 became available (Kim et al. 1998). This may give confidence to the present predictions. Of interest is the finding that the regions comprising strands β_2 and β_6 are hardly conserved. The backbone atoms of these two strands are responsible for the primary monomer-monomer interaction in *M. jannaschii* Hsp16.5. This suggests that dimer formation in most other sHsps may be quite different, in accordance with the variety of sHsp quaternary structures.

7
Alignment and Secondary Structure Prediction of the N-Terminal Domain

As appears from the "alignment" of the N-terminal domains of the studied sHsps, no overall sequence similarities can be detected in this region (Fig. 3). Varying degrees of conservation can only be observed amongst the N-terminal domains of sequences found to be closely related in Fig. 2. It has been proposed that the N-terminal domains of sHsps are characterized by the presence of a conserved F/WDPF motif (Ehrnsperger et al. 1997a, p. 538). In the present data set such a motif is only found in *Stigmatella aurantiaca* SP21 and in *P. horikoshii* 172aa. It is remarkable, however, that almost all other N-terminal domains do contain parts of this motif (DPF, PF, FD, WD) or variants thereof (PWF, FP, YD, PW, DF, etc.) (in black in Fig. 3). This suggests that the presence of such short motifs may somehow be of functional importance.

Also, the N-terminal sequences of the clades 1, 9, 15 and 20 were subjected to secondary structure predictions by PHD. There appears to be a pronounced preference for the α-helical structure in the N-terminal domains (Fig. 3), in sharp contrast to the predominantly β-sheet structure in the α-crystallin domain (Fig. 1). This tendency for α-helicity in the N-terminal domains suggests considerable structural flexibility, in contrast to the more rigid β-sheet structure of the α-crystallin domain. This observation supports the assumption that the core functions of the sHsps are dependent on the conserved α-crystallin domain, while the variable N-terminal domains predominantly allow for the wide variation in structural and functional properties of the sHsps.

8
Summary and Concluding Remarks

To gain insight into the early evolution of sHsps, we searched for their presence in 19 completely sequenced genomes (15 bacteria and 4 archaea), and additionally in the protein databases. Surprisingly, no sHsp-like sequences could be recognized in the genomes of eight bacteria, all being pathogens. Other prokaryotes contain up to 12 often very divergent and sometimes plasmid-borne sHsp genes. Alignment of 40 diverse prokaryotic and 8 representative eukaryotic sHsps confirms that strictly conserved sequences are only present in the C-terminal "α-crystallin" domain. Phylogenetic analyses reveal that the better supported clades of sHsps mostly correspond with generally accepted relationships of prokaryotes. The prokaryote sHsps have a complex history of gene duplications and possible lateral transfer as well.

Secondary structures were predicted for the major clades of prokaryotic sHsps, and compared with the known crystal structure of *M. jannaschii* Hsp16.5 (Kim et al. 1998). Five of the ten β-strands present in the C-terminal domain of this Hsp16.5 are consistently predicted in all prokaryote sHsps; the other secondary structure elements may be less conserved. Secondary structure predictions of the variable N-terminal domains indicate in all instances a predominantly α-helical character. The structural variety of prokaryotic sHsps may relate to their considerable functional diversity, ranging from chaperones to spore coat components. This diversity of features like complex size, surface location, developmental regulation and induction (cf. Table 1) apparently varies independently in different prokaryote lineages (cf. Fig. 2).

Analyses of the rapidly increasing number of known prokaryotic genomes will further extend the presented observations. A deeper understanding of the enormous diversity in sequence and occurrence of the prokaryotic sHsps, as components of the various multichaperone networks in the cell, promises to contribute to our insight in the complexity of protein evolution in general.

References

Archibald JM, Logsdon JM Jr, Doolittle WF (2000) Origin and evolution of eukaryotic chaper-onins: phylogenetic evidence for ancient duplication in CCT genes. Mol Biol Evol 17: 1456–1466

Brocchieri L, Karlin S (2000) Conservation among HSP60 sequences in relation to structure, function, and evolution. Protein Sci 9:476–486

Budin K, Philippe H (1998) New insights into the phylogeny of eukaryotes based on ciliate Hsp70 sequences. Mol Biol Evol 15:943–956

Caspers G-J, Leunissen JAM, De Jong WW (1995) The expanding small heat-shock protein family, and structure predictions of the conserved "α-crystallin domain". J Mol Evol 40:238–248

Cunningham AF, Spreadbury CL (1998) Mycobacterial stationary phase induced by low oxygen tension: cell wall thickening and localization of the 16-kilodalton α-crystallin homolog. J Bacteriol 180:801–808

De Jong WW, Caspers G-J, Leunissen JAM (1998) Genealogy of the α-crystallin – small heat-shock protein superfamily. Int J Biol Macromol 22:151–162

Ehrnsperger M, Buchner J, Gaestel M (1997a) Structural and function of small heat-shock pro-teins. In: Fink AL, Goto Y (eds) Molecular chaperones in the life cycle of proteins. Marcel Dekker Inc, New York, pp 533–575

Ehrnsperger M, Graber S, Gaestel M, Buchner J (1997b) Binding of non-native protein to Hsp25 during heat shock creates a reservoir of folding intermediates for reactivation. EMBO J 16: 221–229

Feder ME, Hofmann G (1999) Heat-shock proteins, molecular chaperones, and the stress response: evolutionary and ecological physiology. Annu Rev Physiol 61:243–282

Felsenstein J (1993) PHYLIP (phylogeny inference package). Distributed by the author. Department of Genetics, University of Washington, Seattle

González-Márquez H, Perrin C, Bracquart P, Guimont C, Linden G (1997) A 16 kDa protein family overexpressed by Streptococcus thermophilus PB18 in acid environments. Microbiol 143: 1587–1594

Gupta RS (1995) Evolution of the chaperonin families (Hsp60, Hsp10, Tcp-1) of proteins and the origin of eukaryotic cells. Mol Microbiol 15:1–11

Heidelbach M, Skladny H, Schrairer HU (1993) Heat shock and development induce synthesis of a low-molecular-weight stress-responsive protein in the myxobacterium Stigmatella auranti-aca. J Bacteriol 175:7479–7482

Henriques AO, Beall BW, Moran CP Jr (1997) CotM of Bacillus subtilis, a member of the α-crystallin family of stress proteins, is induced during development and participates in spore outer coat formation. J Bacteriol 179:1887–1897

Horváth I, Glatz A, Varvasovszki V et al. (1998) Membrane physical state controls the signaling mechanism of the heat shock response in Synechocystis PCC 6803: identification of hsp17 as a "fluidity gene". Proc Natl Acad Sci USA 95:3513–3518

Jobin M-P, Delmas F, Garmyn D, Deviès C, Guzzo J (1997) Molecular characterization of the gene encoding an 18-kilodalton small heat shock protein associated with the membrane of Leuconostoc oenos. Appl Environ Microbiol 63:609–614

Karlin S, Brocchieri L (1998) Heat shock protein 70 family: multiple sequence comparisons, func-tion, and evolution. J Mol Evol 47:565–577

Kim KK, Kim R, Kim S-H (1998) Crystal structure of a small heat-shock protein. Nature 394: 595–599

Lawrence JG (1999) Gene transfer, speciation, and the evolution of bacterial genomes. Curr Opin Microbiol 2:519–523

Lee GJ, Roseman AM, Saibil HR, Vierling E (1997) A small heat shock protein stably binds heat-denatured model substrates and can maintain a substrate in a folding-competent state. EMBO J 16:659–671

Macario AJ, Lange M, Ahring BK, De Macario EC (1999) Stress genes and proteins in the archaea. Microbiol Mol Biol Rev 63:923–967

Michelini ET, Flynn GC (1999) The unique chaperone operon of *Thermotoga maritima*: cloning and initial characterization of a functional Hsp70 and a small heat shock protein. J Bacteriol 181:4237–4244

Münchbach M, Nocker A, Narberhaus F (1999) Multiple small heat shock proteins in rhizobia. J Bacteriol 181:83–90

Narberhaus F, Weiglhofer W, Fischer H-M, Hennecke H (1996) The *Bradyrhizobium japonicum* $rpoH_1$ gene encoding a σ^{32}-like protein is part of a unique heat shock gene cluster together with $groESL_1$ and three small heat shock genes. J Bacteriol 178:5337–5346

Plesofsky-Vig N, Vig J, Brambl R (1992) Phylogeny of the alpha-crystallin-related heat-shock proteins. J Mol Evol 35:537–545

Rost B, Sander C (1994) Combining evolutionary information and neural networks to predict protein secondary structure. Proteins 19:55–72

Saitou N, Nei M (1987) The neighbor-joining method: a new method for reconstructing phylogenetic trees. Mol Biol Evol 4:406–425

Strimmer K, von Haeseler A (1996) Quartet puzzling: a quartet maximum-likelihood method for reconstructing tree topologies. Mol Biol Evol 13:964–969

Thompson JD, Higgins DG, Gibson TJ (1994a) Improved sensitivity of profile searches through the use of sequence weights and gap excision. Comput Appl Biosci 10:19–29

Thompson JD, Higgins DG, Gibson TJ (1994b) CLUSTAL W: improving the sensitivity of progressive multiple sequence alignment through sequence weighting, position-specific gap penalties and weight matrix choice. Nucleic Acids Res 22:4673–4680

Veigner L, Diamant S, Buchner J, Goloubinoff P (1998) The small heat-shock protein IbpB from *Escherichia coli* stabilizes stress-denatured proteins for subsequent refolding by a multichaperone network. J Biol Chem 273:11032–11037

Waters ER, Vierling E (1999) The diversion of plant cytosolic small heat shock proteins preceded the divergence of mosses. Mol Biol Evol 16:127–139

Yoshida A, Nakano Y, Yamashita Y, Oho T, Ohishi M, Koga T (1999) A novel dnaK operon from *Porphyromonas gingivalis*. FEBS Lett 12:287–291

Discovery of Two Distinct Small Heat Shock Protein (HSP) Families in the Desert Fish *Poeciliopsis*

Carol E. Norris, and Lawrence E. Hightower[1]

1
Introduction

The availability of a genetically pedigreed fish colony at the University of Connecticut has made possible a detailed investigation of the heat shock response of a poikilothermic vertebrate. During the past 12 years, two tropical species, six desert species and numerous interspecific hybrids of viviparous fishes in the genus *Poeciliopsis* have been studied. In the natural environment, these animals occupy thermally distinct niches that range in latitude from southern Arizona, through the Sonora Desert of northwestern Mexico across tropical southern Mexico and Central America into Columbia in South America. The diverse habitats include: (1) relatively stable tropical rivers with narrow annual and daily temperature ranges, (2) unstable streams that cross subtropical desert where temperatures range from 4 to 40 °C seasonally and undergo rapid daily changes of as much as 22 °C in 3 h, and (3) pristine head-waters at higher elevations in the foothills of the Sierra Madre Occidental. When we began these studies, no detailed analysis of diversity in HSP families had been attempted. In fact, many investigators considered heat shock genes to be so highly conserved evolutionarily, that a search for variation would be fruitless. This view turned out to be overly pessimistic. Our approach to exploring HSP diversity employed high resolution two-dimensional polyacrylamide gels to display protein isoforms at the level of denatured polypeptide chains. This strategy allowed a relatively rapid screening of multiple HSP families compared with the time required to clone and sequence all of the members of even one multigene family. Analyses of protein patterns also had the advantage of screening out nonfunctional genes and allowing a comparison of the levels of expression of different isoforms.

The results of our studies of the small HSPs of *Poeciliopsis* have not been collected into one paper and our intention is to use the following chapter for this purpose. In addition, several unpublished studies from the research thesis of Dr. Carol Norris (1997) have been included. Variation in small HSP isoform patterns among closely related species of desert fishes is discussed along with

[1] Department of Molecular and Cell Biology, University of Connecticut, Storrs, Connecticut 06269-3044, USA

Progress in Molecular and Subcellular Biology, Vol. 28
A.-P. Arrigo and W.E.G. Müller (Eds.)
© Springer-Verlag Berlin Heidelberg 2002

the finding of small HSP variation within species. During these studies, it was discovered that *Poeciliopsis* genomes encode two classes of small HSPs previously known only by their homologues in other animals such as human HSP27 and *Xenopus* HSP30.

2
Variation in Small HSPs Among Desert Species of *Poeciliopsis*

2.1
Results of the Survey Among Species

The desert species and hemiclones of *Poeciliopsis* used in this study were collected from five river systems of northwestern Mexico between 1961 and 1983 (Fig. 1 in White et al. 1994); more detailed information about collection sites is given in Table 2 in Norris (1997) and in a report by Schultz (1989). Each species inhabits a range of river systems, but with the exception of *P. occidentalis*, all lines of an individual species used in this study originated in a single watershed. *Poeciliopsis* from the Río del Fuerte have been the most intensively studied (thus the large number of lines originating from its tributaries); this river is the site of formation of all naturally-occurring hemiclonal *P. monacha-lucida* lines, since only here do the ranges of *P. monacha* and *P. lucida* overlap (hemiclones are discussed in more detail in Sect. 3.1). Based on their ranges, *P. occidentalis* is characterized as a "northern" species, while *P. viriosa* is characterized as a "southern" species; the ranges of the other desert species are more intermediate.

The protein family designated HSP30 has no phosphorylated members in the *Poeciliopsis* species surveyed to date (Fig. 9 in White et al. 1994) and it is more closely related evolutionarily to *Xenopus* HSP30 than to human HSP27 (Fig. 10 in Norris et al. 1997). Detailed qualitative comparisons of the HSP30 families in six desert species of *Poeciliopsis* were made in order to assess biochemical diversity. Radioisotopic labeling and gel analysis of HSPs were performed a minimum of three times for each species. The 30-kDa regions of representative fluorograms were scanned and the optical density images converted into contour images. Electrophoretically identical polypeptides were assigned the same isoform number in the HSP patterns of each of the six species. Electrophoretic identity was defined as comigration of protein spots during two-dimensional gel analysis of a mixture of ^{35}S-methionine-labeled proteins of two species. The 30-kDa family in six species of *Poeciliopsis* exhibited a high degree of biochemical variability even among these closely related species (Fig. 1). Due to the number of 30-kDa isoforms and the presence of many elongated spots, it was sometimes difficult to establish isoform identities from mixing experiments. Uncertain identifications are indicated by a "?" following the spot number. For example, the spot labeled 14? in the *P. prolifica*

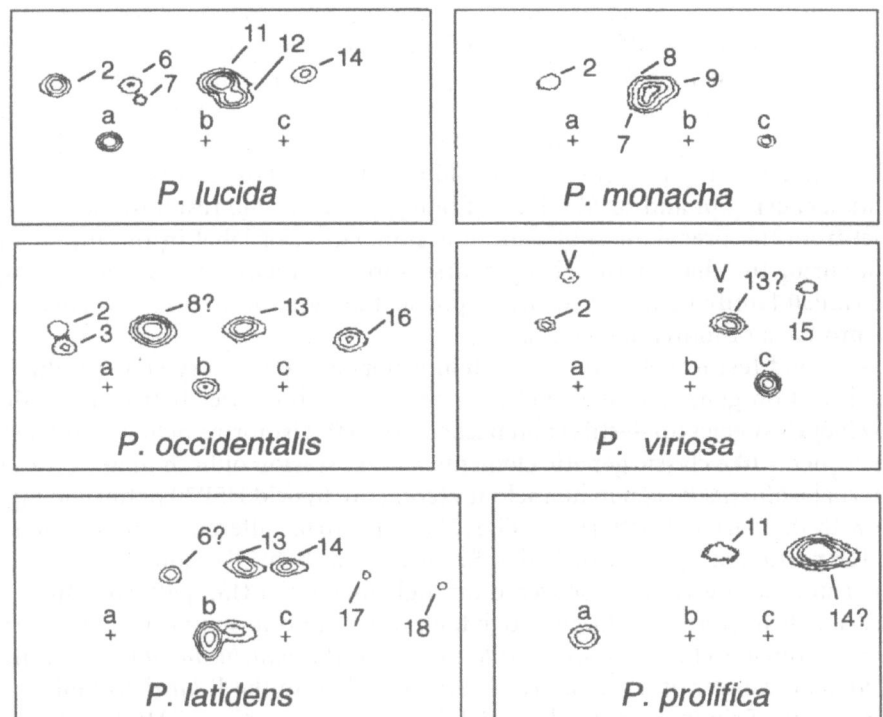

Fig. 1. Comparison of HSP30 isoform patterns among desert species. The 30-kDa region of two-dimensional gels of proteins labeled in heat shocked primary hepatocytes was analyzed as described in the text. Isoforms with the same isoelectric point were aligned vertically. Isoform numbers were assigned based on the co-migration of isoforms in mixed samples. Ambiguous designations were indicated by a "?". Spots *a–c* were constitutively synthesized polypeptides used as marker spots. The positions of each marker spot was indicated in each pattern to aid in comparison of isoforms. Other constitutively expressed proteins were marked with a "V". This figure was originally published in White et al. (1994) and is used with permission

pattern was probably two overlapping isoforms. Spot 14 of *P. lucida* migrates somewhere within the 14? region, but it was not possible to locate it exactly. Comparison of the 30-kDa patterns of three species with identical 70-kDa patterns indicated that there were spots in common (spots 11 and 14 in *P. lucida* and *P. prolifica*, spot 2 in *P. lucida* and as a minor spot in *P. occidentalis*), but none of the isoforms were present in all three species.

2.2
HSP Isoforms as Primary Gene Products

A limitation to the analysis of variation at the level of protein isoforms is that some of the isoforms may be created by posttranscriptional processes.

Pulse-chase analysis was performed in order to assess precursor-product relationships among the different HSP isoforms. Incorporation of ^{35}S-methionine
into proteins during a 5-min labeling period was terminated by the replacement of the labeling medium with medium containing 25 times the normal
amount of unlabeled methionine. During the chase period, the fate of polypeptides synthesized during the pulse could be followed. Posttranslational modification causing pI and/or Mr shifts of polypeptides would result in the loss of
label from the precursor spot and an accumulation of label in the modified
spot during the chase period. No precursor-product relationships exist among
the HSP30 family members, which supports the hypothesis that isoforms are
the products of individual genes.

A second test to determine if protein isoforms are primary gene products
was based on genetic crosses between species (such crosses between certain
Poeciliopsis species are fertile) and backcrosses. HSP isoforms behaved in these
experiments like classic genetic elements. Co-expression of both maternal and
paternal isoforms is seen in hemiclones (compare hybrid HSP30 pattern in Fig.
2D with the parental patterns in Figs. 2C,E). Isoform differences are diagnostic for particular species, such that an accidental hybridization event in the
aquarium room was first discovered through analysis of HSP patterns. On this
occasion, fish from a *P. latidens* stock tank were found to synthesize a mixture
of HSP isoforms characteristic of *P. latidens* and *P. monacha-lucida*, indicating
accidental transfer of a *P. monacha-lucida* female into the *P. latidens* tank. (C.
Norris and J. Schultz, unpubl. data). Additionally, segregation of HSP70, HSP60
and HSP30 isoforms was seen in a backcross experiment performed with *P.
monacha* and *P. viriosa* (two inbred species which interbreed to form fertile
hybrids). The protein patterns seen in the offspring indicated that within the
HSP30 family, isoforms were genetically linked. In other words, offspring
inherited the complete *monacha* or *viriosa* cluster rather than a random
mixture of the two. No evidence of linkage between HSP70 and HSP30 or
between each of these and HSP60 was seen (Norris 1997).

2.3
Conclusions

We concluded that the HSP30 family has been much less highly conserved
during evolution of *Poeciliopsis* species than the HSP70 family, which also has
been studied in detail (White et al. 1994; Norris et al. 1995). Most of the HSP30
isoforms which are characteristic of a particular species were unique to that
species. It is not known whether the HSP30 isoforms are qualitatively different
in function. The diversity of HSP30 isoforms among species of *Poeciliopsis* may
offer a selective advantage but could also be due to fewer selective constraints
on HSP30 than on other HSPs. It is known that the HSP30 family in *Poeciliopsis*
is induced only near the extreme upper thermal limit for these fishes (37–
40°C), and therefore, it is reserved for responses to the most extreme thermal
stress encountered by these animals in their natural environments. Fish placed

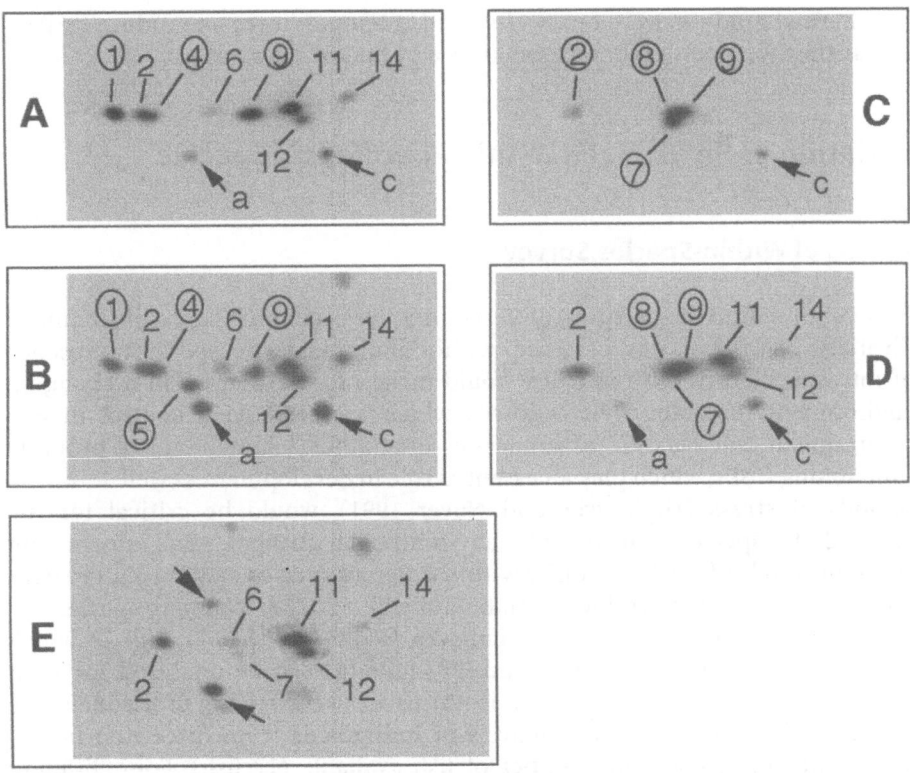

Fig. 2A–E. Example of within-species variation: HSP30 isoforms synthesized in *P. monacha* strains and *P. monacha-lucida* hemiclones. HSP isoform patterns of individuals were characterized on two-dimensional gels. Constitutively expressed proteins are marked by *arrows. Circled numbers* indicate isoforms attributable to the *P. monacha* genome; in the hemiclonal patterns the paternal *P. lucida* isoforms are numbered as in panel E. **A** Pattern in hybrids containing an inbred male *lucida* M61–31 genome and female *monacha* genomes found at Arroyos San Pedro, Cuchujaqui, and Guirocoba. **B** Pattern in hybrids containing an inbred male *lucida* M61–31 genome and female *monacha* genomes found at Arroyo Jaguari. **C** Pattern in two homozygous, inbred *monacha* strains (S68–4 and S68–5) developed from fish originally collected at Arroyo Jaguari. **D** Pattern in laboratory-synthesized hybrids (Syn-4 and Syn-5) containing an inbred male *lucida* M61–31 genome and either inbred *monacha* S68–4 or S68–5. **E** Pattern from homozygous, inbred *lucida* M61–31 strain developed from fish originally collected at Arroyo San Pedro. Originally published in a different arrangement of Figs. 5 and 6 in Norris et al. 1995, with permission

in a thermal gradient were rarely observed at temperatures that induce HSP30, unless they were foraging for food after a period of deprivation.

3
Variation in Small HSPs Within *Poeciliopsis* Species

3.1
Results of Within-Species Survey

As a continuation of the study of variability in *Poeciliopsis* HSPs, biochemical diversity of HSP30 was assessed within individual species. The presence of intraspecific diversity in HSPs would make rapid evolution in a changing environment possible: even isoforms which are selectively neutral in one environment may become advantageous in another. Maintenance of properly functioning HSPs, which play an essential role in development as well as during periods of stress (Hightower and Nover 1991), would be critical for the survival of a species confronted by environmental changes. Also, information about the levels of HSP variability within different species may shed more light on the processes that produce or maintain HSP variability among species.

HSP30 isoforms were analyzed in seven stocks of *P. lucida*, four of which were inbred strains. Between 3 and 20 individuals were analyzed for each inbred strain; no variability in HSP isoforms was seen among individuals of a particular inbred strain. The analysis of individuals from three non-inbred strains was limited by the number of fish available (12 fish). Four different HSP30 patterns were obtained from these non-inbred fish (Fig. 5 in Norris et al. 1995). The four patterns contained identical constitutive proteins used as markers but differed in which subset of the six inducible isoforms was present. One of these patterns is shown as an example in Fig. 2E. Some of these inducible protein isoforms also differed in intensity among the four patterns. Only isoform 11 was synthesized at a high rate in all four patterns, while isoform 14 was always synthesized at a lower rate. Isoform 2 was a major component only in one pattern (Fig. 2E), isoform 6/8? (which was probably two isoforms that did not resolve well from each other) was a major component of two other patterns, and isoform 7 was a major component of two other patterns as well. These isoforms were also found at low levels in the other patterns. Isoform 12 was found only in the pattern shown in Fig. 2E, a pattern characteristic of genomes that were homozygous at HSP30 loci of inbred strains.

HSPs synthesized by inbred strains of *P. monacha* were analyzed, as well as the HSPs contributed by maternal "wild" *monacha* genomes to the HSP patterns of *P. monacha-lucida* hemiclones (Fig. 2A–D). Hybridization in the natural environment between a *P. monacha* female and a *P. lucida* male produces *monacha-lucida* hybrid females. Each of the hybrid females contains a randomly assorted haploid *monacha* genome which will be passed clonally to all offspring by a genetic mechanism called hybridogenesis, a novel method of reproduction (Schultz 1961, 1969). Briefly, this mechanism consists of the loss of the paternal *lucida* genome entirely during premeiotic divisions of oogenesis in the hybrid, thus precluding any recombination and reassortment between maternal and paternal genomes. Upon fertilization of the hybrid female by a *P. lucida* male, the clonally derived *monacha* genome and a new paternal genome are incorporated into the zygote. Thus, individuals derived from a single hybrid female share identical maternal genomes but unique paternal genomes in the wild. When virgin offspring of captured wild females are mated in the laboratory to a homozygous, inbred strain of *P. lucida* such as the one with the HSP30 pattern shown in Fig. 2E, each female becomes the progenitor of a strain of genetically identical individuals, and strains established from different females contain different *monacha* alleles but share a common paternal genome. This is how the different hemiclonal lines of *monacha-lucida* used in this study were established. Each hemiclonal line contains a unique, randomly assorted haploid *monacha* genome which is passed clonally to all offspring. Analysis of HSP30 expression in parental inbred strains of *P. monacha* (Fig. 2C) and *P. lucida* (Fig. 2E), and hybrids synthesized in the laboratory from them, showed that maternal and paternal HSP30 isoforms were co-expressed in the hybrid (Fig. 2D). Thus the contribution of the *monacha* genome to the HSP pattern of wild hemiclones (Fig. 2A,B) mated to inbred *P. lucida* M61–31 (Fig. 2E) could be deduced by subtracting isoforms synthesized by the paternal strain. This was indicated in Fig. 2 by circling only those isoforms not seen in the *P. lucida* pattern in Fig. 2E.

3.2
Conclusions

The geographical distribution of HSP30 diversity in *P. monacha* separates the fishes of the Arroyo Jaguari from those found in the rest of the Río del Fuerte. This grouping parallels extensive work on allozyme frequencies in both hemiclonal (Vrijenhoek et al. 1978) and sexual (Vrijenhoek 1979) *monacha* genomes. Only two allozyme haplotypes were found in hemiclonal *monacha* genomes sampled from the Jaguari over a 15-year period; each allozyme haplotype was linked with a specific histocompatibility locus (Angus and Schultz 1979) and mitochondrial DNA haplotype (Quattro et al. 1992). These haplotypes represented a subset of the diversity found in the sexually reproducing *monacha* population at this site, and were not found elsewhere in the Fuerte, while haplotypes found in the other three tributaries were not found in the Jaguari. This is similar to the HSP30 isoform distribution: patterns B and C were found only in Jaguari fishes while pattern A was common to the other three tributaries and was not found in the Jaguari. Isolation of the Jaguari population from the rest of the Fuerte is thought to account for these differences; hybridization events occurring in headwaters of the Jaguari where populations of *P. monacha* and *P. lucida* overlap would fix *monacha* genomes unique to this locale. The barriers restricting *monacha* gene flow would presumably also restrict the spread of the resulting hemiclones to the rest of the Río del Fuerte.

P. lucida collected from the Río del Fuerte exhibited lower levels of allozyme polymorphism than *P. monacha* in the same sites (Leslie 1979). Again, the Jaguari fishes exhibited differential traits when compared to the rest of the Fuerte. Differences in caudal fin ray number (Angus and Schultz 1983) and the monomorphic nature of the allozyme complement of *P. lucida* collected from the headwaters of the Jaguari over many years (Vrijenhoek, personal communication) indicates isolation of the Jaguari *lucida* population. However, heterozygosity in histocompatibility loci was seen when tissue graft analysis was performed on this population (Eileen Fielding, personal communication). The HSP30 complements of strains of *P. lucida* from the Jaguari were not unique to that site; each pattern was also found in at least two other sites. The pattern in Fig. 2E showed reduced levels of isoform 7, and both this and a second pattern showed reduced expression of isoform 6 relative to some of the other patterns. Both the pattern shown in Fig. 2E and a second one were patterns from homozygous (inbred) fish; thus it appears that stable, heritable regulatory changes can also govern expression of these proteins and contribute to diversity. Inbred and partially inbred strains of *P. lucida* were analyzed, as opposed to the *monacha* genomes assayed mainly in hemiclonal form. In both cases a minimum estimate of the variability present in wild populations was obtained, but the different selective and/or random events that fix genes during inbreeding compared with those governing the formation of hemiclones may give us different snapshots of the diversity present in wild species.

4
Discovery of Two Classes of Small HSPs in a Single Taxon

4.1
Differential Synthesis

Analyses of newly synthesized HSPs on one-dimensional gels revealed several major size classes of small HSPs in *Poeciliopsis*. In order to get a clearer picture of changes in the synthesis of sHSPs over the course of a prolonged heat shock, extracts of ^{35}S-methionine-labeled cells were analyzed on two-dimensional gels by non-equilibrium pH gradient electrophoresis (Fig. 3). HSP30 appeared as a cluster of four spots whose synthesis was greatest at 2 h; HSP27 synthesis was detectable at 2 h but increased with longer incubation at heat shock temperature; band 1 was synthesized mainly after prolonged heat shock (7 h). Quantification of fluorograms like those in Fig. 3 allowed visualization of the kinetic changes in HSP synthesis during prolonged heat shock (Fig. 4). The transient nature of HSP30 induction in comparison to HSP27 synthesis was evident along with the delayed induction of Band 1 synthesis, perhaps in response to a different signal than that which induced the initial heat shock

Fig. 3. Kinetics of sHSP synthesis. Extracts of cultures labeled with ^{35}S-methionine for 1 h at 30°C (*Co*) or after 2, 4 and 7 h at 39.5°C were run on two-dimensional NEPHGE gels in order to visualize individual small HSPs. Time of incubation at heat shock temperature is indicated on the *left*. *Arrows* indicate constitutively synthesized proteins; isoforms of HSP30, HSP27 and band 1 are labeled. Only the low molecular weight regions of the gels are shown

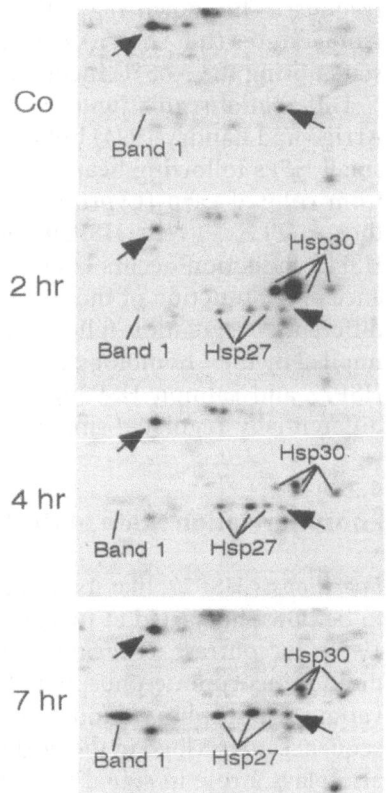

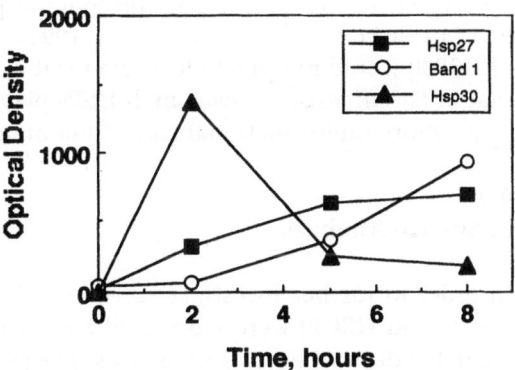

Fig. 4. Quantification of HSP synthesis during a prolonged heat shock. The fluorographs in Fig. 3 were quantified and total optical density in each sHSP was calculated (the three isoforms of HSP27 were summed). For graphing purposes, the HSP30 values were divided by 2

response. This down-regulation was not seen during incubations at higher temperatures (not shown), consistent with a partial recovery from the stressed state during the 39.5 °C incubation.

Differential regulation of small HSPs occurs in other systems (reviewed in Arrigo and Landry 1994) but differential expression of two different classes of small HSPs following heat shock has been reported only in cell lines derived from colored carp (*Cyprinus carpio*). These cells exhibit changes similar to those in PLHC-1 cells: HSP30 synthesis decreases in parallel with HSP70, while HSP27 induction occurs later and persists longer (Ku et al. 1992a,b). A difference in the function of the two small HSPs of *Poeciliopsis* is suggested by their differential regulation following heat shock, and by the sequence differences implied by their homology to divergent small HSPs from other species (human HSP27 and salmon HSP30); band 1 apparently represents another class of differentially-regulated small HSP.

4.2
Phosphorylation State of the Small HSPs

Poeciliopsis HSP27, like its human counterpart (Landry et al. 1992), appeared to be phosphorylated at two sites following heat shock (Fig. 7 in Norris et al. 1997). In contrast, *P. lucida* HSP30 was a cluster of protein isoforms which did not incorporate phosphate label (Fig. 9 in White et al. 1994). The conservation of phosphorylation sites between the human and *Poeciliopsis* HSP27 sequences, described in the next section, makes it likely that *Poeciliopsis* HSP27 also plays a role in signal transduction to the actin cytoskeleton. In contrast, the ability of mammalian HSP27 to function as a molecular chaperone in vitro is independent of phosphorylation. Recombinant murine HSP27 prevents aggregation of unfolded proteins and assists refolding regardless of phosphorylation state. The recombinant protein was phosphorylated in vitro with purified MAPKAP kinase-2 (Knauf et al. 1994). Thus, the lack of phosphorylation of HSP30 would not preclude a role for it as a molecular chaperone. It may be that in *Poeciliopsis*, the two small HSPs play complementary roles, in contrast to the more multifunctional role of mammalian HSP27.

4.3
Sequence Analysis

In order to further investigate these relationships, cDNA clones for *P. lucida* HSP27 and HSP30 were sequenced and evolutionary analysis was performed using the derived protein sequences. The sequence of PLHSP27 (Fig. 5; Norris et al. 1997) included an uninterrupted reading frame of 603 nt, along with 20 nt of 5' UTR and a 3' UTR of 435 nt. The predicted amino acid sequence contained three consensus phosphorylation sites for MAPKAP kinase-2 (-RXXS-) which are conserved in human HSP27 (Landry et al. 1992; Arrigo and Landry 1994). This is consistent with the incorporation of labeled phosphate into the two most acidic isoforms of HSP27 in heat-shocked cell cultures.

The sequences of two independent *HSP30* clones, PLHSP30A and PLHSP30B, were aligned (Fig. 6 in Norris et al. 1997). Sequence differences between these two clones and an additional isolate (PLHSP30C) are diagrammed in Fig. 5. All of the 5' UTRs were identical. The inferred coding regions of the three *HSP30* clones were identical in length (615 nt) but differed at six (clone C) or seven (clone B) positions when compared to clone A (asterisks in Fig. 5). The 3' UTRs were highly divergent, except for a region of 38 nucleotides immediately after the termination codon that was conserved in clones A and B. The nucleotide substitutions among the three clones were mostly synonymous; only four amino acid differences between clone B and the other two clones were predicted. These clones probably represent separate genes rather than allelic variants, since the PLHC-1 cell line from which the cDNAs were derived originated from an inbred strain of *P. lucida* which was homozygous as judged by allozyme analysis (Vrijenhoek et al. 1978) and by histocompatibility analysis (Angus and Schultz 1979; Schultz 1989). In clones A and B, the loss of sequence similarity after the first 38 nt of the 3' UTRs could

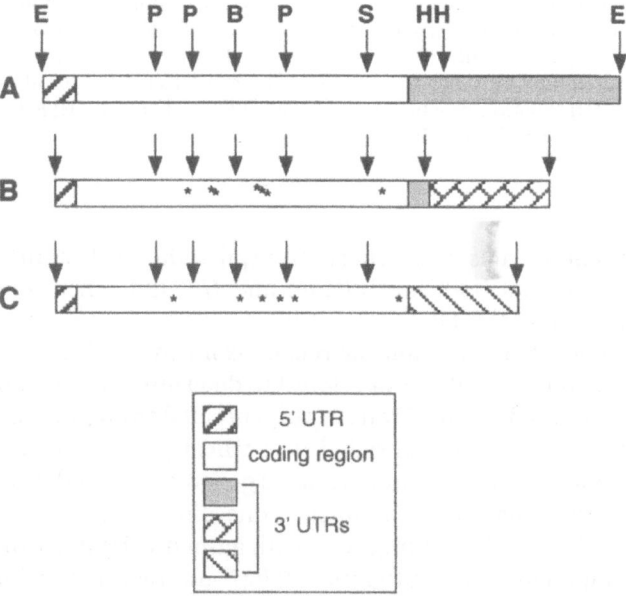

Fig. 5. Diagrammatic representation of 3' UTR variability among *HSP30* cDNA clones. A, B, and C represent PLHSP30 clones A, B and C, respectively. Regions where the sequence was identical (e.g., the 5' UTRs) were shaded identically; coding regions differed only at the sites marked by *asterisks*. 3' UTRs were highly divergent apart from a short sequence shared between clones A and B. Each sequence was represented by at least two independent isolates from the cDNA library. Restriction sites are indicated as follows: *E* Eco RI cloning site; *P* Pst I; *B* Bgl II; *S* Sst I; *H* Hinc II

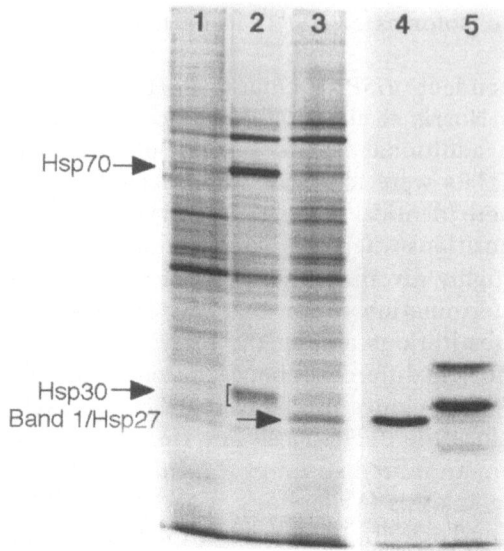

Fig. 6. In vitro transcription/translation products of PLHSP27 and PLHSP30B. RNA generated from linearized plasmids using T7 polymerase (RiboMAX, Ambion) was translated in a wheat germ extract in the presence of ³⁵S-methionine. Protein products were run on one-dimensional gels in parallel with ³⁵S-methionine-labeled extracts from control cells (*lane 1*) or cells heat shocked for 1.5 h (*lane 2*) or 6 h (*lane 3*). *Lane 4* Translation products from the PLHSP27 clone; *lane 5* translation products from the PLHSP30B clone. The HSP30 region is bracketed. On this gel HSP27 and band 1 did not resolve

be generated using an alternative splice site (-GT- at nt 717 in PLHSP30A). This predicts that both the *HSP30A* and *HSP30B* genes could produce transcripts with either 3′ UTR.

Since there are multiple related isoforms of HSP30 in *P. lucida*, which differ in pI and/or M_r, it was of interest to determine which of the isoforms is encoded by the HSP30 clone. Transcripts generated from linearized PLHSP30B using T7 polymerase were translated in a wheat germ extract in the presence of ³⁵S-methionine. Translation products generated from the *HSP27* cDNA in the same way were analyzed in parallel on one-dimensional polyacrylamide gels (Fig. 6). The HSP27 product migrated with authentic HSP27 from labeled cells and also comigrated with authentic HSP27 on two-dimensional gels. The protein pattern from the HSP30 transcripts was more complex, and contained a band which migrated above the HSP30 region as well as one that comigrated with HSP30 from a labeled cell extract. This may have been due to the production of aberrant transcripts in the in vitro transcription reaction. If the band in the HSP30 region of the gel is taken as the true product of PLHSP30B cDNA, then it appears that this clone codes for the lower molecular weight HSP30 isoform (the lower, acidic isoform of the HSP30 cluster; see Fig. 3). A comparison was

Table 1. Comparison of predicted and measured values of pI and molecular mass of *P. lucida* HSP30 and HSP27. Predicted values were calculated from the protein sequences using the Protein Analysis Toolbox in the program MacVector (International Biotechnologies, Inc.). Measurement of mass was performed by comparing migration of protein bands in an 11% SDS-PAGE gel with the migration of protein standards. The pI of isoform 12 of the HSP30 cluster (see Fig. 1) was approximated from its migration in isoelectric focusing (IEF) on two-dimensional gels relative to the measured pH gradient across similar gels, as well as through comparison with the migration of IEF marker proteins. The pI of the unphosphorylated (basic) isoform of HSP27 was approximated from its migration in NEPHGE two-dimensional gels relative to IEF marker proteins, as well as through comparison with *P. lucida* proteins whose pIs had been determined on IEF gels

	Predicted pI	Measured pI	Predicted mass (kDa)	Measured mass (kDa)
HSP27	6.26	~7	22.6	25
HSP30A	5.32		23.6	
HSP30B	5.31	~5.7	23.5	26.5

then made of predicted vs. measured pI and molecular mass for the HSP30 and HSP27 clones (Table 1). For both proteins, the measured pIs were more basic than those predicted, and the measured molecular masses were higher. However, both methods supported the relative pI and mass relationships: HSP27 was smaller and more basic than HSP30. In addition, based on the predicted values, it appeared that the protein products of HSP30 clones A and B would co-migrate on two-dimensional gels, in spite of the differences in their amino acid sequences.

Based on an analysis of HSP30 isoform diversity, it has been proposed that *Poeciliopsis HSP30* is a multigene family derived by duplication and divergence (White et al. 1994; Norris et al. 1995). Since the three clones which have been characterized here probably code for electrophoretically identical products, the minimum estimate of diversity in this gene family (based on the diversity of HSP30 protein isoforms) will have to be expanded to include multiple genes that code for a single HSP30 isoform.

4.4
Evolutionary Analysis

Amino acid sequences for HSP27 and HSP30 were used in searches of the Non-redundant SwissProt database using FastA. Search results indicated that the two *Poeciliopsis* small HSPs belonged to the α-crystallin/small HSP superfamily. However, the two proteins belonged to different groups within this family: *P. lucida* HSP27 was most similar to mammalian small HSPs, while *P. lucida* HSP30 was most similar to salmon and *Xenopus* HSP30s. The highest scoring sequences for each protein were used separately to generate multiple sequence alignments. As a group, the HSP27s showed a higher degree of conservation than the HSP30s. Pairwise comparisons of *P. lucida* and individual mammalian or avian HSP27 sequences yielded values of 53–55% identity and 30–31% similarity across the entire protein sequence. The overlapping regions of *P. lucida* HSP30 and salmon HSP30 showed 68.3% identity and 23.0% similarity, but this alignment excluded a region of 40 residues at the C-terminus that was highly divergent. Only ~30% identity was seen when *P. lucida* HSP30 was compared with *Xenopus* HSP30s.

The HSP30 and HSP27 and four α-crystallins (αA- and αB-crystallin from human and from chicken) were aligned. Phylogenetic trees were generated by neighbor-joining (Clustal W) and minimum error (Fitch-Margoliash) methods as well as by protein parsimony analysis. All methods gave consensus trees identical to that shown in Fig. 7, which was generated by neighbor-joining. Bootstrap analysis indicated high levels of support for all branches in both distance matrix (lowest value 799 out of 1000) and maximum parsimony (lowest value 75 out of 100) trees. HSP27 is most similar to a group of mammalian and avian small HSPs, with which it shares induction patterns that differ from those of HSP70, stress-inducible phosphorylation (Arrigo and

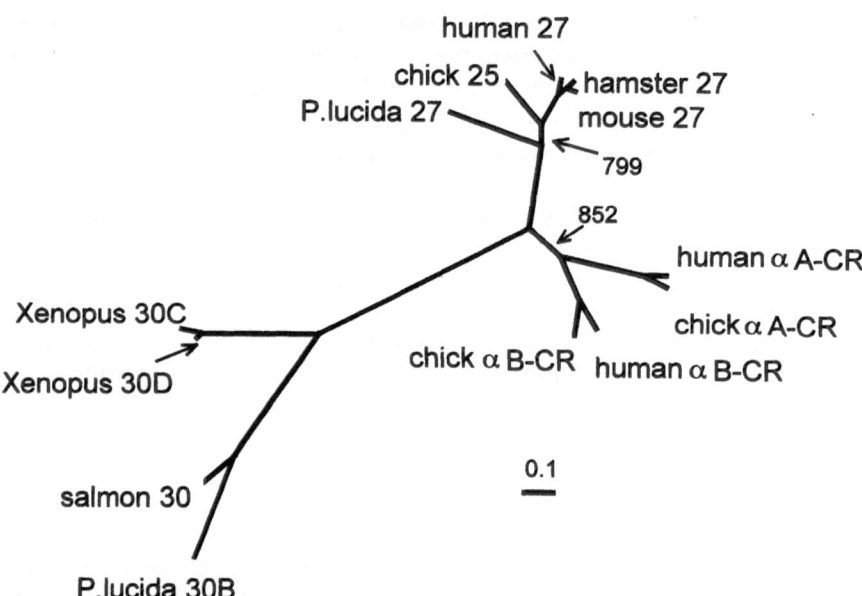

Fig. 7. Evolutionary relationships among members of the α-crystallin/sHSP superfamily. The distance matrix tree generated in Clustal W (neighbor-joining) for members of the α-crystallin/small HSP family. Bootstrap values (out of 1000 replicates) are shown for branches with <99.5% support. The *scale bar* represents the average number of substitutions per position. Identical consensus trees were derived by Fitch-Margoliash distance matrix (minimum error) analysis and by parsimony analysis, using the programs PROTDIST, FITCH and PROTPARS as implemented in Phylip3.5c (Felsenstein 1993). Originally published as Fig. 10 in Norris et al. 1997, and used with permission

Landry 1994), and sequence similarity . The HSP30 sequence is most similar to that of *Xenopus* and salmon HSP30s, as previously intimated by the cross-reactivity of an anti-salmon HSP30 antibody with the *Poeciliopsis* protein (Norris 1997).

Our favored explanation for the high degree of divergence of HSP30s from other vertebrate members of the superfamily is that the HSP30 lineage has evolved at a more rapid rate. This is supported by an examination of the tree in Fig. 7, which shows that the distance separating salmon and *Poeciliopsis* HSP30 is the same as that separating human and *Poeciliopsis* HSP27. If there is an orthologous relationship between human and *Poeciliopsis* HSP27, and between salmon and *Poeciliopsis* HSP30, then each of these pairs of genes has been diverging since the speciation events separating the corresponding lineages. However, the time over which the ancestral HSP27 sequence has been diverging (since the most recent common ancestor of fish and humans) is much greater than the time over which the ancestral HSP30 sequence has been diverging (since the speciation event separating the two fish lineages). That HSP30 and HSP27 are paralogs (derived by gene duplication in an ancestral species followed by sequence divergence) is clearly supported by our characterization of members of both lineages in a single taxon, but the relative positioning of the various vertebrate branches could be re-examined in light of the possibility of unequal evolutionary rates.

4.5
Conclusions

It is interesting that the more highly conserved of the two *Poeciliopsis* small HSPs also displayed an unusual pattern of induction during heat shock; this may indicate the presence of evolutionary constraints on HSP27 due to a conserved function in signal transduction pathways, during development, or in response to a stressor other than heat. If the non-phosphorylated *Poeciliopsis* HSP30s function only as molecular chaperones, interacting in a relatively non-specific way with hydrophobic regions of unfolded proteins, they may not be as evolutionarily constrained as small HSPs involved in actin dynamics or in regulation of development. This would allow for a more rapid rate of evolutionary change in the HSP30 lineage.

The characterization of HSP27 and HSP30 homologues in *Poeciliopsis lucida* helps delineate the complex evolutionary history of this family, but raises additional questions. These questions derive from the ancient origin of the two lineages, and concern the fate of the HSP27 lineage in *Xenopus* and salmon, and that of the HSP30 lineage in mammals. So far, attempts to find HSP27 homologues in salmon and *Xenopus* have not been successful (L. Weber and J. Heikkila, pers. comm.), and no HSP30 homologues have been identified in mammals. It will be interesting to determine if these genes have been lost or inactivated, or have diverged and been recruited to play another role in the cell.

Acknowledgements. We wish to thank R. Jack and Mary E. Schultz for more than 30 years of work in establishing and maintaining a genetically pedigreed fish colony and initial studies by Jack Schultz of intrinsic and acquired thermotolerance in *Poeciliopsis* species. Former graduate students whose work contributed directly to the studies discussed in this chapter include Mary A. Brown, Philip J. diIorio and Eileen Fielding. We also acknowledge the contributions of our collaborators and friends, Lee Weber and Eileen Hickey, on the small heat shock proteins.

This work was supported by a Marine/Freshwater Biomedical Sciences Center grant (ES03848) from the National Institute of Environmental Health Sciences and by the University of Connecticut Research Foundation. Carol Norris was a Howard Hughes Medical Institute Predoctoral Fellow.

References

Angus RA, Schultz RJ (1979) Clonal diversity in the unisexual fish *Poeciliopsis monacha-lucida*: a tissue graft analysis. Evol 33:27–40

Angus RA, Schultz RJ (1983) Meristic variation in homozygous and heterozygous fish. Copeia 1983:287–299

Arrigo A-P, Landry J (1994) Expression and function of the low-molecular-weight heat shock proteins. In: Morimoto RI, Tissiéres A, Georgopoulos C (eds) The biology of heat shock proteins and molecular chaperones. Cold Spring Harbor Laboratory Press, Plainville, NY, pp 335–374

Felsenstein J (1993) Phylogeny inference package, version 3.5c. Distributed by the author. Dept of Genetics, Univ of Washington, Seattle

Hightower LE, Nover L (ed) (1991) Heat shock and development. Springer, Berlin Heidelberg New York

Knauf U, Jakob U, Engel K, Buchner J, Gaestel M (1994) Stress- and mitogen-induced phosphorylation of the small heat shock protein Hsp25 by MAPKAP kinase 2 is not essential for chaperone properties and cellular thermoresistance. EMBO J 13:54–60

Ku CC, Chen SN, Kou GH (1992a) Dynamic changes in the localization and quantity of heat shock proteins in a cultured fin cell line of color carp *Cyprinus carpio*. Bull Inst Zool Acad Sin 31: 276–289

Ku CC, Lu CH, Kou GH, Chen SN (1992b) Purification and immunological characterization of color carp (*Cyprinus carpio*) fibroblast heat shock proteins. Comp Biochem Physiol 107B: 147–159

Landry J, Lambert H, Zhou M, Lavoie J, Hickey E, Weber LA, Anderson CW (1992) Human hsp27 is phosphorylated at serines 78 and 82 by heat shock and mitogen-activated kinases that recognize the same amino acid motif as S6 kinase. J Biol Chem 267:794–803

Leslie JF (1979) Genetic studies on sexually and clonally inherited genomes of *Poeciliopsis* (Pisces: Poeciliidae). Rutgers University, New Brunswick, NJ

Norris CE (1997) Analysis of diversity in the heat shock response of the desert topminnow *Poeciliopsis*. Molecular and cell biology. University of Connecticut, Storrs, 153 pp

Norris CE, diIorio PJ, Schultz RJ, Hightower LE (1995) Variation in heat shock proteins within tropical and desert species of poeciliid fishes. Mol Biol Evol 12:1048–1062

Norris CE, Brown MA, Hickey E, Weber LA, Hightower LE (1997) Low-molecular weight heat shock proteins in a desert fish (*Poeciliopsis lucida*): homologs of human Hsp27 and *Xenopus* Hsp30. Mol Biol Evol 14:1050–1061

Quattro JM, Avise JC, Vrijenhoek RC (1992) An ancient clonal lineage in the fish genus *Poeciliopsis* (Atheriniformes: Poeciliidae). Proc Natl Acad Sci USA 89:348–352

Schultz RJ (1961) Reproductive mechanism of unisexual and bisexual stains of the viviparous fish *Poeciliopsis*. Evol 15:302–325

Schultz RJ (1969) Hybridization, unisexuality, and polyploidy in the teleost *Poeciliopsis* (Poeciliidae) and other vertebrates. Am Nat 103:605–619

Schultz RJ (1989) Origins and relationships of unisexual poeciliids. In: Meffe GK, Snelson FF (eds) Ecology and evolution of poeciliid fishes. Prentice-Hall, Englewood Cliffs, pp 69–87

Vrijenhoek RC (1979) Genetics of a sexually reproducing fish in a highly fluctuating environment. Am Nat 113:17–29

Vrijenhoek RC, Angus RA, Schultz RJ (1978) Variation and clonal structure in a unisexual fish. Am Nat 112:41–55

White CN, Hightower LE, Schultz RJ (1994) Variation in heat-shock proteins among species of desert fishes (Poeciliidae, *Poeciliopsis*). Mol Biol Evol 11:106–119

Chaperone Function of sHsps

Martin Haslbeck and Johannes Buchner[1]

1
Introduction

Small heat shock proteins (sHsps) are a widespread but diverse class of proteins. In contrast to other families of Hsps, they contain certain conserved
sequence motifs only in the C-terminal part of the protein, the so called α-
crystallin domain. These are low molecular mass proteins (15–42 kDa) which
form oligomeric structures ranging from 9 to 50 subunits. sHsps display a
chaperone function in vitro. In addition, it has been suggested that they are
involved in the inhibition of apoptosis, organization of the cytoskeleton and
contribute to the optical properties of the eye lens in the case of α-crystallin.
How these different functions can be explained by a common mechanism is
unclear at the present state of investigations. However, as most of the observed
phenomena involve non-native protein, the repeatedly reported chaperone
properties of sHsps seem to be of key importance for understanding their
function.

2
General Aspects of Chaperone Function

2.1
Chaperones

Not surprisingly, most classes of heat-shock proteins have been identified as
molecular chaperones in recent years (Buchner 1996). Molecular chaperones
support the correct folding of proteins during translation and translocation
as well as during stress periods when cellular proteins are in danger of irre-
versibly aggregating. To achieve this aim, chaperones must be able to differentiate between native and non-native forms of target proteins. The binding of
unfolded polypeptide chains to the chaperone reduces the probability of irreversible interactions of exposed hydrophobic sequence stretches and thus

[1] Institut für Organische Chemie und Biochemie, Technische Universität München, 85747
Garching, Germany

Progress in Molecular and Subcellular Biology, Vol. 28
A.-P. Arrigo and W.E.G. Müller (Eds.)
© Springer-Verlag Berlin Heidelberg 2002

allows folding or maintenance of protein structure under non-permissive conditions.

The molecular chaperone concept is not a contradiction to the principle of spontaneous folding (Anfinsen 1973), as molecular chaperones do not convey any specific information for folding but rather support correct structure by preventing improper or unproductive reactions that could lead to protein mis-folding and/or aggregation. Without becoming part of the native structure, molecular chaperones therefore ensure high fidelity in the protein folding and assembly process.

It is rewarding to summarize and contrast briefly the major classes of these functionally related proteins before discussing the chaperone function of sHsps specifically. Hsp70 proteins bind hydrophobic regions in linear sequences of polypeptides (Flynn et al. 1991; Blond-Elguindi et al. 1993; Rüdiger et al. 1997). The affinity for the substrate is modulated by the binding of ATP and by co-chaperones (Bukau and Horwich 1998). This is also true for the role model chaperone GroE from *E. coli*. GroEL, the main component of the GroE complex, forms a cylindrical two ring structure of seven subunits each. Non-native proteins are bound in the central cavity of this cylinder. The co-chaperone GroES forms a heptameric ring. Binding of GroES to the ends of the cylinder (in the presence of ATP) leads to the encapsulation of the GroE-bound protein. This step limits the size range of the substrates to 10–55 kDa (Ewalt et al. 1997). The substrate interaction is regulated by the binding and hydrolysis of ATP. In an iterative mechanism, several cycles of binding and release lead to the folding of the substrate (Todd et al. 1996; Bukau and Horwich 1998; Sigler et al. 1998; Beissinger et al. 1999). In contrast to the detailed mechanistic picture that can be drawn for GroE, the mode of action of the chaperone Hsp90 in concert with its numerous co-factors is still largely unknown. Hsp90 seems to chaperone many signal transducing proteins with native-like structures in an ATP-dependent way. The binding of Hsp90 stabi-lizes these structures. In some cases, this is the prerequisite for the binding of ligands to the substrate proteins (Bohen et al. 1995; Sullivan et al. 1997; Buchner 1999). Finally, members of the Hsp100 family are ATP-dependent chaperones (Kruger et al. 1994; Schirmer et al. 1994) capable of dissolving pre-formed protein aggregates (Parsell et al. 1994; Schirmer and Lindquist 1998; Goloubinoff et al. 1999).

sHsps are sometimes termed junior chaperones because out of the major classes of Hsps, they have been most recently identified as molecular chaper-ones. In contrast to other chaperone families, sHsps bind several non-native proteins per oligomeric complex, thus representing the most efficient chaperone family in terms of the quantity of substrate binding (Jaenicke and Creighton 1993; Ehrnsperger et al. 1997; Lee et al. 1997). In some cases, the release of substrate proteins from the sHsp complex is achieved in coopera-tion with Hsp70 in an ATP-dependent reaction (Ehrnsperger et al. 1997; Lee et al. 1997), suggesting that the role of sHsps in the network of chaperones discussed above is to create a reservoir of non-native refoldable protein (see below).

2.2
Analysis of Chaperone Function

Progress in understanding the mechanism of chaperone function has been achieved mainly by a reductionist biochemical approach using purified chaperones and "model substrate proteins". This term implies that the substrate proteins do not necessarily represent natural, in vivo substrates. In many cases, the underlying rationale was that proteins which are recalcitrant folders in vitro may also have problems folding in the cell and are therefore good representatives of the unknown natural targets.

A number of different substrate proteins has been used to study the function of chaperones including RuBisCO (Goloubinoff et al. 1989), Rhodanese (Mendoza et al. 1991), malate dehydrogenase (MDH) (Chen et al. 1994; Ranson et al. 1997; Ben-Zvi et al. 1998), and citrate synthase (Buchner et al. 1991, 1998b). These substrates differ in their quaternary structure, their rate of folding, and their tendency to undergo irreversible side reactions during folding and unfolding (cf. Jaenicke 1993). Citrate synthase is a commercially available homodimeric enzyme, which allows a comparison of the mode of action of different chaperones since the influence of four of the five major classes of molecular chaperones on citrate synthase (CS) folding has been determined already (Buchner 1996; Buchner et al. 1991; Ehrnsperger et al. 1997; Jakob et al. 1995). The use of CS as a chaperone substrate allows the question to be addressed whether chaperones suppress aggregation by monitoring changes in light scattering after thermal or chemical denaturation. This is a first test for the ability of a protein to discriminate between native and non-native proteins. Using this assay, it turned out that sHsps are very efficient in binding non-native proteins. When CS is unfolded by incubation at temperatures above 40 °C, both mammalian 25-kDa murine heat shock protein (Hsp25) and 26-kDa yeast heat shock protein (yeast Hsp26) were able to suppress aggregation significantly at equimolar ratios (Fig. 1).

Aggregation is a late step in the thermal unfolding pathway of CS. Early steps are manifested in the loss of enzymatic activity. Following the inactivation of CS over time in the presence or absence of chaperones allows their ability to influence this process to be defined and to differentiate between stable or transient interactions during these early steps (Fig. 2). In the presence of Hsp25, the inactivation kinetics of CS are identical to the reaction in the absence of Hsps (Jakob et al. 1993; Jakob and Buchner 1994; Buchner et al. 1998a). Thus the complexes between Hsp25 and early unfolding intermediates seem to be stable. This is similar to results obtained for the chaperone GroEL (Grallert et al. 1998). In contrast, Hsp26 apparently slows down the inactivation process, seemingly stabilizing CS. This thermoprotective effect may be due to more

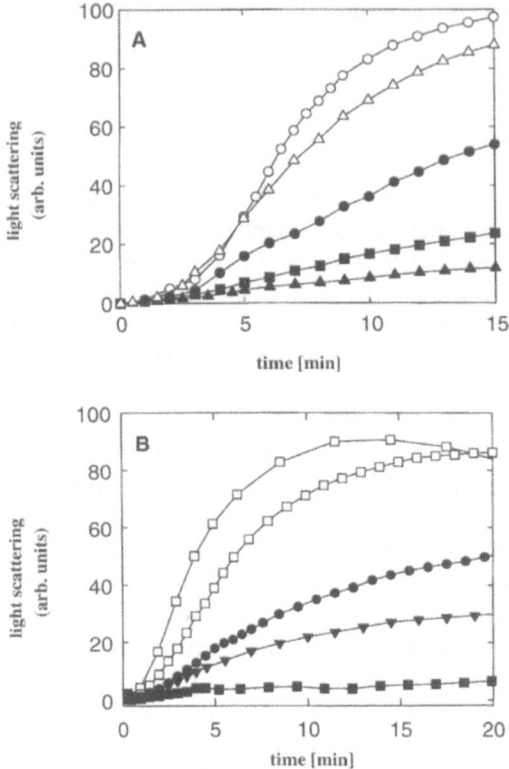

Fig. 1. A Influence of Hsp25 on the thermal aggregation of citrate synthase (CS). CS (final concentration: 75 nM) was diluted into a thermostatted solution of 30 nM (●) (), 60 nM (■), 200 nM Hsp25 (▲) and 0.5 mg/ml IgG (▲). *Open circles* (○) represent the spontaneous aggregation of CS at 43 °C. **B** Influence of Hsp26 on the thermal aggregation of CS. CS (final concentration: 75 nM) was diluted into a thermostatted solution of 37.5 nM (□), 55 nM (●), 75 nM (▼) and 150 nM Hsp26 complex (■). *Open circles* represent the spontaneous aggregation of CS at 43 °C. The kinetics of aggregation were determined by measuring the light scattering of the samples

transient interactions between Hsp26 and CS conformers which are native-like and able to refold to the native state (Haslbeck et al. 1999).

Binding of the substrate oxaloacetic acid to CS shifts the midpoint of the thermal transition from 48 to 66.5 °C (Srere 1966; Zhi et al. 1991). This stabilization was exploited to analyze the unfolding and refolding pathway of CS in more detail. Two defined inactive dimeric unfolding intermediates were detected (see Jakob et al. 1995). Subsequently, the protein dissociates and aggregates. This dissection of the unfolding pathway allows one to determine to what extent the intermediates are stabilized and populated by different sHsps. The stability of CS-sHsp complexes can be determined by performing a temperature shift to 25 °C. If the CS stays stably bound to the Hsp, no increase in activ-

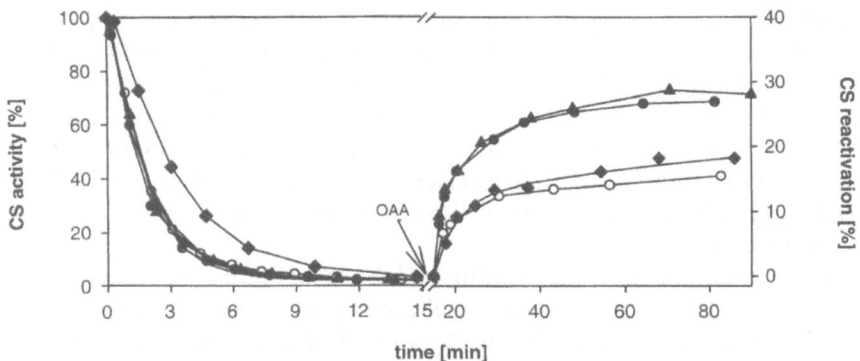

Fig. 2. Thermal inactivation and reactivation of CS. CS (75 nM) was inactivated at 43 °C for 15 min in the absence (○) and presence of 0.3 μM Hsp25 (▲), 0.3 μM Hsp27 (●) and 0.3 μM authentic bovine lens α-crystallin (◆). CS activity was measured at the time points indicated as described in the text. To start reactivation, at time 15 min OAA was diluted 1:100 into the samples (final concentration: 1 mM). The tubes were then transferred to the folding permissive temperature of 25 °C and reactivation was monitored by measuring CS activity. The concentrations of Hsp25 and Hsp27 refer to 16-mers. α-crystallin concentration was calculated for a 25-mer

ity is observed after the temperature shift. This is the case for Hsp25-CS complexes (Fig. 2). Here, reactivation by addition of OAA can be triggered at time points as late as several hours after the temperature shift. For an effective release of bound intermediates from Hsp25 in the absence of OAA, the addition of molecular chaperones is required (see below; Ehrnsperger et al. 1997).

3
The Chaperone Properties of sHsps

3.1
In Vivo Functions of sHsps

Under physiological conditions, the abundance of sHsps varies according to cell type and organism studied (Ciocca et al. 1993). In addition, the expression of sHsps is dependent upon development, growth cycle, differentiation and the oncogenic status of the cell (Bond and Schlessinger 1987; Gaestel et al. 1989; Crête and Landry 1990; Pauli et al. 1990; Ciocca et al. 1993; Klemenz et al. 1993; Arrigo 2000; Wehmeyer and Vierling 2000). In this context, a well-characterized example of developmental induction is the accumulation of sHsps during seed maturation (Vierling and Sun 1989; Almoguera and Jordano 1992). With the exception of the eye lens, under physiological conditions, the levels of sHsps in different cell types range from non-existent to about 1% in the case of human Hsp20 in striated muscle, heart and diaphragm cells (Niwa et al. 2000; Sugiyama et al. 2000). Studies on the expression of sHsps under heat

shock and other stress conditions have shown that this group of Hsps is among
the most strongly induced heat shock proteins (Klemenz et al. 1991, 1993;
Inaguma et al. 1992; Arrigo and Landry 1994). Final yields can be up to 1% of
total protein depending on the cells and tissue analyzed (Arrigo and Landry
1994; Waters et al. 1996). In addition to the quantity, the number of sHsps also
varies significantly in different species, with up to 30 sHsps in plants, where
sHsp appear in the cytosol as well as in all organelles in multiple isoforms
(Vierling and Sun 1989).

The regulation of αB-crystallin and 27-kDa human heat shock protein
(Hsp27) expression in the same cellular compartment can be either independent or co-ordinated, according to the stress factors applied (Head et al. 1994).
sHsps thus seem to be regulated by several independent systems, indicating a
wide set of functions, which need further characterization in the future.

The overexpression of sHsps has been reported to convey thermoresistance
in a number of organisms and cell types. This phenomenon has been shown
in mammalian cells for human Hsp27 (Landry et al. 1989), murine Hsp25
(Knauf et al. 1992), *Drosophila* Hsp27 (Rollet et al. 1992; Mehlen et al. 1993),
αB-crystallin (Aoyama et al. 1993) and αA-crystallin (van den Ijssel et al. 1994),
indicating a general thermoprotective function of these proteins. Plant sHsps
are strongly induced during heat shock according to the developmental status
of the cells (DeRocher and Vierling 1994). Protection against heat and cold
stress by sHsps was demonstrated in chestnut seeds (Collada et al. 1997). In
cyanobacteria, the photosynthetic apparatus is protected against heat shock by
the constitutive expression of sHsps (Nakamoto et al. 2000). Surprisingly,
the yeast homologue Hsp26 does not seem to have a demonstrable function in
thermoprotection or during development (Petko and Lindquist 1986; Susek
and Lindquist 1989).

Overexpression of Hsp27 has not only been associated with thermotolerance but also with conferring resistance to oxidative stress (Mehlen et al. 1995;
Huot et al. 1995, 1996) as well as protection against several cytotoxic
agents including cytochalasin D (Lavoie et al. 1993, 1995), tumor necrosis
factor (Mehlen et al. 1995) and some anticancer drugs (Oesterreich et al.
1993).

A clear picture of the mechanism of sHsps in protecting cells against stress
has not emerged so far. On the one hand, sHsps have repeatedly been reported
to act as promiscuous molecular chaperones. On the other hand, there is
evidence that sHsps can specifically interact with certain components of the
cytoskeleton such as actin (Lavoie et al. 1993; Benndorf et al. 1994; Zhu et al.
1994a,b) or intermediate filaments (Nicholl and Quinlan 1994). In this context,
myopathy patients carrying an αB-crystallin mutation revealed aggregates of
desmin associated with αB-crystallin, confirming previous findings that sHsps
interact with intermediate filaments (Carter et al. 1995; Djabali et al. 1997; van
den Ijssel et al. 1999).

It could well be that the general chaperone properties leading to the
stabilization of thermolabile intracellular proteins under stress conditions and
the specific interaction with the cytoskeleton, possibly preserving cell shape

and integrity, are two related properties contributing to protection against stress.

An increasing number of diseases have been described where mutations and modifications of sHsps exert their pathological effect by altering protein folding or assembly. Increased expression of sHsps and especially αB-crystallin has been observed in several neurodegenerative disorders, such as Alzheimer's disease (Shinohara et al. 1993; Renkawek et al. 1994), Alexander disease (Iwaki et al. 1993; Head et al. 1993), Creutzfeld Jakob syndrome (Kato et al. 1992; Renkawek et al. 1992) and others. The phosphorylation patterns of Hsp25 and αB crystallin in brains of patients with Alexander disease have been reported to be distinct from control brains, indicating that post-translational modification might play a role in the progress of the disease (Head et al. 1993). A common feature of all these neurodegenerative disorders is the deposition of improperly folded protein in fibers, inclusion bodies or plaques in the nervous system (cf. Thomas et al. 1995). sHsps are typically found in association with these insoluble protein-aggregates and ubiquitin (Lowe et al. 1993; Cooper et al. 1995). Increased expression of both α-crystallin and Hsp27 was observed in the context of these diseases. Interestingly, in scrapie-infected mouse neuroblastoma cells, induction of Hsp27 and Hsp72 by metabolic stress is blocked, but Hsp73 and Hsp90 expression is enhanced (Tatzelt et al. 1995). This observation could indicate that Hsp27, together with Hsp72, is specifically involved in the pathogenesis of the prion disease.

The elevated expression may thus not only be considered as a strategy for survival during unfavorable times but also as a potential mechanism of defense against diseases, which makes sHsps an interesting target for novel therapeutic approaches (cf. Clark and Muchowski 2000; Ehrnsperger et al. 2000). Preliminary experiments have already shown the feasibility of gene therapy for high level expression of sHsps (Wagstaff et al. 1999).

In summary, the function of altered sHsp expression in the course of several seemingly unrelated diseases is still enigmatic. A common feature of many of the described disorders is the accumulation of misfolded protein (Cheetham 1995; Thomas et al. 1995).

However, an understanding of the precise cellular role of sHsps is still lacking. Interestingly, a *Synechocystis* sp. mutant deleted for its single sHsp gene, Hsp16.6, showed a defect in thermotolerance, indicating the importance of sHsps at high temperatures in this organism (E. Vierling, pers. comm.). Other prokaryotic sHsps were found to be tightly associated with intracellular aggregated endogenous proteins (Laskowska et al. 1996) and with inclusion bodies which are formed during high level protein expression. In *E. coli*, the two endogenous sHsps were therefore termed inclusion body proteins (IbpA and IbpB; Allen et al. 1992). As in many other organisms, the deletion of sHsps in *E. coli* is not lethal, nor does it show a phenotype (Thomas and Baneyx 1998) The deletion of Ibp had only major effects on a strain also mutated in DnaK function (Thomas and Baneyx 1998).

Taken together, these results suggest a biological function of sHsps in intracellular protein aggregation.

3.2
sHsps Bind Non-Native Protein

All sHsps investigated so far share conserved regions in the C-terminal part of the protein, while the N-terminal part is quite divergent in sequence and length (Arrigo and Landry 1994; Ehrnsperger et al. 1997). The conserved C-terminal domain exhibits high sequence homology to the family of α-crystallins (de Jong et al. 1993). In addition to the sequence homology, both sHsps and α-crystallins form large oligomeric complexes, although complex size and subunit stoichiometry vary substantially (de Jong et al. 1993). For Hsp16.5 from *Methanococcus jannaschii* and α-crystallin, the three-dimensional structures have been described as hollow shells (Haley et al. 1998; K.K. Kim et al. 1998). While α-crystallin seems to be dynamic and exchanges subunits, Hsp16.5 is a precisely defined complex.

The interaction of sHsps with non-native protein was first demonstrated for α-crystallin, murine Hsp25 and human Hsp27 (Horwitz 1992; Jakob et al. 1993; Merck et al. 1993). Meanwhile this was extended to sHsps from a number of organisms. In contrast to other chaperone families, sHsps bind several non-native proteins per oligomeric complex, comprising the most effective chaperone family concerning substrate binding (Ehrnsperger et al. 1997; Lee et al. 1997; Jaenicke and Creighton 1993).

One of the first indications of sHsps being implicated in chaperoning protein folding was the finding that an ammonium sulfate fraction, enriched in plant sHsps, could prevent soluble proteins from thermal denaturation in vitro (Jinn et al. 1989). Mammalian sHsps and α-crystallin have subsequently been shown to be able to suppress aggregation of thermally denaturing model substrates. In addition, the in vivo relevance of these interactions was demonstrated by the suppression of aggregation of the eye lens proteins β- and γ-crystallin in the presence of α-crystallin (Horwitz 1992). Generally, sHsps seem to be rather promiscuous in their binding behavior, as complex formation with a variety of heat denatured substrates was reported. Interaction with heat-inactivated CS was shown for Hsp27, Hsp25, α-crystallin (Jakob et al. 1993; Ehrnsperger et al. 1997), Hsp26 from yeast (Haslbeck et al. 1999), cytosolic sHsps from Pea (Lee et al. 1995), a sHsp from chestnut seeds (Collada et al. 1997), Hsp16.5 from *Methanococcus jannaschii* (R. Kim et al. 1998), Hsp16.3 from *Mycobacterium tuberculosis* (Chang et al. 1996) and lipocortin (Kim et al. 1997). In addition Hsp25, Hsp26 and Hsp 16.3 from *Mycobacterium tuberculosis* interact with the insulin B-chain (Haslbeck et al. 1999; Yang et al. 1999). Promiscuity in binding was also shown for the cytosolic plant sHsp 18.1 (Lee et al. 1997) and for α-crystallin which forms stable complexes with a number of non-native proteins such as carbonic anhydrase (Rao et al. 1993), Rhodanese (Das et al. 1996) and γ-, β_L-, β_H-crystallin (Raman et al. 1995a,b; Wang and Spector 1995; Das et al. 1996). Interestingly, α-crystallin was more efficient in preventing aggregation of other lens crystallins than that of model enzymes (Horwitz 1992).

The chaperone function of sHsps is not restricted to the prevention of thermal protein aggregation since renaturation of chemically denatured protein was reported as well. Renaturation of γ-crystallin was supported by α-crystallin (Horwitz 1992) and the yield of reactivation of different model enzymes was significantly increased in the presence of either Hsp25, Hsp27 or α-crystallin (Jakob et al. 1993). The mode of action was almost indistinguishable among the different sHsp species (Jakob et al. 1993; Merck et al. 1993).

Recently, an influence of ATP on the chaperone functions of αB-crystallin has been described (Muchowski and Clark 1998), suggesting structural rearrangements upon ATP binding (Reddy et al. 1992; Palmisano et al. 1995). In α-crystallin, ATP induced a conformational change that was accompanied by a concomitant internalization of a previously exposed hydrophobic surface, as measured by 8-anilino-1-naphtalene sulfonate (ANS) binding (Rawat and Rao 1998). In this context, four conserved residues in the core domain of α-crystallin seem to be protected against proteolytic cleavage in the presence of ATP. This may indicate ATP-dependent structural modifications and an ATP-binding motif in the core domain of sHsps (Muchowski et al. 1999). In contrast to these findings, the chaperone activity of sHsps in vitro was so far found to be independent of ATP-binding or hydrolysis (Horwitz 1992; Jakob et al. 1993; Merck et al. 1993; Lee et al. 1997; Haslbeck et al. 1999). Detailed studies on the specificity of ATP binding and hydrolysis of sHsps and their relation to the binding of non-native proteins will be required to resolve this issue.

3.3
Complex Formation and Substrate Range

The underlying mechanism of sHsps chaperoning protein folding is still enigmatic. Most of the sHsps studied form very stable complexes with non-native substrate proteins (Rao et al. 1994; Lee et al. 1995; Das et al. 1996; Ehrnsperger et al. 1998). For example, the interaction between α-crystallin and carbonic anhydrase is stable at 4 °C as well as at room temperature for at least 18 h (Rao et al. 1994). The stoichiometry of the complexes between sHsps and non-native protein seems to be dependent on the substrate investigated (Lee et al. 1997). However, several reports suggest a maximum binding capacity of one denatured protein molecule per subunit of the sHsp oligomers (Farahbakhsh et al. 1995; Wang and Spector 1995; Lee et al. 1997; Haslbeck et al. 1999). In the case of Hsp26 and Hsp25, the formation of the complex is a highly co-operative process (Haslbeck et al. and M. Ehrnsperger, unpubl. results). The structural requirements for proteins to be recognized and bound by sHsps are not entirely clear yet. In the case of Rhodanese bound to α-crystallin, tryptophane fluorescence suggests that the protein is more native than when complexed with GroEL (Das et al. 1996). Similar binding patterns were observed when α-crystallin was complexed with alcohol dehydrogenase and γ-crystallin (Carver et al. 1995b). In contrast to other chaperones like GroEL, DnaK and SecB, α-crystallin does not interact with stable unfolded, hydrophobic proteins such as carboxymethylated α-lactalbumin and α-casein (Carver et al. 1995b). Based on these data, it was suggested that sHsps preferentially recognize early labile intermediates which are most sensitive to aggregation (Carver et al. 1995b; Das et al. 1996). In contrast to other chaperones like Hsp70 (Stege et al. 1994) or Hsp90 (Jakob et al. 1995), some sHsps investigated do not seem to alter the inactivation kinetics of proteins (Horwitz et al. 1992; Rao et al. 1994, 1995; Ehrnsperger et al. 1997). On the other hand, Hsp26, lipocortin and several members of plant sHsps are able to decelerate the inactivation kinetics of substrate proteins (Lee et al. 1995; Kim et al. 1997; Collada et al. 1997; Haslbeck et al. 1999). While the general property of interacting with non-native proteins seems to be conserved, differences in detail seem to exist between different members of the sHsp family.

Another interesting feature of sHsps is their wide substrate range. sHsps can support the folding of thermally and chemically denatured proteins ranging from at least 4 to 100 kDa (Horwitz et al. 1992; Jakob et al. 1993; Rao et al. 1993; Ehrnsperger et al. 1997, 1998). For example, at physiological temperatures, α-crystallin interacts with the reduced and aggregation-prone B-chains of insulin, and melittin a peptide from bee venom which adopts an unordered conformation (Farahbakhsh et al. 1995). Interestingly, sHsp also seem to bind structured folding intermediates which are prone to aggregation (Carver et al. 1995; Das et al. 1996; Ehrnsperger et al. 1997; Treweek et al. 2000). The broad substrate range and the stable binding of sHsps can be exploited to stabilize labile antigens, as demonstrated for Hsp25 in diagnostic immunological assays (Ehrnsperger et al. 1998).

3.4
The Substrate Binding Site

The location of the binding sites of non-native proteins in the large oligomeric structures of sHsps is not fully understood yet. Boyle et al. (1993) suggested binding to a central cavity as in the case of GroEL. However, the ability of binding stoichiometric amounts of substrate per sHsp monomer argues against this idea (see above). Furthermore, recent electron microscope studies strongly suggest that the substrate protein is bound on the surface of the oligomeric sHsp structure (Ehrnsperger et al. 1997; Lee et al. 1997; Haslbeck et al. 1999). Theoretical studies supporting an open micellar structure and surface binding of protein as a working model for α-crystallin (Singh et al. 1995) agree with experimental data (Wang and Spector 1994, 1995; Farahbakhsh et al. 1995).

The conserved α-crystallin domain of sHsps was suggested to be involved in the chaperone function of the proteins. Removal of parts of the C-terminal domains of α-crystallin and Hsp27 from *Drosophila* had been reported to lead to a marked reduction of the ability of the protein to suppress thermal aggregation and to protect against thermal stress in vivo (Mehlen et al. 1993; Takemoto et al. 1993). In agreement with these findings, Lee et al. (1997) could show by cross-linking that bound substrates were in contact with a conserved region in the C-terminal part of the protein. This consensus sequence is found in all members of the sHsp family (Plesofsky-Vig et al. 1992; de Jong et al. 1993; Waters 1995). Cross-linking studies on αB-crystallin also revealed two sequence stretches belonging to the conserved α-crystallin domain (Sharma et al. 1997). In addition, site-directed mutations within the α-crystallin domain of αB-crystallin decreased the chaperone activity of the protein (Muchowsiki et al. 1999). However, studies with the isolated α-crystallin domain of Hsp25 did not provide any evidence for in vitro chaperone activity of this region (Merck et al. 1992). The flexible C-terminal tail, which is found in virtually all sHsps with variable sequence and length (Carver et al. 1992, 1995a; Wistow 1993), has been recently proposed to be involved in substrate binding (Carver and Lindner 1998). However, the truncation of the C-terminal 18 residues of Hsp25 has no effect on protective chaperone function using CS as substrate. The truncated Hsp25 was not able to stabilize α-lactalbumin against precipitation following reduction with dithiothreitol (Lindner et al. 2000).

It should be noted that non-native protein could not only be cross-linked to C-terminal consensus sequences but also to a hydrophobic N-terminal region, indicating that, together with the C-terminal domain, conserved N-terminal sequence stretches might be involved in substrate binding (Lee et al. 1997). These findings indicate that the conserved α-crystallin domain might be important but is not sufficient for the chaperone activity of sHsps. Apparently, additional parts of the proteins are required.

3.5
Mechanism of Substrate Binding

Not only is the binding site on sHsp oligomers still unknown, the mechanism of interaction with denatured substrate molecules remains enigmatic as well. It is widely believed that hydrophobic contacts contribute to binding. This is mostly based on data indicating that the hydrophobicity profiles of some sHsps change under stress conditions. α-crystallin and Hsp26 bind increasing amounts of the hydrophobic dyes 8-anilino-1-naphtalene sulfonate (ANS) (Carver et al. 1995; Raman et al. 1995a,b; Haslbeck et al. 1999) and bis-ANS (Das and Surewicz 1995a,b) at temperatures above 30°C, suggesting that α-crystallin undergoes a temperature-dependent structural change that increases surface hydrophobicity. However, data on the two possible binding sites of plant sHsps and site-directed mutagenesis seem to contradict the hypothesis that hydrophobic interactions are exclusively involved in substrate binding (Lee et al. 1997; van den Ijssel et al. 1999). The putative binding site identified by cross-linking (Lee et al. 1997; Sharma et al. 1997) contains several highly conserved aliphatic residues with intervening charged amino acids. This alternating polar-non-polar sequence suggests that efficient binding of denatured protein is mediated by hydrophobic residues interacting with exposed non-polar amino acids of the substrate while charged amino acids might serve to optimally space the residues and aid in hydration of the region (Lee et al. 1997). Previous studies on the binding of spin-labeled insulin B-chain and melittin to α-crystallin support the notion that polar residues are involved in substrate binding to sHsps (Farahbakhsh et al. 1995; van den Ijssel et al. 1999). Electron spin resonance (EPR) analysis of the labeled peptides showed that the nitroxide side chains are immobilized in a polar environment on α-crystallin. The bound insulin chains are not in a fully extended conformation and melittin does not appear to bind to hydrophobic surfaces on α-crystallin as an amphiphatic helix, as it does to membranes and some other proteins (Farahbakhsh et al. 1995). Smulders et al (1995) could show that the D69S mutant of αA-crystallin shows decreased chaperone activity indicating that the charged Asp69 might be involved in protein binding. Furthermore, mutations at position R116 in αA-crystallin and R120 in αB-crystallin proved to be responsible for genetic disorders (Litt et al. 1998; Vicart et al. 1998). These mutations affected the residue, equivalent to arginine 107 in the α-crystallin domain of *M. jannaschii* Hsp16.5. In the Hsp16.5 crystal structure (Kim et al. 1998a), arginine 107 plays an unusual role because this highly charged residue is buried in the hydrophobic core of the protein core. It appears that the position of the arginine 107 side chain is rigidly defined within the structure. The arginine 107 side chain is directed inwards from a β-sheet towards the hydrophobic core of the protein, but the charged end group reaches the surface by pointing sideways to the edge of the sheet, making about half of the arginine charged end group accessible on the surface. This would make the arginine an ideal contributor to substrate binding (van den Ijssel et al. 1999).

3.6
Regulation of Functional Properties

Chaperones are often regulated by cofactors or ATP. In the case of sHsps other mechanisms seem to exist. An intrinsic mechanism of regulation was shown for yeast Hsp26. Under physiological conditions, oligomeric Hsp26 complexes consisting of 24 monomers exist as a largely active storage form. Heat shock leads to the dissociation of these complexes into dimers. This species specifically recognizes and binds non-native polypeptide chains resulting in the cooperative formation of large, well-defined Hsp26-substrate complexes (Haslbeck et al. 1999).

It is tempting to speculate that the heat-induced dissociation and activation of Hsp26 represents an early and simple mechanism, which developed into a more sophisticated system of functional regulation during the evolution of higher eukaryotes. Instead of a direct temperature sensing system, additional signals could now be integrated into the activation process of the respective sHsp. Phosphorylation is a widespread modification of mammalian sHsps, which seems to have functional consequences. Like heat in the case of Hsp26, the phosphorylation of sHsps seems to be associated with changes in the oligomeric structure of sHsp complexes. Depending on the cell type, the extra-

cellular stimulus and the time after stimulation, phosphorylation of Hsp27 correlates with either the dissociation or further increase in size of the oligomers
(Kato et al. 1994; Mehlen and Arrigo 1994; Lambert et al. 1999). The phosphorylated, dissociated form of Hsp27 shows no chaperone activity (Rogalla et al.
1999) and thermoprotection is decreased (Martin et al. 1999). Interestingly,
larger oligomeric complexes of Hsp 25 were observed upon heat treatment and,
for a non-phosphorylatable mutant of Hsp25, after TNFα treatment (Preville
et al. 1998). In addition, Hsp25 showed chaperone activity as a hexadecamer
and as a large mega-dalton complex (Ehrnsperger et al. 1999). The mammalian
sHsp-related protein Hsp20 dissociates when phosphorylated. This may
mediate cellular signaling processes that lead to vasorelaxation (Brophy et al.
1999). Furthermore, phosphorylation of sHsps seems to alter the interaction
with actin (Lavoie et al. 1993, 1995; Benndorf et al. 1994; Guay et al. 1997).

From the data above one may speculate that phosphorylation is necessary
but not sufficient for dissociation of mammalian sHsp oligomers. At present,
only a few members of the diverse family of sHsps have been investigated in
detail. It could well be that, in addition to dissociation and phosphorylation,
other factors could contribute to the regulation of the oligomeric size of sHsps
and thus modulate function via structural changes.

3.7
sHsps in the Context of the Cellular Chaperone Machinery

Taken together, sHsps seem to be potent and efficient chaperones in that
they are able to selectively bind non-native proteins in large quantities per
oligomeric sHsp complex. However, in general, no release of bound substrate
has been observed. For α-crystallin, it was suggested that irreversible binding
of other lens crystallins, which might occur in the aging lens, prevents aggregation and light scattering (Smulders et al. 1995; Horwitz et al. 1999; Horwitz
2000). Along these lines, Lee et al. (1995) suggested that in the case of plant
sHsps, bound substrate was held in a soluble form for subsequent degradation.
In the living cell it seems detrimental to entertain a chaperone system which
traps non-native proteins without allowing reactivation once a stress period is
over. A first hint that sHsp substrate complexes were not dead-end but productive intermediates was the discovery that CS could be dissociated and reactivated from Hsp25CS complexes by oxaloacetate (OAA), a stabilizing ligand
of CS (Ehrnsperger et al. 1997). Thus, Hsp25 is able to bind to thermally
inactivated proteins and trap it in a soluble and folding-competent state until
permissive folding conditions are restored. While the above-mentioned experiments used the ligand OAA, in the cell a physiological release factor would be
required.

A clue for the contribution of different chaperones to this reaction came
from in vivo experiments. Kampinga et al. (1994, 1995) and Stege et al. (1995)
could demonstrate that the over expression of Hsp70 led to a deceleration
of the inactivation of marker proteins in transfected cells. Whereas in the

presence of Hsp27, the inactivation of the marker protein was not influenced and both the substrate and the chaperone, were found to be insolubilized. However, when these cells were transferred to permissive conditions, both proteins became soluble again and the cells recovered much faster than in the absence of Hsp27. This is reminiscent of the finding that prokaryotic sHsps are found specifically associated with inclusion bodies, intracellular aggregates of overexpressed proteins (Allen et al. 1992).

In agreement with these findings, it was shown in vitro that the chaperone Hsp70 allowed reactivation of the sHsp-bound substrate in the presence of ATP (Ehrnsperger et al. 1997). In addition, for Hsp18.1 from pea, the ATP-dependent release and refolding of bound luciferase by different Hsp70 and Hsp40 was demonstrated (Lee et al. 1997; Lee and Vierling 2000). Evidence for the release of substrate proteins bound to sHsp was also shown for *E. coli* IbpB. Here, bound malate dehydrogenase and lactate dehydrogenase were specifically transferred to the DnaK–DnaJ–GrpE system and subsequently reactivated (Veinger et al. 1998).

Sequencing of the genome of the thermophilic eubacterium *T. maritima* revealed another piece of evidence for the cooperation of sHsps with Hsp70. Here, the genes for Hsp70 and a 17.0-kDa sHsp were found in one operon (Michelini and Flynn 1999).

Based on these data, a model that allows sHsps to be integrated into the cellular chaperone machinery was proposed (Figs. 3, 4). Under stress conditions, Hsp25 binds non-native proteins, prevents their irreversible aggregation and complexes them in a refoldable state. Upon restoration of physiological temperatures, the non-native protein will either dissociate from the complex spontaneously, or ATP-dependent chaperones such as Hsp70 will complete the refolding. Whether other factors such as Hsp70 co-chaperones or additional Hsp families are involved in this process remains to be seen. Another open question is whether the proposed model (Fig. 4) can be generalized for all organisms.

4
Conclusions

With good reason, sHsps have been called "the forgotten chaperones", since, although they had been implicated to be involved in a number of fundamental cellular processes in addition to the heat shock response, their function remained enigmatic. Recent experiments using in vitro protein folding assays have allowed the functional properties of sHsps under stress conditions to be defined. These results showed that sHsps belong to the class of molecular chaperones. The current evidence suggests that a specific function of sHsps is to effectively trap aggregation-prone folding intermediates and maintain them in a refoldable conformation. At the moment, structural information of sHsps is rather poor. Thus the underlying conformational features responsible for the

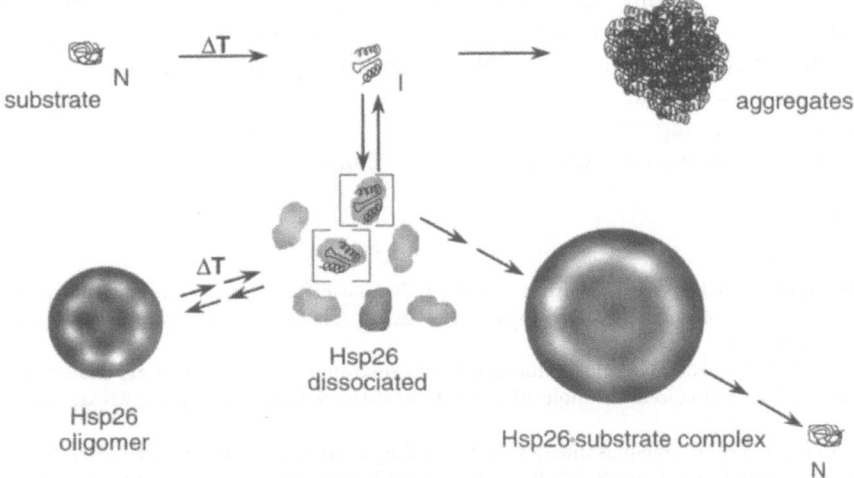

Fig. 3. Model of the chaperone function of Hsp26. Under physiological conditions Hsp26 exists in an inactive oligomeric storage form. Upon heat shock (ΔT) unfolding of native protein (N) leads to the formation of non-native, aggregation-prone intermediates (I). Hsp26 dissociates into smaller complexes which are able to bind non-native proteins (I), thus preventing their aggregation. Well-defined Hsp26·non-native protein complexes assemble subsequently from the dissociated species. After release of the substrate protein, Hsp26 reassociates to the inactive storage form under physiological conditions

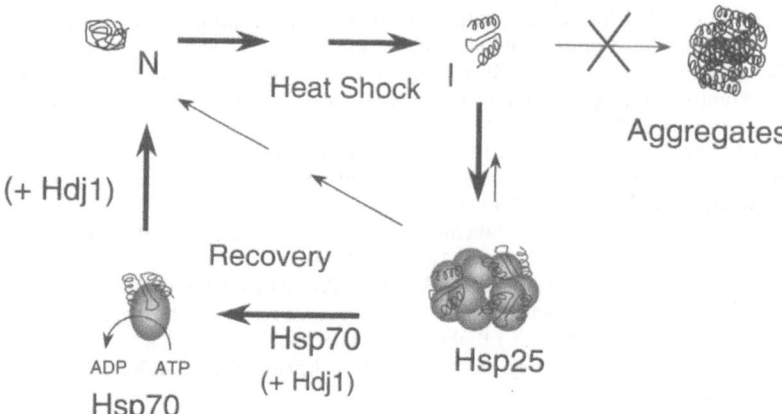

Fig. 4. Model for the chaperone function of Hsp25 under heat shock conditions. Unfolding of native protein (N) leads to the formation of non-native, aggregation-prone intermediates (I). In the presence of Hsp25, these unfolding intermediates can be bound in high quantity, in a soluble and refoldable state. Even at permissive temperatures the non-native protein stays bound. However, in the presence of Hsp70, ATP, and possibly cofactors like Hdj-1, sHsp-bound protein can be released and refold into its native state

stable binding of non-native proteins remain unknown. In the eye lens, stable binding of α-crystallin to other crystallins seems to be sufficient to protect the lens efficiently from irreversible aggregation processes which could cause turbidity. However, in the cytosol and potentially in organelles, cooperation with other ATP-dependent chaperone systems seems to be required to achieve productive folding, even under permissive conditions.

References

Allen SP, Polazzi JO, Gierse GK, Easton AM (1992) Two novel heat shock genes encoding proteins produced in response to heterologous protein expression in *Escherichia coli*. J Bacteriol 174:6938–6947

Almoguera C, Jordano J (1992) Developmental and environmental concurrent expression of sunflower dry-seed-stored low-molecular-weight heat-shock protein and Lea mRNAs. Plant Mol Biol 19:781–792

Anfinsen CB (1973) Principles that govern the folding of protein chains. Science 181:223–230

Aoyama A, Fröhli E, Schäfer R, Klemenz R (1993) α B-crystallin expression in mouse NIH 3T3 fibroblasts: glucocorticoid responsiveness and involvement in thermal protection. Mol Cell Biol 13:1824–1835

Arrigo AP (2000) sHsp as novel regulators of programmed cell death and tumorigenicity. Pathol Biol (Paris) 48:280–288

Arrigo AP, Landry J (1994) Expression and function of the low-molecular-weight heat shock proteins. In: Morimoto RI (ed) The biology of heat shock proteins and molecular chaperones. Cold Spring Harbor Laboratory Press, Plainview, NY, 335 pp

Beissinger M, Buchner J (1998) How chaperones fold proteins. Biol Chem 379:245–259

Benndorf R, Hayess K, Ryazantsev S, Wieske M, Behlke J, Lutsch G (1994) Phosphorylation and supramolecular organization of murine small heat shock protein HSP25 abolish its actin polymerization-inhibiting activity. J Biol Chem 269:20780–20784

Ben-Zvi AP, Chatellier J, Fersht AR, Goloubinoff P (1998) Minimal and optimal mechanisms for GroE-mediated protein folding. Proc Natl Acad Sci USA 95:15275–15280

Blond-Elguindi S, Cwirla SE, Dower WJ, Lipshutz RJ, Sprang SR, Sambrook JF, Gething MJ (1993) Affinity panning of a library of peptides displayed on bacteriophages reveals the binding specificity of BiP. Cell 75:717–728

Bond U, Schlesinger MJ (1987) Heat-shock proteins and development. Adv Genet 24:1–29

Bohen SP, Kralli A, Yamamoto KR (1995) Hold 'em and fold 'em: chaperones and signal transduction. Science 268:1303–1304

Boyle D, Gopalakrishnan S, Takemoto L (1993) Localization of the chaperone binding site. Biochem Biophys Res Commun 192:1147–1154

Brophy CM, Dickinson M, Woodrum D (1999) Phosphorylation of the small heat shock-related protein, HSP20, in vascular smooth muscles is associated with changes in the macromolecular associations of HSP20. J Biol Chem 274:6324–6329

Buchner J (1996) Supervising the fold: functional principles of molecular chaperones. FASEB J 10:10–19

Buchner J (1999) Hsp90 & Co. - a holding for folding. Trends Biochem Sci 24:136–141

Buchner J, Schmidt M, Fuchs M, Jaenicke R, Rudolph R, Schmid FX, Kiefhaber T (1991) GroE facilitates refolding of citrate synthase by suppressing aggregation. Biochemistry 30:1586–1591

Buchner J, Ehrnsperger M, Gaestel M, Walke S (1998a) Purification and characterization of small heat shock proteins. Methods Enzymol 290:339–349

Buchner J, Grallert H, Jakob U (1998b) Analysis of chaperone function using citrate synthase as non-native substrate protein. Methods Enzymol 290:323–338

Bukau B, Horwich AL (1998) The Hsp70 and Hsp60 chaperone machines. Cell. 92:351–366

Carter JM, Hutcheson AM, Quinlan RA (1995) In vitro studies on the assembly properties of the lens proteins CP49, CP115: coassembly with alpha-crystallin but not with vimentin. Exp Eye Res 60:181–192

Carver JA, Lindner RA (1998) NMR spectroscopy of alpha-crystallin. Insights into the structure, interactions and chaperone action of small heat-shock proteins. Int J Biol Macromol 22: 197–209

Carver JA, Aquilina JA, Truscott RJW, Ralston GB (1992) Identification by ^{1}H NMR spectroscopy of flexible C-terminal extensions in bovine lens α-crystallin. FEBS Lett 311:143–149

Carver JA, Esposito G, Schwedersky G, Gaestel M (1995a) ^{1}H NMR spectroscopy reveals that mouse Hsp25 has a flexible C-terminal extension of 18 amino acids. FEBS Lett 369:305–310

Carver JA, Guerreiro N, Nicholls KA, Truscott RJW (1995b) On the interaction of α-crystallin with unfolded protein. Biochim Biophys Acta 1252:251–260

Chang Z, Primm TP, Jakana J, Lee IH, Serysheva I, Chiu W, Gilbert HF, Quiocho FA (1996) *Mycobacterium tuberculosis* 16-kDa antigen (Hsp16.3) functions as an oligomeric structure in vitro to suppress thermal aggregation. J Biol Chem 271:7218–7223

Cheetham ME (1995) Cell stress genes and chronic neurodegenerative disorders. Neuropathol Appl Neurobiol 21:486–468

Chen S, Roseman AM, Hunter AS, Wood SP, Burston SG, Ranson NA, Clarke AR, Saibil HR (1994) Location of a folding protein and shape changes in GroEL-GroES complexes imaged by cryo-electron microscopy. Nature 371:261–264

Ciocca DR, Oesterreich S, Chamness GC, McGuire WL, Fuqua SA (1993) Biological and clinical implications of heat shock protein 27,000 (Hsp27): a review. J Natl Cancer Inst 85:1558–1570

Clark JI, Muchowski PJ (2000) Small heat-shock proteins and their potential role in human disease. Curr Opin Struct Biol 10:52–59

Collada C, Gomez L, Casado R, Aragoncillo C (1997) Purification and in vitro chaperone activity of a class I small heat-shock protein abundant in recalcitrant chestnut seeds. Plant Physiol 115:71–77

Cooper PN, Jackson M, Lennox G, Lowe J, Mann DM (1995) Tau, ubiquitin and α B-crystallin immunohistochemistry define the principal causes of degenerative frontotemporal dementia. Arch Neurol 52:1011–1015

Crête P, Landry J (1990) Induction of HSP27 phosphorylation and thermoresistance in Chinese hamster cells by arsenite, cycloheximide, A23187, and EGTA. Radiat Res 121:320–327

Das KP, Surewicz WK (1995a) Temperature-induced exposure of hydrophobic surfaces and its effect on the chaperone activity of α-crystallin. FEBS Lett 369:321–325

Das KP, Surewicz WK (1995b) On the substrate specificity of α-crystallin as a molecular chaperone. Biochem J 311:367–370

Das KP, Petrash JM, Surewicz WK (1996) Conformational properties of substrate proteins bound to a molecular chaperone α-crystallin. J Biol Chem 271:10449–10452

De Jong WW, Leunissen JA, Voorter CE (1993) Evolution of the α-crystallin/small heat shock protein family. Mol Biol Evol 10:103–126

DeRocher AE, Vierling E (1994) Developmental control of small heat shock protein expression during pea seed maturation. Plant J 5:93–104

Djabali K, de Nechaud B, Landon F, Portier MM (1997) AlphaB-crystallin interacts with intermediate filaments in response to stress. J Cell Sci 110:2759–2769

Ehrnsperger M, Graber S, Gaestel M, Buchner J (1997) Binding of non-native protein to Hsp25 during heat shock creates a reservoir of folding intermediates for reactivation. EMBO J 16: 221–229

Ehrnsperger M, Hergersberg C, Wienhues U, Nichtl A, Buchner J (1998) Stabilization of proteins and peptides in diagnostic immunological assays by the molecular chaperone Hsp25. Anal Biochem 259:218–225

Ehrnsperger M, Lilie H, Gaestel M, Buchner J (1999) The dynamics of Hsp25 quaternary structure. Structure and function of different oligomeric species. J Biol Chem 274:14867–14874

Ehrnsperger M, Gaestel M, Buchner J (2000) Analysis of chaperone properties of small Hsp's. Methods Mol Biol 99:421–429

Ewalt KL, Hendrick JP, Houry WA, Hartl FU (1997) In vivo observation of polypeptide flux through the bacterial chaperonin system. Cell. 90:491–500

Farahbakhsh ZT, Huang QL, Ding LL, Altenbach C, Steinhoff HJ, Horwitz J, Hubbell WL (1995) Interaction of α-crystallin with spin-labeled peptides. Biochemistry 34:509–516

Flynn GC, Pohl J, Flocco MT, Rothman JE (1991) Peptide-binding specificity of the molecular chaperone BiP. Nature 353:726–730

Gaestel M, Gross B, Benndorf R, Strauss M, Schunk WH, Kraft R, Otto A, Böhm H, Stahl J, Drabsch H et al. (1989) Molecular cloning, sequencing and expression in Escherichia coli of the 25-kDa growth-related protein of Ehrlich ascites tumor and its homology to mammalian stress proteins. Eur J Biochem 179:209–213

Goloubinoff P, Christeller JT, Gatenby AA, Lorimer GH (1989) Reconstitution of active dimeric ribulose bisphosphate carboxylase (RuBisCo) from an unfolded state depends on two chaperonin proteins and Mg-ATP. Nature 342:884–889

Goloubinoff P, Mogk A, Zvi AP, Tomoyasu T, Bukau B (1999) Sequential mechanism of solubilization and refolding of stable protein aggregates by a bichaperone network. Proc Natl Acad Sci USA 96:13732–13737

Grallert H, Rutkat K, Buchner J (1998) GroEL traps dimeric and monomeric unfolding intermediates of citrate synthase. J Biol Chem 273:33305–33310

Guay J, Lambert H, Gingras-Breton G, Lavoie JN, Huot J, Landry J (1997) Regulation of actin filament dynamics by p38 map kinase-mediated phosphorylation of heat shock protein 27. J Cell Sci 110:357–368

Haley DA, Horwitz J, Stewart PL (1998) The small heat-shock protein, αB-crystallin, has a variable quarternary structure. J Mol Biol 277:27–35

Haslbeck M, Walke S, Stromer T, Ehrnsperger M, White HE, Chen S, Saibil HR, Buchner J (1999) Hsp26: a temperature-regulated chaperone. EMBO J 18:6744–6751

Head MW, Corbin E, Goldman JE (1993) Overexpression and abnormal modifications of the stress proteins α B-crystallin and HSP27 in Alexander disease. Am J Pathol 143:1743–1753

Head MW, Corbin E, Goldman JE (1994) Coordinate and independent regulation of αB-crystallin and HSP27 expression in response to physiological stress. J Cell Physiol 159:41–50

Horwitz J (1992) α-crystallin can function as a molecular chaperone. Proc Natl Acad Sci USA 89:10449–10453

Horwitz J (2000) The function of alpha-crystallin in vision. Semin Cell Dev Biol 11:53–60

Horwitz J, Bova MP, Ding LL, Haley DA, Stewart PL (1999) Lens alpha-crystallin: function and structure. Eye 13:403–408

Huot J, Lambert H, Lavoie JN, Guimond A, Houle F, Landry J (1995) Characterization of 45-kDa / 54-kDa HSP27 kinase, a stress-sensitive kinase which may activate the phosphorylation-dependent protective function of mammalian 27-kDa heat-shock protein HSP27. Eur J Biochem 227:416–427

Huot J, Houle F, Spitz DR, Landry J (1996) HSP27 phosphorylation-mediated resistance against actin fragmentation and cell death induced by oxidative stress. Cancer Res 56: 273–279

Inaguma Y, Shinohara H, Goto S, Kato K (1992) Translocation and induction of α B crystallin by heat shock in rat glioma (GA-1) cells. Biochem Biophys Res Commun 182:844–850

Iwaki T, Iwaki A, Tateishi J, Sakaki Y, Goldman JE (1993) αB-crystallin and 27-kd heat shock protein are regulated by stress conditions in the central nervous system and accumulate in Rosenthal fibers. Am J Pathol 143:487–495

Jaenicke R (1993) What does protein refolding in vitro tell us about protein folding in the cell? Philos Trans R Soc Lond B Biol Sci 339:287–294

Jaenicke R, Creighton TE (1993) Junior chaperones. Curr Biol 3:234–235

Jakob U, Buchner J (1994) Assisting spontaneity: the role of Hsp90 and small Hsps as molecular chaperones. Trends Biochem Sci 19:205–211

Jakob U, Gaestel M, Engel K, Buchner J (1993) Small heat shock proteins are molecular chaperones. J Biol Chem 268:1517–1520

Jakob U, Meyer I, Bugl H, Andre S, Bardwell JC, Buchner J (1995) Structural organization of procaryotic and eucaryotic Hsp90. Influence of divalent cations on structure and function. J Biol Chem 270:14412–14419

Jinn TL, Yeh YC, Chen YM, Lin CY (1989) Stabilization of soluble proteins in vitro by heat shock proteins-enriched ammonium sulfate fraction from soybean seedlings. Plant Cell Physiol 30:463–467

Kampinga HH, Brunsting GF, Stege GJJ, Konings AWT, Landry J (1994) Cells overexpressing Hsp27 show accelerated recovery from heat-induced nuclear protein aggregation. Biochem Biophys Res Commun 204:1170–1177

Kampinga HH, Brunsting JF, Stege GJJ, Burgman PWJJ, Konings AWT (1995) Thermal protein denaturation and protein aggregation in cells made thermotolerant by various chemicals: role of heat shock proteins. Exp Cell Res 219:536–546

Kato K, Hasegawa K, Goto S, Inaguma Y (1994) Dissociation as a result of phosphorylation of an aggregated form of the small stress protein, hsp27. J Biol Chem 269:11274–11278

Kato S, Hiranov A, Umahara T, Llena JF, Herz F, Ohama E (1992) Ultrastructural and immuno-histochemical studies on ballooned cortical neurons in Creutzfeldt-Jakob disease: expression of α B-crystallin, ubiquitin and stress-response protein 27. Acta Neuropathol Berl 84:443–448

Kim GY, Lee HB, Lee SO, Rhee HJ, Na DS (1997) Chaperone-like function of lipocortin 1. Biochem Mol Biol Int 43:521–528

Kim KK, Yokota H, Santoso S, Lerner D, Kim R, Kim SH (1998) Purification, crystallization, and preliminary X-ray crystallographic data analysis of small heat shock protein homolog from *Methanococcus jannaschii*, a hyperthermophile. J Struct Biol 121:76–80

Kim R, Kim KK, Yokota H, Kim SH (1998) Small heat shock protein of *Methanococcus jannaschii*, a hyperthermophile. Proc Natl Acad Sci USA 95:9129–9133

Klemenz R, Frohli E, Steiger RH, Schafer R, Aoyama A (1991) α B-crystallin is a small heat shock protein. Proc Natl Acad Sci USA 88:3652–3656

Klemenz R, Andres AC, Frohli E, Schafer R, Aoyama A (1993) Expression of the murine small heat shock protein hsp25 and α B crystallin in the absence of stress. J Cell Biol 120:639–645

Knauf U, Bielka H, Gaestel M (1992) Over-expression of the small heat shock protein hsp25 inhibits growth of Ehrlich ascites tumor cells. FEBS Lett 309:297–302

Lambert H, Charette SJ, Bernier AF, Guimond A, Landry J (1999) HSP27 multimerization mediated by phosphorylation-sensitive intermolecular interactions at the amino terminus. J Biol Chem 274:9378–9385

Landry J, Chrétien P, Lambert H, Hickey E, Weber LA (1989) Heat shock resistance conferred by expression of the human HSP27 gene in rodent cells. J Cell Biol 109:7–15

Laskowska E, Wawrzynow A, Taylor A (1996) IbpA and IbpB, the new heat-shock proteins, bind to endogenous *Escherichia coli* proteins aggregated intracellularly by heat shock. Biochimie 78:117–122

Lavoie JN, Hickey E, Weber LA, Landry J (1993) Modulation of actin microfilament dynamics and fluid phase pinocytosis by phosphorylation of heat shock protein 27. J Biol Chem 268:24210–24214

Lavoie JN, Lambert H, Hickey E, Weber LA, Landry J (1995) Modulation of cellular thermoresistance and actin filament stability accompanies phosphorylation-induced changes in the oligomeric structure of heat shock protein 27. Mol Cell Biol 15:505–516

Lee GJ, Vierling E (2000) A small heat shock protein cooperates with heat shock protein 70 systems to reactivate a heat-denatured protein. Plant Physiol 122:189–198

Lee GJ, Pokala N, Vierling E (1995) Structure and in vitro molecular chaperone activity of cytosolic small heat shock proteins from pea. J Biol Chem 270:10432–10438

Lee GJ, Roseman AM, Saibil HR, Vierling E (1997) A small heat shock protein stably binds heat-denatured model substrates and can maintain a substrate in a folding-competent state. EMBO J 16:659–671

Lindner RA, Carver JA, Ehrnsperger M, Buchner J, Esposito G, Behlke J, Lutsch G, Kotlyarov A, Gaestel M (2000) Mouse Hsp25, a small shock protein. The role of its C-terminal extension in oligomerization and chaperone action. Eur J Biochem 267:1923–1932

Litt M, Kramer P, LaMorticella DM, Murphey W, Lovrien EW, Weleber RG (1998) Autosomal dominant congenital cataract associated with a missense mutation in the human alpha crystallin gene CRYAA. Hum Mol Genet 7:471–474

Lowe, Mayer RJ, Landon M (1993) Ubiquitin in neurodegenerative diseases. Brain Pathol 3:55–65

Martin JL, Hickey E, Weber LA, Dillmann WH, Mestril R (1999) Influence of phosphorylation and oligomerization on the protective role of the small heat shock protein 27 in rat adult cardiomyocytes. Gene Expr 7:349–355

Mehlen P, Arrigo AP (1994) The serum-induced phosphorylation of mammalian hsp27 correlates with changes in its intracellular localization and levels of oligomerization. Eur J Biochem 221:327–334

Mehlen P, Briolay J, Smith L, Diaz-latoud C, Fabre N, Pauli D, Arrigo AP (1993) Analysis of the resistance to heat and hydrogen peroxide stresses in COS cells transiently expressing wild type or deletion mutants of the Drosophila 27-kDa heat-shock protein. Eur J Biochem 215:277–284

Mehlen P, Preville X, Chareyron P, Briolay J, Klemenz R, Arrigo AP (1995) Constitutive expression of human hsp27, Drosophila hsp27, or human αB-crystallin confers resistance to TNF- and oxidative stress-induced cytotoxicity in stably transfected murine L929 fibroblasts. J Immunol 154:363–374

Mendoza JA, Rogers E, Lorimer GH, Horowitz PM (1991) Chaperonins facilitate the in vitro folding of monomeric mitochondrial rhodanese. J Biol Chem 266:13044–13049

Merck KB, de Haard-Hoekman WA, Oude-Essink BB, Bloemendal H, de Jong WW (1992) Expression and aggregation of recombinant αA -crystallin and its two domains. Biochim Biophys Acta 1130:267–276

Merck KB, Groenen PJ, Voorter CE, de Haard-Hoekman WA, Horwitz J, Bloemendal H, de Jong WW (1993) Structural and functional similarities of bovine α-crystallin and mouse small heat-shock protein. A family of chaperones. J Biol Chem 268:1046–1052

Michelini ET, Flynn GC (1999) The unique chaperone operon of Thermotoga maritima: cloning and initial characterization of a functional Hsp70 and small heat shock protein. J Bacteriol 181:4237–4244

Muchowski PJ, Clark JI (1998) ATP-enhanced molecular chaperone functions of the small heat shock protein human alphaB crystallin. Proc Natl Acad Sci USA 95:1004–1009

Muchowski PJ, Wu GJ, Liang JJ, Adman ET, Clark JI (1999) Site-directed mutations within the core "alpha-crystallin" domain of the small heat-shock protein, human alphaB-crystallin, decrease molecular chaperone functions. J Mol Biol 289:397–411

Nakamoto H, Suzuki N, Roy SK (2000) Constitutive expression of a small heat-shock protein confers cellular thermotolerance and thermal protection to the photosynthetic apparatus in cyanobacteria. FEBS Lett 483:169–174

Nicholl ID, Quinlan RA (1994) Chaperone activity of alpha-crystallins modulates intermediate filament assembly. EMBO J 13:945–953

Niwa M, Kozawa O, Matsuno H, Kato K, Uematsu T (2000) Small molecular weight heat shock-related protein, HSP20, exhibits an anti-platelet activity by inhibiting receptor-mediated calcium influx. Life Sci 66:PL7–PL12

Oesterreich S, Hilsenbeck SG, Ciocca DR, Allred DC, Clark GM, Chamness GC, Osborne CK, Fuqua SA (1993) The small heat shock protein HSP27 is not an independent prognostic marker in axillary lymph node-negative breast cancer patients. Clin Cancer Res 2:1199–1206

Palmisano DV, Groth-Vasselli B, Farnsworth PN, Reddy MC (1995) Interaction of ATP and lens alpha crystallin characterized by equilibrium binding studies and intrinsic tryptophan fluorescence spectroscopy. Biochim Biophys Acta 1246:91–97

Parsell DA, Kowal AS, Singer MA, Lindquist S (1994) Protein disaggregation mediated by heat-shock protein Hsp104. Nature 372:475–478

Pauli D, Tonka CH, Tissières A, Arrigo AP (1990) Tissue-specific expression of the heat shock protein HSP27 during Drosophila melanogaster development. J Cell Biol 111:817–820

Petko L, Lindquist S (1986) Hsp26 is not required for growth at high temperatures, nor for thermotolerance, spore development, or germination. Cell 45:885–894

Plesofsky-Vig N, Vig J, Brambl R (1992) Phylogeny of the α-crystallin-related heat-shock proteins. J Mol Evol 35:537–545

Preville X, Schultz H, Knauf U, Gaestel M, Arrigo AP (1998) Analysis of the role of Hsp25 phosphorylation reveals the importance of the oligomerization state of this small heat shock protein in its protective function against TNFalpha- and hydrogen peroxide-induced cell death. J Cell Biochem 69:436–452

Raman B, Ramakrishna T, Rao CM (1995a) Rapid refolding studies on the chaperone-like α-crystallin. Effect of α-crystallin on refolding of β- and γ-crystallins. J Biol Chem 270:19888–19892

Raman B, Ramakrishna T, Rao CM (1995b) Temperature dependent chaperone activity of α-crystallin. FEBS Lett 365:133–136

Ranson NA, Burston SG, Clarke AR (1997) Binding, encapsulation and ejection: substrate dynamics during a chaperonin-assisted folding reaction. J Mol Biol 266:656–664

Rao PV, Horwitz J, Zigler JS (1993) α-Crystallin, a molecular chaperone, forms a stable complex with carbonic anhydrase upon heat denaturation. Biochem Biophys Res Commun 190:786–793

Rao PV, Horwitz J, Zigler JS (1994) Chaperone-like activity of α-crystallin. The effect of NADPH on its interaction with zeta-crystallin. J Biol Chem 269:13266–13272

Rao PV, Huang QL, Horwitz J, Zigler JS (1995) Evidence that α-Crystallin prevents non-specific protein aggregation in the intact eye lens. Biochim Biophys Acta 1245:439–447

Rawat U, Rao M (1998) Interactions of chaperone alpha-crystallin with the molten globule state of xylose reductase. Implications for reconstitution of the active enzyme. J Biol Chem 273: 9415–9423

Reddy MC, Palmisano DV, Groth-Vasselli B, Farnsworth PN (1992) 31P NMR studies of the ATP/alpha-crystallin complex: functional implications. Biochem Biophys Res Commun 189: 1578–1584

Renkawek K, de Jong WW, Merck KB, Frenken CW, van Workum FP, Bosman GJ (1992) α B-crystallin is present in reactive glia in Creutzfeldt-Jakob disease. Acta Neuropathol Berl 83:324–327

Renkawek K, Voorter CE, Bosman GJ, van Workum FP, de Jong WW (1994) Expression of α B-crystallin in Alzheimer's disease. Acta Neuropathol Berl 87:155–160

Rogalla T, Ehrnsperger M, Preville X, Kotlyarov A, Lutsch G, Ducasse C, Paul C, Wieske M, Arrigo AP, Buchner J, Gaestel M (1999) Regulation of Hsp27 oligomerization, chaperone function, and protective activity against oxidative stress/tumor necrosis factor alpha by phosphorylation. J Biol Chem 274:18947–18956

Rollet E, Lavoie JN, Landry J, Tanguay RM (1992) Expression of Drosophila's 27 kDa heat shock protein into rodent cells confers thermal resistance. Biochem Biophys Res Commun 185: 116–120

Rüdiger S, Buchberger A, Bukau B (1997) Interaction of Hsp70 chaperones with substrates. Nat Struct Biol 4:342–349

Schirmer EC, Lindquist S (1998) Purification and properties of Hsp104 from yeast. Methods Enzymol 290:430–444

Schirmer EC, Lindquist S, Vierling E (1994) An Arabidopsis heat shock protein complements a thermotolerance defect in yeast. Plant Cell 6:1899–1909

Sharma KK, Kaur H, Kester K (1997) Functional elements in molecular chaperone alpha-crystallin: identification of binding sites in alpha B-crystallin. Biochem Biophys Res Commun 239:217–222

Shinohara H, Inaguma Y, Goto S, Inagaki T, Kato K (1993) α B-crystallin and HSP28 are enhanced in the cerebral cortex of patients with Alzheimer's disease. J Neurol Sci 119:203–209

Sigler PB, Xu Z, Rye HS, Burston SG, Fenton WA, Horwich AL (1998) Structure and function in GroEL-mediated protein folding. Annu Rev Biochem 67:581–608

Singh K, Groth-Vasselli B, Kumosinski TF, Farnsworth PN (1995) α-Crystallin quaternary structure: molecular basis for its chaperone activity. FEBS Lett 372:283–287

Smulders RHPH, Merck KB, Aendekerk J, Horwitz J, Takemoto L, Slingsby C, Bloemendal H, de Jong WW (1995) The mutation Asp69δSer affects the chaperone-like activity of αB-crystallin. Eur J Biochem 232:834–838

Srere PA (1966) Citrate-condensing enzyme-oxalacetate binary complex. Studies on its physical and chemical properties. J Biol Chem 241:2157–2165

Stege GJJ, Li GC, Li L, Kampinga HH, Konings AWT (1994) On the role of hsp72 in heat-induced intranuclease protein aggregation. Int J Hyperthermia 10:659–674

Stege GJJ, Brunsting JF, Kampinga HH, Konings AWT (1995) Thermotolerance and nuclear protein aggregation: protection against initial damage or better recovery? J Cell Physiol 164:579–586

Sugiyama Y, Suzuki A, Kishikawa M, Akutsu R, Hirose T, Waye MM, Tsui SK, Yoshida S, Ohno S (2000) Muscle develops a specific form of small heat shock protein complex composed of MKBP/HSPB2 and HSPB3 during myogenic differentiation. J Biol Chem 275:1095–1104

Sullivan W, Stensgard B, Caucutt G, Bartha B, McMahon N, Alnemri ES, Litwack G, Toft D (1997) Nucleotides and two functional states of hsp90. J Biol Chem 272:8007–8012

Susek RE, Lindquist SL (1989) hsp26 of Saccharomyces cerevisiae is related to the superfamily of small heat shock proteins but is without a demonstrable function. Mol Cell Biol 9:5265–5271

Takemoto L, Emmons T, Horwitz J (1993) The C-terminal region of α-crystallin: involvement in protection against heat induced denaturation. Biochem J 294:435–438

Tatzelt J, Zuo J, Voellmy R, Scott M, Hartl U, Prusiner SB, Welch WJ (1995) Scrapie prions selectively modify the stress response in neuroblastoma cells. Proc Natl Acad Sci USA 92:2944–2948

Thomas JG, Baneyx F (1998) Roles of the Escherichia coli small heat shock proteins IbpA and IbpB in thermal stress management: comparison with ClpA, ClpB, and HtpG in vivo. J Bacteriol 180:5165–5172

Thomas PJ, Qu BH, Pedersen PL (1995) Defective protein folding as a basis of human disease. Trends Biochem Sci 20:456–459

Todd MJ, Lorimer GH, Thirumalai D (1996) Chaperonin-facilitated protein folding: optimization of rate and yield by an iterative annealing mechanism. Proc Natl Acad Sci USA 93:4030–4035

Treweek TM, Lindner RA, Mariani M, Carver JA (2000) The small heat-shock chaperone protein, alpha-crystallin, does not recognise stable molten globule states of cytosolic proteins. Biochim Biophys Acta 1481:175–188

Van den Ijssel PR, Overkamp P, Knauf U, Gaestel M, de Jong WW (1994) αA-crystallin confers cellular thermoresistance. FEBS Lett 355:54–56

Van den Ijssel PR, Norman DG, Quinlan RA (1999) Molecular chaperones: small heat shock proteins in the limelight. Curr Biol 9:R103–R105

Veinger L, Diamant S, Buchner J, Goloubinoff P (1998) The small heat-shock protein IbpB from Escherichia coli stabilizes stress-denatured proteins for subsequent refolding by a multichaperone network. J Biol Chem 273:11032–11037

Vierling E, Sun A (1989) Developmental expression of heat shock protein in higher plants. In Cherry J (ed) Environmental stress in plants. Springer, Berlin Heidelberg New York, pp 343–354

Vicart P, Caron A, Guicheney P, Li Z, Prevost MC, Faure A, Chateau D, Chapon F, Tome F, Dupret JM, Paulin D, Fardeau M (1998) A missense mutation in the alphaB-crystallin chaperone gene causes a desmin-related myopathy. Nat Genet 20:92–95

Wagstaff MJ, Collaco-Moraes Y, Smith J, de Belleroche JS, Coffin RS, Latchman DS (1999) Protection of neuronal cells from apoptosis by Hsp27 delivered with a herpes simplex virus-based vector. J Biol Chem 274:5061–5069

Wang K, Spector A (1994) The chaperone activity of bovine α-crystallin. Interaction with other lens crystallins in native and denatured states. J Biol Chem 269:13601–13608

Wang K, Spector A (1995) α-crystallin can act as a chaperone under conditions of oxidative stress. Invest Ophthalmol Vis Sci 36:311–321

Waters ER (1995) The molecular evolution of the small heat-shock proteins in plants. Genetics 141:785–795

Waters ER, Lee JL, Vierling E (1996) Evolution, structure and function of the small heat shock proteins in plants. J Exp Bot 47:325–338

Wehmeyer N, Vierling E (2000) The expression of small heat shock proteins in seeds responds to discrete developmental signals and suggests a general protective role in desiccation tolerance. Plant Physiol 122:1099–1108

Wistow G (1993) Identification of lens crystallins: a model system for gene recruitment. Methods Enzymol 224:563–575

Yang H, Huang S, Dai H, Gong Y, Zheng C, Chang Z (1999) The *Mycobacterium tuberculosis* small heat shock protein Hsp16.3 exposes hydrophobic surfaces at mild conditions: Conformational flexibility and molecular chaperone activity. Protein Sci 8:174–179

Zhi W, Srere PA, Evans CT (1991) Conformational stability of pig citrate synthase and some active-site mutants. Biochemistry 30:9281–9286

Zhu Y, O'Neill S, Saklatvala J, Tassi L, Mendelsohn ME (1994a) Phosphorylated HSP27 associates with the activation-dependent cytoskeleton in human platelets. Blood 84:3715–3723

Zhu Y, Tassi L, Lane W, Mendelsohn ME (1994b) Specific binding of the transglutaminase, platelet factor XIII, to Hsp27. J Biol Chem 269:22379–22384

The Small Heat Shock Proteins of the Nematode *Caenorhabditis elegans*: Structure, Regulation and Biology

E. Peter M. Candido[1]

1
Introduction

The first investigations of the heat shock response in *C. elegans* were carried out by analyzing the patterns of proteins synthesized by ^{35}S-labeled nematodes cultured at various temperatures (Snutch and Baillie 1983). These experiments established a temperature profile for the heat shock response, identified the major molecular weight classes of heat shock proteins (Hsps) and noted the homology between *Drosophila* and *C. elegans* Hsp70 genes, based on cross-hybridization experiments. In these studies, a prominent, labeled protein with an apparent molecular mass of 16 kDa was noted in heat shocked animals. Partial cDNAs encoding Hsp16 species were subsequently cloned, and these were the first small Hsp (sHsp) sequences determined after those of *Drosophila*, to which they were found to be related (Russnak et al. 1983). The Hsp16 sequences of *C. elegans* thus provided the first indication that small Hsps of varying molecular masses from distantly related organisms were similar to each other and to the mammalian α-crystallins. The completion of the *C. elegans* genome sequence (The *C. elegans* Sequencing Consortium 1998) provides an opportunity to examine the entire complement of sHsps from this metazoan. This review will provide an overview of the structures, expression, developmental and tissue specificities and biochemical properties of the *C. elegans* sHsps which have been studied to date.

2
The Small Heat Shock Protein Family of *C. elegans*

A BLAST (Basic Local Alignment Search Tool; Altschul et al. 1990) search of the *C. elegans* genome using an Hsp16 sequence as query yields 16 genes (Fig. 1). However, this number includes two identical copies of the Hsp16–1 and Hsp16–48 genes (Hsp16–1A and Hsp16–1B; Hsp16–48A and Hsp16–48B; see Sect. 3.1 below). Thus the nematode genome encodes 14 distinct, clearly identifiable small heat shock proteins. These sHsp genes are located on four of the

---

[1] Department of Biochemistry and Molecular Biology, University of British Columbia, 2146 Health Sciences Mall, Vancouver V6T 1Z3, Canada

Progress in Molecular and Subcellular Biology, Vol. 28
A.-P. Arrigo and W.E.G. Müller (Eds.)
© Springer-Verlag Berlin Heidelberg 2002

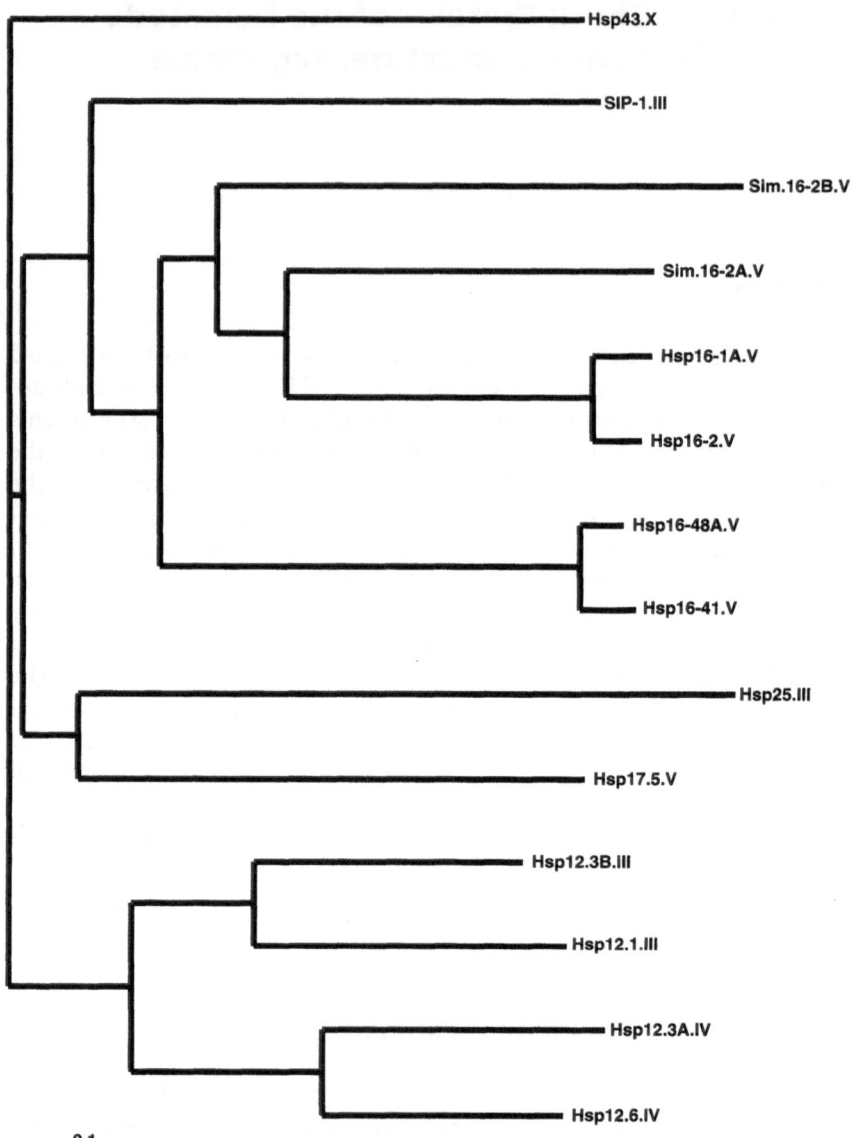

Fig. 1. Dendrogram of the small heat shock proteins of *C. elegans*. The *roman numeral* following the name of each Hsp indicates the chromosomal location of the corresponding gene. The analysis was carried out using CLUSTAL W (Thompson et al. 1994) and displayed using TreeView (Page 1996). Note that the *C. elegans* genome contains two identical copies (designated *A* and *B*) of the genes encoding Hsp16-1 and Hsp16-48, and only one of each is shown. GenBank accession numbers: Hsp43, AAA80366; SEC-1, CAA84703; Sim.16-2B, CAB01146; Sim.16-2A, CAB01147; Hsp16-1A, AAB04839; Hsp16-2, AAF60615; Hsp16-48A, AAB04840; Hsp16-41, AAF60616; Hsp25, AAA68804; Hsp17.5, AAB37035; Hsp12.3B.III, AAA27953; Hsp12.3A.IV, CAA92771; Hsp12.1, CAB03380; Hsp12.6, CAA92771. *Scale bar* 0.10 nucleotide substitutions/site

six nematode chromosomes, i.e., every linkage group except chromosomes I and II. Interestingly, 9 of the 16 sHsp genes are found on chromosome V, including all of the Hsp16 genes, as well as the Hsp17.5 gene.

3
The 16-kDa Stress Proteins (Hsp16s)

Hsp16 cDNAs were cloned on the basis of their very high differential expression following a heat shock (Russnak et al. 1983). At the normal *C. elegans* growth temperature of 20°C, Hsp16 mRNA is undetectable by northern blot analysis, while high levels of mRNA are produced following a 1–2 h heat shock at temperatures ranging from 27–35°C. Based on S1 nuclease analyses of transcription initiation sites and the locations of polyadenylation sites, the Hsp16 mRNAs range from approximately 560–620 nucleotides in length, excluding the poly-A tail (Russnak and Candido 1985; Jones et al. 1986).

3.1
Hsp16 Gene Organization

The genes encoding the four well-studied Hsp16s (Hsp16-1, Hsp16-2, Hsp16-41, Hsp16-48) are arranged in divergently transcribed pairs (Russnak and Candido 1985; Jones et al. 1986) on chromosome V (Fig. 2B). The individual genes within the pairs all exhibit the same basic organization: (1) a canonical heat shock transcription element (HSE) followed within 15 nucleotides or so by a TATA element; (2) a short exon encoding 42–46 amino acids followed by a short intron (46–58 bp); (3) a second exon encoding the conserved α-crystallin domain, and (4) a short 3′ non-coding region containing a polyadenylation signal. The amino acid sequences encoded by the first exons are of two types: those of Hsp16-1 and Hsp16-2 are 93% identical and 42 amino acids in length, while those of Hsp16-41 and Hsp16-48 are 93% identical and 46 residues in length (Fig. 2A). There is little similarity between the first exons of the two groups (Jones et al. 1986), raising the question of potential functional differences between the two classes of Hsp16s.

A remarkable feature of the Hsp16-1/48 pair is its exact duplication in an inverted orientation (Fig. 2B), the duplicated segments of 1.9 kb being separated by only 416 bp (Russnak and Candido 1985). The duplicated genes are identical in sequence, including introns, 5′ non-coding and 3′ non-coding sequences, suggestive of either an evolutionarily recent duplication event or of frequent gene conversion between the two arms of the inverted regions.

Interestingly, each Hsp16 "locus" contains one of each type of Hsp16, as defined by the exon 1-encoded amino acid sequence: Hsp16-1 is paired with Hsp16-48, and Hsp16-2 is paired with Hsp16-41. The members of the divergently arranged pairs are very closely linked. For instance, only 346 base pairs separate the ATG initiation codons of Hsp16-2 and Hsp16-41 (Jones et al.

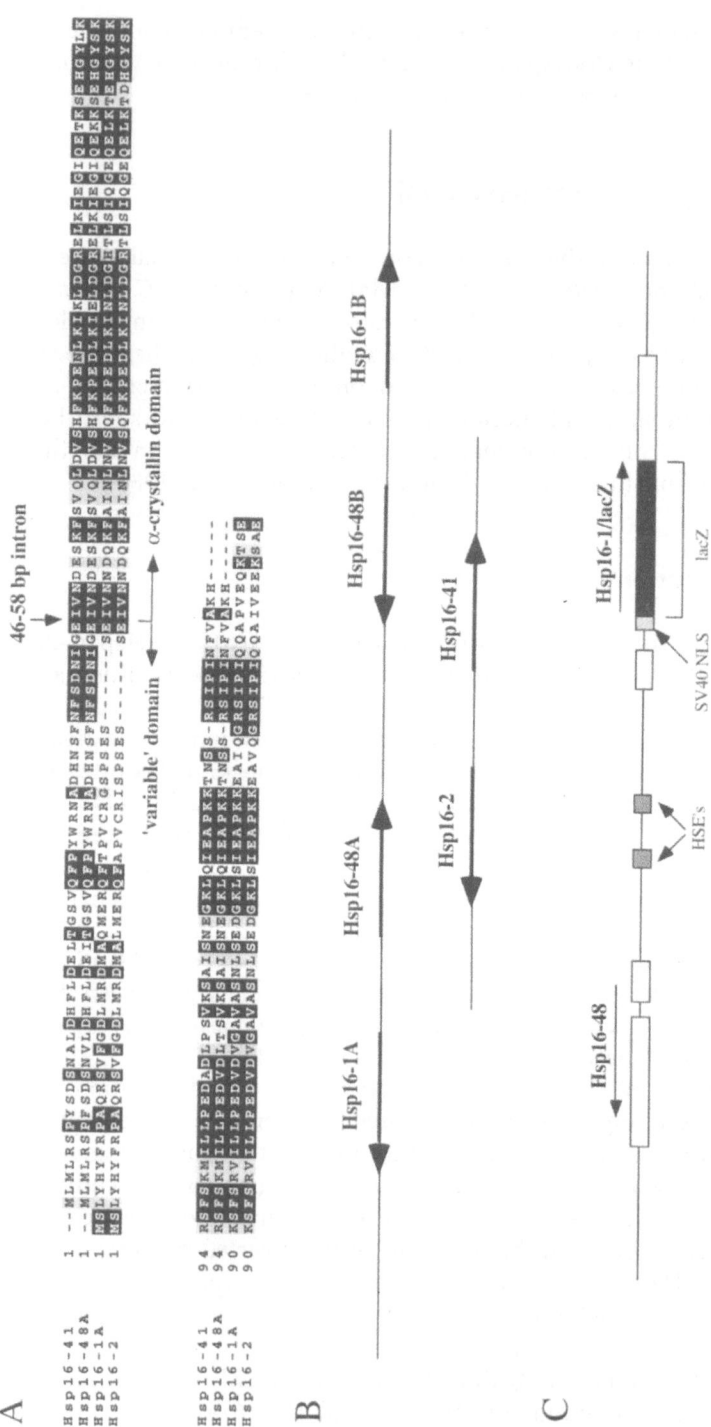

Fig. 2. **A** A comparison of amino acid sequences of stress-inducible 16 kDa sHsps. *Dashes* Gaps introduced to maximize alignment. *Light shading* Similar residues; *black shading* identical residues. The position of the intron relative to the coding sequence in each of the genes is indicated. Note that Hsp16-41 and Hsp16-48 have nearly identical N-terminal amino acid sequences and total lengths, as do Hsp16-1 and Hsp16-2. **B** Genomic arrangement of the *hsp16* genes. Each *hsp16* gene is represented by an *arrow* indicating the direction of transcription; introns are not shown. The *hsp16-1/48* locus contains an inverted duplication of the *hsp16-1/hsp16-48* gene pair, i.e., 16-1A and B are identical to each other, and 16-48A and B are identical to each other, including introns and flanking sequences. The *hsp16-2/41* locus contains a single pair of genes. Both loci are located on chromosome V. **C** Structure of the reporter gene construct in strain PC72 (see text and Fig. 3). *Open boxes* represent exons, *lines* represent non-coding sequences. HSEs, heat shock promoter elements; SV40 NLS, nuclear localization signal from SV40; lacZ, *E. coli lacZ* coding region, fused in-frame with the second exon of *hsp16-1*. The two *hsp16* genes are transcribed in opposite directions under the control of the heat shock elements in the intergenic region. Flanking sequences of the fusion gene, including the 3' untranslated region, are derived from the *hsp16-1* gene

1986), and the HSEs are separated by only about 150 base pairs. Given the proximity of the HSEs within a gene pair and the symmetry of the alternating TTC and GAA motifs in the HSEs, it was suspected that the Hsp16 HSEs might act bi-directionally to activate transcription of both genes, and this was found to be the case when various Hsp16 mutant promoter combinations were tested by transfection into cultured mouse fibroblasts (Kay et al. 1986). Thus the activity of each Hsp16 gene in a pair seems to be influenced by both sets of HSEs.

Two other sHsp genes, most similar to Hsp16-2, are found on chromosome V; these are designated Sim.16-2A and Sim.16-2B in Fig. 1. These genes lie in the same orientation, 820 bp apart. They have not been studied, and it is not known if they are stress-inducible; between the end of gene A and the start of gene B a putative heat shock element is discernible, preceding a canonical TATA box upstream of the initiation codon, but HSE and TATA elements are not readily identifiable upstream of gene A. Like the other Hsp16 genes, however, Sim.16-2A and Sim.16-2B each contain a single intron.

3.2
Hsp16 Regulation

The Hsp16 HSEs consist of three perfect nGAAn motifs of alternating polarity, with the Hsp16-1 and Hsp16-2 sequences being identical (CtcGAAtg TTCtaGAAa), while those of Hsp16-41 and Hsp16-48 are also identical (CtaGGAccTTCtaGAAcaTTCt). The last TTC or GAA triplet is separated by 15–21 bp from a downstream TATA element. Unlike many genes in *C. elegans*, the Hsp16 genes do not undergo trans-splicing, and transcription start sites were determined to lie 15–20 bp downstream of the TATA element (Russnak and Candido 1985; Jones et al. 1986). Thus the Hsp16 genes exhibit an extremely compact organization, each divergently oriented pair being encompassed within 1435–1508 bp, including their polyadenylation signals, introns and the shared intergenic region.

Studies of *hsp16* expression at the RNA and protein levels indicate that these genes are virtually inactive in unstressed animals (Russnak et al. 1983; Jones et al. 1986). Following a heat shock, the HSP16s are expressed at all developmental stages; the levels of the mRNAs are greatest in embryos and L1 (first larval stage), and decline gradually from L1 to adults (Jones et al. 1989). The Hsp16-2/41 locus produces up to seven-fold more mRNA than does the Hsp16-1/48 locus, or a ratio of up to 14:1 on a per gene basis. This difference is not attributable to differences in mRNA stability, but to the longer time period of activation of the 16-2/41 locus following a heat shock (Candido et al 1989; Jones et al. 1989). The basis for this difference in promoter efficiency between the two loci is unknown. In unshocked embryos, nuclease hypersensitive sites are evident in the chromatin upstream of the HSEs in both Hsp16 loci, and during activation of the genes these give way to an overall nuclease sensitivity over each gene pair (Dixon et al. 1990).

The tissue distribution of Hsp16s has been investigated using transgenic strains carrying in-frame fusions of *hsp16* genes with the *E. coli lacZ* coding region (Stringham et al. 1992). Consistent with earlier measurements of Hsp16 mRNA, in situ staining of these transgenic nematodes showed the presence of β-galactosidase activity only in heat-shocked animals. Constructs which included the entire intergenic region of an *hsp16* gene pair, e.g., Hsp16–1/48, were expressed from late embryos (gastrulation) onward, and in most if not all somatic tissues of larvae and adults, with expression being highest in intestine and pharynx. Recent experiments in which early embryonic blastomeres were reprogrammed by expression of an endoderm-specific GATA factor under control of an *hsp16* promoter (Zhu et al. 1998) indicate that *hsp16* promoters can be activated as early as the 12-cell stage in early *C. elegans* embryos. This biological assay therefore seems to be much more sensitive than histochemical staining for β-galactosidase.

In transgenic reporter strains, no expression was seen in the germ line. However, the lack of germ-line expression is a general characteristic of transgenes in *C. elegans*, and results from a specific repression mechanism operating in meiotic cells (Seydoux et al. 1996). Indeed, immunohistochemical analysis using anti-Hsp16 polyclonal antibodies confirms the ubiquitous expression of Hsp16s in somatic tissues, and shows that they are also made in sperm cells, following heat shock (Ding and Candido 2000c).

Transgenic *Hsp16-lacZ* strains provide a facile readout of *hsp16* expression, and these strains have been used to investigate the effects of chemical stressors (Fig. 3; Stringham and Candido 1994; Jones et al. 1996). As is the case with Hsp genes in mammalian cells, *Drosophila* and other systems, certain chemical agents which are known to cause protein damage also activate the *C. elegans* *hsp16* genes. These include heavy metal ions such as Cd^{2+}, Hg^{2+}, Pb^{2+}, Cu^{2+}, arsenite, the oxidizing agent, paraquat (Stringham and Candido 1994) and certain fungicides (Jones et al. 1996). In contrast to the general pattern of *hsp16* activation seen with heat shock, however, chemical inducers show distinct tissue specificity. Most chemical inducers activate Hsp16 synthesis in the pharynx, a muscular organ which serves to pump in and grind bacterial cells, the nematode's main food source. Intestinal cells are the next most common target of induction by chemical stressors. These patterns of induction suggest that the degree of tissue exposure to exogenous substances is the major factor determining the induction of Hsp16s by chemical agents, as the nematode cuticle is impermeable to most small molecules. Thus the pharynx and intestine are the first tissues to become exposed to dissolved substances in the medium, and show the greatest induction of the Hsp16 transgenes. Interestingly, chemical agents which induce a stress response in the nematode also tend to inhibit feeding, which in turn limits the exposure of internal tissues to chemical stressors (Jones and Candido 1999). Such a response would seem to have obvious survival value.

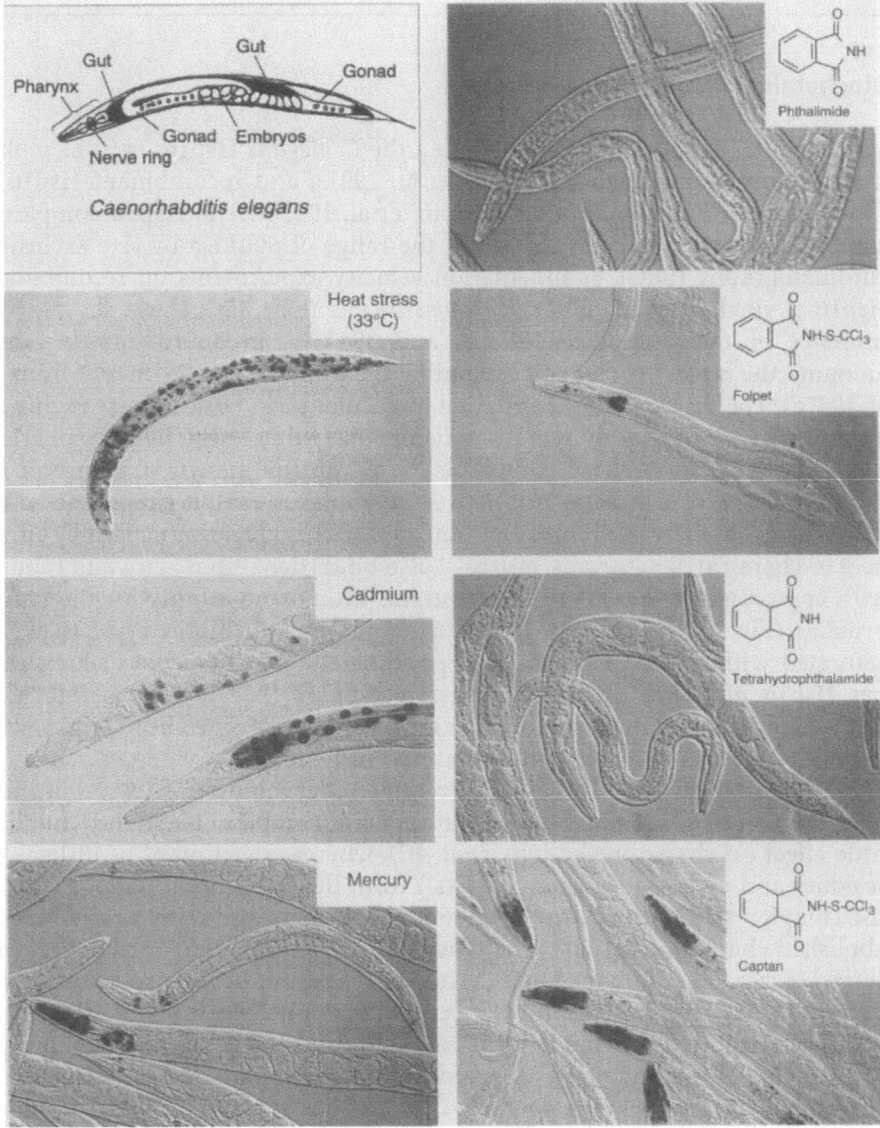

Fig. 3. Histochemical detection of β-galactosidase activity in transgenic *C. elegans*. Strain PC72 nematodes carrying integrated *hsp16–1/lacZ* transgenes (Fig. 2C) were exposed to solutions of the indicated compounds for 5 h, then stained for β-galactosidase activity. Darkly staining nuclei indicate the presence of active β-galactosidase. The enzyme is sequestered in cell nuclei due to the presence of a nuclear localization sequence in the enzyme. Heat stress (33 °C for 1 h followed by a 30-min recovery period) induces a stress response in most cells of the nematode. Cadmium (5 ppm in the presence of bacteria as food) induces a stress response in the gut cells; in the absence of bacteria the response to cadmium is largely restricted to the pharynx. Mercury (2 ppm) induces a stress response in the pharynx and in the nerve ring. The right panels show the effects of four related compounds, all tested at 50 ppm. Phthalimide and tetrahydrophthalimide do not induce a stress response while the fungicides captan and folpet are effective inducers. Stress induction by captan occurs primarily in the anterior portion of the pharynx while stress induction by folpet is primarily in the posterior pharynx. (Reproduced with permission of Elsevier Science Ltd. from Candido and Jones 1996)

3.3
Biochemical Properties of Hsp16s

Like small Hsps studied in other species, the *C. elegans* Hsp16s exist as multimeric complexes in vivo (Hockertz et al. 1991), and recombinant Hsp16–2
shows similar behavior in vitro (Leroux et al. 1997a). The Hsp16 complexes
exhibit apparent molecular masses in the range of 500 kDa by size exclusion
chromatography (SEC). Sedimentation velocity experiments on recombinant
Hsp16–2 yielded lower molecular mass values than SEC; these showed the
presence of two complexes of 239 and 394 kDa, in approximately equal
amounts, the larger species corresponding theoretically to a 24-mer (Leroux et
al. 1997a). The discrepancy in apparent molecular mass between size exclusion
chromatography and sedimentation data is likely due to the hollow spherical
nature of these quaternary complexes, based on the known structure of an
archaeal sHsp (Kim et al. 1998). Hsp16–2 complexes exhibit chaperone activity, as determined by inhibition of citrate synthase aggregation caused by either
heat or chemical denaturants. Hsp16–2 also binds denatured actin and tubulin
with approximately equal affinities, but has little or no affinity for the native
proteins (Fig. 4; Leroux et al. 1997a). The binding stoichiometry of Hsp16–2
saturated with firefly luciferase is approximately one molecule of luciferase
per Hsp16 monomer in the oligomeric complex (L. Ding, M. Leroux and
E.P.M. Candido, unpubl. results). This small Hsp therefore seems to bind a wide
variety of denatured or partially denatured proteins.

Deletion experiments on Hsp16–2 revealed that removal of 15 amino acids
from the C-terminal end increased the apparent complex size slightly, but had
little effect on chaperone activity (Fig. 4C), whereas deletion of as little as 15
residues and up to 44 residues (Fig. 4D) from the N-terminal region reduced
the complex size to a mixture of monomers, dimers and trimers and largely
abolished chaperone activity. The N-terminal deletions were outside the conserved α-crystallin domain. Since (1) loss of chaperone activity correlated with
decreased oligomeric size, and (2) experiments on N-terminally His-tagged
Hsp16–2 suggested that the N-terminal region was buried in the oligomers, it
was concluded that the non-conserved N-terminal region of the Hsp16s likely
contributes to intersubunit contacts in the complexes, while the C-terminal
was likely accessible on the surface (Leroux et al. 1997a). This conclusion is
supported by tryptic digestion experiments on native Hsp16–2 complexes,
showing that the major cleavage site corresponds to Lys123 or Lys124 (i.e.,
21–22 residues from the carboxy-terminal end; see Fig. 2A), and that there are
no cleavages in the N-terminal region (L. Ding, M. Leroux and E.P.M. Candido,
unpubl. results). The deduced orientation of Hsp16–2 subunits within
oligomeric complexes is consistent with the determined three-dimensional
structure of *Methanococcus jannaschii* Hsp16.5 (Kim et al. 1998).

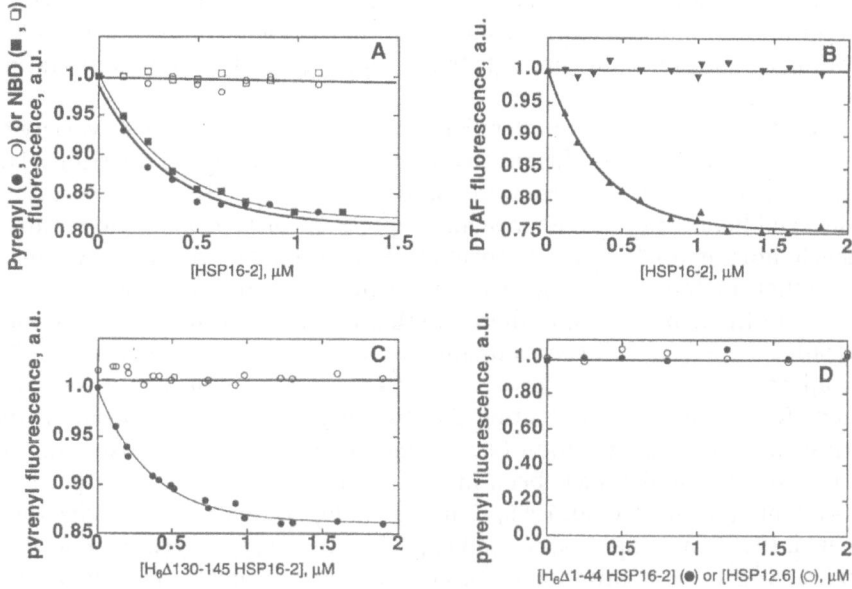

Fig. 4. Protein binding of sHsps, as measured by quenching of fluorescently labeled native or denatured actin and tubulin. **A** Increasing amounts of Hsp16-2 were added to solutions of native (O) or urea-denatured (●) pyrenyl-actin and native (□) or denatured (■) NBD-actin in 5 mM Tris–HCl, pH 7.8, 0.2 mM DTT, 0.2 mM ATP, 0.1 mM CaCl₂. **B** Increasing amounts of Hsp16-2 were added to solutions of native (▼) or urea-denatured (▲) DTAF-tubulin in 0.1 M PIPES, pH 6.9, 1 mM EGTA, 0.5 mM MgCl₂ and 1 mM GTP. **C** increasing amounts of H₆Δ130–145 Hsp16-2 were added to a solution of native (O) or urea-denatured (●) pyrenyl-actin. **D** Increasing amounts of H₆Δ1–44, Hsp16-2 (●) or Hsp12.6 (O) were added to a solution of urea-denatured pyrenyl-actin. The concentration of actin and tubulin solutions was 1 µM. The fluorescence was recorded after each addition. All measurements were made at 20 °C in a Spex fluorolog 2 spectrofluorometer. The excitation monochromator was set at 345, 468, and 492 nm and the emission recorded at 386, 535, and 517 nm for pyrenyl-actin, NBD-actin and DTAF-tubulin, respectively. Quenching of fluorescence indicates binding to the labeled protein. Note that truncation of the last 15 amino acid residues of histidine-tagged Hsp16-2 (H₆Δ130–145) has no effect on binding to denatured actin (**C**), whereas truncation of the first 44 residues (H₆Δ1–44) abolishes binding (**D**). (Reproduced with permission of The American Society for Biochemistry and Molecular Biology, Inc., from Leroux et al. 1997a)

3.4
Uses of Hsp16 Promoters in *C. elegans* Research

The strict dependence of *hsp16* gene expression on stress signals has rendered their promoters useful in a variety of *C. elegans* studies. A transgenic strain (PC72) carrying a stably integrated *hsp16-lacZ* fusion (Fig. 2C) was used to show that β-galactosidase activity can be induced in individual cells of living animals using a laser microbeam (Stringham and Candido 1993). This laser activation technique was subsequently used on transgenic strains carrying *hsp16* promoter/*mab-5* fusions to show that the Hox gene, *mab-5*, acts within migrating Q neuroblast descendants to determine their direction of migration (Harris et al. 1996).

More generally, *hsp16* promoters have been used to drive a wide variety of genes under controlled conditions to answer specific biological questions. For instance, *mec-3/hsp16* fusions were used to identify sequences necessary for the establishment, maintenance and repression of *mec-3*, a gene required for mechanoreceptor function (Way et al. 1991). An *hsp16* promoter was used to show that the human *bcl-2* gene can prevent the programmed death of specific cells in the nematode, and therefore that the mechanism of programmed cell death is conserved between humans and nematodes (Vaux et al. 1992). The same approach was used to express a putative transposase of the transposable element Tc3, thereby demonstrating that the availability of transposase is the limiting factor for transposition (van Luenen et al. 1993). The most common uses of *hsp16* promoters have been in expression of transgenes to rescue specific mutants (Herman et al. 1995; Cali et al 1999; Herman and Horvitz 1999; Sym et al. 1999), in ectopic expression experiments, i.e., to determine the effect of expressing a gene product in tissues where it is not normally present (Duggan et al. 1998; Hong et al. 1998; Zhu et al. 1998; Kim and Wadsworth 2000), or to overexpress a gene product to elucidate regulatory mechanisms (Fraser et al. 1999; Mehra et al. 1999). A particularly striking application of ectopic expression is that of Zhu et al. 1998, who reprogrammed most of the cells in early embryos into intestinal cells by heat shock-dependent expression of a specific GATA transcription factor. The variety of biological assays employed in the above studies further confirms the very tight, stress-dependent activation of *hsp16* promoters.

The inducibility of Hsp16s by stressors other than heat shock in transgenic *hsp16*-reporter strains such as PC72 has led to the proposed use of such strains as biomonitors of environmental stress (Candido and Jones 1996; Mutwakil et al. 1997). Such strains can also be powerful tools in elucidating intracellular mechanisms of the stress response; for instance, PC72 nematodes were used to show that polyglutamine repeats, such as those found in several neurodegenerative diseases, activate the heat shock response in individual cells (Satyal et al. 2000).

4
The 12-kDa Small Heat Shock Proteins (Hsp12s)

To date the smallest member of the α-crystallin family of sHsps is a 12-kDa protein first identified from genome sequence data in *C. elegans* (Caspers et al. 1995). Three other members of this group, all of similar size, were later identified as genomic sequences (Leroux et al. 1997b). Hsp12.6 and Hsp12.3A are encoded by adjacent genes on chromosome IV, while Hsp 12.1 and Hsp12.3B (accession no. AAA27953; designated Hsp12.2 in Kokke et al. 1998) are encoded on chromosome III, but are not closely linked. Hsp12.3A and Hsp12.3B are distinct sequences, with 42% identity to each other; the closely linked Hsp12.3A and Hsp12.6 are most closely related, at 67% identity (Fig. 1).

The Hsp12s possess a conserved α-crystallin domain; however, relative to other sHsps, they are truncated at both ends. The N-terminal "domain" of the Hsp12s consists of only 24–25 residues, while the C-terminal extension is essentially lacking, making them even smaller than prokaryotic sHsps such as IbpA from *E. coli*. Interestingly, despite considerable sequence variation among these Hsp12s, they are either 109 or 110 residues in length, suggesting that this may be close to a minimum size for a stable α-crystallin domain.

By western blot analysis, Hsp12.6 is expressed predominantly at the L1 or first larval stage in *C. elegans*. However, in situ immunofluorescence studies show that although it is present throughout most tissues of the animal at L1, it is distinctly localized to the vulva and spermatheca in maturing L4 animals and adult hermaphrodites (Ding and Candido 2000c; Fig. 5A–D). This pattern likely reflects the combined distributions of at least three of the four Hsp12s, due to the cross-reactivity of the antibody. The expression level of Hsp12.6 (and likely of the other Hsp12s) was not significantly altered by a variety of stressors tested (Leroux et al. 1997b).

Recombinant Hsp12s exist as monomers to tetramers in solution, and do not form large multimeric complexes (Leroux et al. 1997b; Kokke et al. 1998); the size of Hsp12.6 in nematode extracts was also consistent with that of a dimer (Leroux et al. 1997b). These proteins fail to interact with denatured actin (Fig. 4D), citrate synthase, or insulin in classic chaperone assays (Leroux et al. 1997a, 1997b; Kokke et al. 1998). The lack of both a significant N-terminal domain and chaperone activity in the Hsp12s supports suggestions from various studies that the N-terminal regions are important for multimerization of sHsps (Merck et al. 1992; Leroux et al. 1997a; Lambert et al. 1999). At the same time, these properties of the Hsp12s present an interesting puzzle regarding their possible functions, since they are clearly expressed in vivo at significant levels. The possibility that the Hsp12s might be capable of forming mixed complexes with Hsp16s and thereby modulating their activity was tested by renaturing Hsp12.6 and Hsp16-2 together from 8 M urea (M. Leroux, L. Ding and E.P.M. Candido, unpubl. results). Mixed complexes were not seen, nor was the chaperone activity of Hsp16-2 affected relative to the same protein renatured in the absence of Hsp12.6. Perhaps the Hsp12s act on specific substrate proteins, and/or require specific co-factors or post-synthetic modification for their activation. If so, the low molecular weight of Hsp12.6 seen in nematode extracts suggests such interactions are transient or involve only a small proportion of the Hsp12 pool.

Hsp12

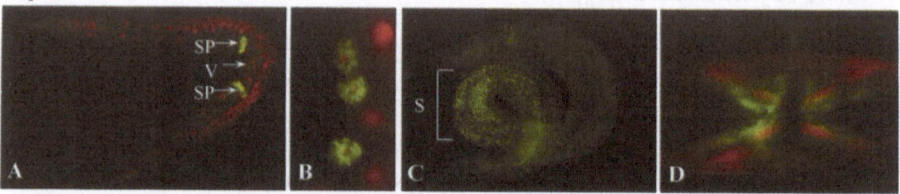

Hsp25

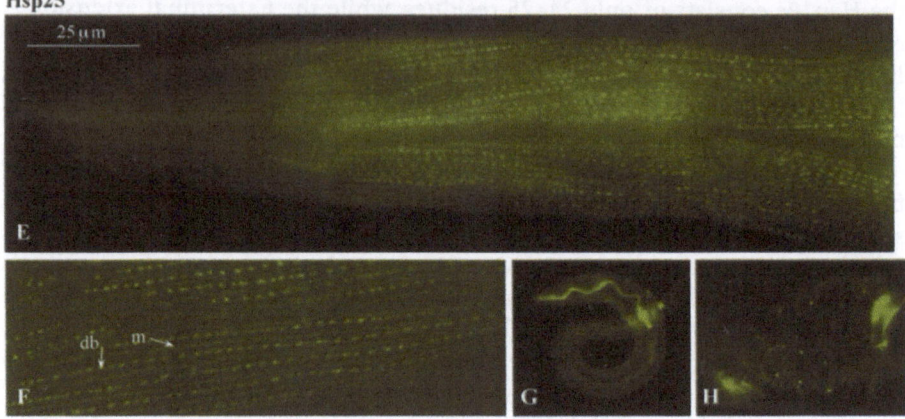

Hsp43

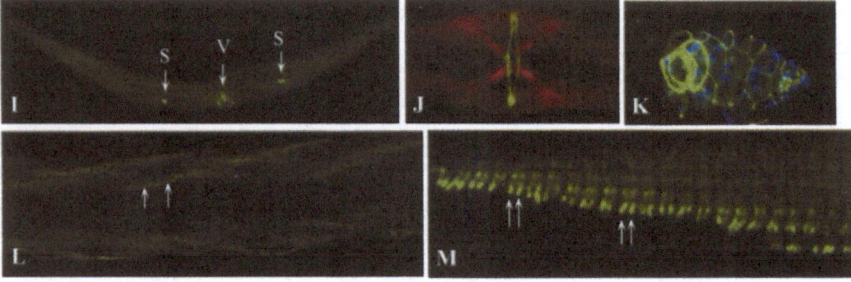

Fig. 5. Immunolocalization of Hsp12, Hsp25 and Hsp43 in *C. elegans*. Anterior is to the *left* in all images. **A** Overview of Hsp12 staining in an L4 hermaphrodite; spermathecae (*SP*) are strongly labeled and vulva (*V*) is weakly labeled. **B** Localization of Hsp12 in sperm: Hsp12 staining in *green*, nuclei (4,6-diamidino-2-phenylindole (DAPI) stain for DNA) in *red*. **C** Adult male showing Hsp12 staining in sperm of testis (*S*). **D** Staining of Hsp12 in a subset of vulva muscles is shown in *green*, and anti-myosin staining in *red*. **E** View of anterior of adult hermaphrodite stained for Hsp25, showing striated pattern in body wall muscle. **F** Higher magnification view of Hsp25 in body wall muscle, showing localization to dense bodies (*db*) and M-lines (*m*). **G** Pharyngeal staining of Hsp25 in L2 larva. **H** Staining of Hsp25 at junctions between cells of the spermathecal wall. **I** Overview of Hsp43 staining in adult hermaphrodite, showing spermathecal (*SP*) and vulval (*V*) localization. **J** Ventral view of Hsp43 in vulval cells (*green*) compared to vulval muscle (*red*, phalloidin stain). **K** Staining of Hsp43 at junctions between cells of spermathecal cage (*green*), with DAPI-stained nuclei shown in *blue*. **L** Overview of Hsp43 staining over body wall muscle of L3 larva; signal is more intense where muscle cells contact one another (*arrows*). **M** Magnified region of body wall muscle in L3 larva, showing punctate pattern of Hsp43 staining forming double columns of elongated spots (*paired arrows*) running circumferentially over muscle cells, probably at sites of contact with the overlying hypodermis. (A–D and I–M reproduced with permission of The Biochemical Society from Ding and Candido 2000b; Ding and Candido 2000c, and E–H with permission of The American Society for Biochemistry and Molecular Biology, Inc. from Ding and Candido 2000a)

5
SIP-1

SIP-1, originally named SEC-1 (Linder et al. 1996) is a 17.8-kDa sHsp with a calculated pI of 8.2, as opposed to isoelectric points of 5.0–5.9 for the Hsp16s. The internal pH of *C. elegans* is approximately 6.3 (Wadsworth and Riddle 1988), so SIP-1 will be positively charged in vivo. Encoded on chromosome III, SIP-1 is not up-regulated by cadmium exposure or heat shock; indeed, levels of its mRNA decline almost two-fold after heat shock.

SIP-1 mRNA is strictly regulated with respect to the developmental stage, being present in embryos but undetectable in L1-L4 stages; it is also seen in mature adults, which carry developing embryos. Western analysis of SIP-1 protein confirms the developmental expression pattern seen by northern blotting. Thus SIP-1 seems to be embryo-specific. Analysis of SIP-1 mRNA in situ shows it to be abundant throughout embryos in early and middle stages, and it declines precipitously as embryos undergo morphogenesis. In adults, SIP-1 mRNA is highly expressed in the gonadal arms, which contain developing oocytes, but is absent from other cells. SIP-1 appears to carry out essential functions in the early embryo, since the introduction of antisense oligonucleotides into oocytes and subsequent fertilization results in disorganized embryos which abort development (Linder et al. 1996).

SIP-1 is able to function as a chaperone, since *E. coli* cells expressing the protein survive an intense thermal stress, while control cells carrying plasmids lacking the SIP-1 gene do not (Linder et al. 1996). SIP-1 thus carries out some essential function(s) in early embryos; the requirement for SIP-1 cannot be met by other sHsps, presumably either because the latter fail to be expressed at the appropriate time and place, or because SIP-1 possesses unique properties. As a working hypothesis, Linder et al. postulated that SIP-1 may be needed to facilitate the folding of large quantities of nascent polypeptides produced during the intense biosynthetic phase of early embryonic development.

6
Hsp17.5

Hsp17.5, encoded on chromosome V, has not been studied to date; its response to stress and its tissue specificity are unknown.

7
Hsp25

Hsp25, first identified from sequencing of the *C. elegans* genome, has been studied using biochemical and immunological techniques (Ding and Candido 2000a). Hsp25 shares approximately 65% sequence identity with p27, a 27-kDa protein from the parasitic nematode *Dirofilaria immitis* (dog heartworm;

Lillibridge et al. 1996). This is the closest match between p27 and the other *C. elegans* sHsps, suggesting that these proteins are orthologs. Recombinant Hsp25 shows a polydisperse profile on size exclusion columns, indicative of the formation of predominantly trimers and tetramers, and these complexes exhibit chaperone activity. Hsp25 levels are not affected by heat shock, and it is expressed at all developmental stages. In situ immunohistochemical staining of whole animals with a polyclonal anti-Hsp25 antibody (Ding and Candido 2000a) revealed specific localization to body wall muscle and the lining of the pharynx (Fig. 5E–G), and to junctions between cells forming the spermathecal wall (Fig. 5H). The immunostaining in muscle cells was localized to a series of thin lines parallel to the long axis of the muscle fiber, alternating with thicker lines consisting of discrete spots (Fig. 5E-F). This pattern is consistent with the localization of Hsp25 to the dense bodies and M lines of the myofibrils. In nematode muscle the dense bodies are the sites of attachment of actin filaments, i.e., they correspond to the Z lines in vertebrate muscle, while the M lines are analogous to those of vertebrate muscle, and arise from the stacking of the central portions of the myosin filaments (Waterston 1988). An affinity column using immobilized, recombinant Hsp25 bound vinculin and α-actinin, components of the dense bodies, but not actin, from *C. elegans* extracts. The distribution of Hsp25 thus suggests an association with the dense bodies in nematode muscle cells, which are analogous to the focal adhesion plaques of vertebrate non-muscle cells and the dense plaques of smooth muscle (Waterston 1988). It was therefore postulated that Hsp25 might function in the maintenance, turnover or assembly of focal adhesion structures (Ding and Candido 2000a).

In the dog heartworm, *D. immitis*, p27 was also found to be constitutively expressed, and localized adjacent to the hypodermal membrane in L3 and L4 larvae (Lillibridge et al. 1996). Since dense bodies of muscle cells attach through integrins to the extracellular matrix and thence to the overlying hypodermis, p27 and Hsp25 seem to have similar localizations in *C. elegans* and *D. immitis*, at least at the current level of resolution.

Double-stranded RNA mediated gene interference (RNAi) was used to examine the possible consequences of blocking Hsp25 expression. RNAi is a powerful approach to the study of gene function in *C. elegans*, and derives from the finding that double-stranded RNA complementary to a gene sequence can often produce a phenocopy of a true genetic deletion (Fire 1999). In this technique, double-stranded RNA, prepared from the sequence of the mature gene transcript, is injected into hermaphrodites, and the progeny of the injected individual are examined for phenotypic abnormalities. The effect is both highly specific for the gene of interest and highly penetrant, but does not generally persist in the germ line. RNAi experiments using double-stranded RNA complementary to Hsp25 yielded no obvious phenotype (Ding and Candido 2000a), suggesting that either this gene is non-essential for development and viability, or that other gene products may be able to compensate for its function (see below).

8
Hsp43

Hsp43, the largest member of the sHsp family in the *C. elegans* genome, is the only one encoded on the X chromosome. Interestingly, despite its length of 368 residues, the only known domain identifiable in GenBank is the α-crystallin domain (Candido, unpubl. results). Hsp43 is constitutively expressed at all developmental stages, and its levels are unaffected by heat shock (Ding and Candido 2000b). Although its physical properties have not been extensively studied due to the poor solubility of the bacterially expressed protein, Hsp43 seems to form complexes with molecular masses of at least 670 kDa, suggesting that it forms multimers of at least 16 subunits.

Immunohistochemical staining (Ding and Candido 2000b) for Hsp43 in adult and L4 animals shows very strong, specific staining in the vulva and spermatheca (Fig. 5I, J). In the vulval region, Hsp43 is localized in the utse cells, which attach the uterus to the seam cells of the hypodermis and in the uv1 cells, which form the junction between the vulva and uterus. At the same developmental stages, Hsp43 staining is seen in desmosomes of cells which make up the spermathecal cage and the spermathecal valve (Fig. 5K), reminiscent of patterns seen with Hsp25. In all postembryonic stages, Hsp43 staining is observed in a very regular punctate pattern over body wall muscle, thought to represent labeling of hemidesmosomes associated with the hypodermis, where it contacts the muscle cells (Fig. 5L, M). Interestingly, the monoclonal antibody IFA, which defines an epitope conserved on all classes of mammalian intermediate filaments (Pruss et al. 1981), yields a very similar immunofluorescence pattern (Francis and Waterston 1991).

As in the case of Hsp25, RNAi experiments on Hsp43 yield no obvious phenotype. Given the similarities in tissue distribution of several sHsps, and in particular the expression patterns of Hsp25 and Hsp43, it is possible that these proteins can compensate for each other under such conditions.

9
Conclusions and Prospects

The availability of whole genome sequences provides a new perspective on the evolution of protein families. Multicellular eukaryotes, with their larger genomes, are likely to have expanded and diversified protein families, complicating the identification of orthologous genes between distantly related species. A comparison of the small Hsp genes in the completed genomes of *C. elegans* and *D. melanogaster* is instructive in this regard. A Clustal W analysis (Thompson et al. 1994) of the 14 distinct sHsp sequences from *C. elegans* together with seven *D. melanogaster* sHsps results in clustering of six of the *Drosophila* sequences together, while the nematode Hsp43 and fly Hsp67Ba (accession no. AAA28634.1) emerge as nearest neighbors (data not shown).

In general, BLAST searches using a given *C. elegans* sHsp as query sequence yield the highest scores with sequences (either genomic or cDNA) from other nematode species, while matches with any of the *Drosophila* sequences are always somewhat lower. For instance, *C. elegans* Hsp17.5 shows 66% identity over 149 residues with a cDNA from dog hookworm (*Ancylostoma caninum*), while the best match with *Drosophila* is 43%, over the core α-crystallin domain (41/96 residues), with EST #AW942187. In another example, a *Brugia malayi* cDNA with 60% identity to *C. elegans* Hsp12 has been isolated, suggesting that it is an ortholog of Hsp12, while the best *Drosophila* match is 35% over a small region (34/97 residues, with cDNA #AI390000).

Such comparisons suggest that the family of α-crystallin domain proteins diversified after the divergence of insects and nematodes from a common ancestor. If so, do the diverse members of this family carry out a common set of functions in metazoans, or have some sHsps evolved to carry out different functions in distantly related species? In the near future, it seems likely that a focus on the basic biochemical mechanisms of sHsp chaperone activity, as well as on the structure, cell biology and genetics of the tissue-specific, constitutively expressed members of the family will yield important new insights into the breadth of small Hsp function in biological systems.

References

Altschul SF, Gish W, Miller W, Myers EW, Lipman DJ (1990) Basic local alignment search tool. J Mol Biol 215:403–410

Cali BM, Kuchma SL, Latham J, Anderson P (1999) *smg-7* is required for mRNA surveillance in *Caenorhabditis elegans*. Genetics 151:605–616

Candido EPM, Jones D (1996) Transgenic *Caenorhabditis elegans* strains as biosensors. Trends Biotechnol 14:125–129

Candido EPM, Jones D, Dixon DK, Graham RW, Russnak RH, Kay RJ (1989) Structure, organization, and expression of the 16-kDa heat shock gene family of *Caenorhabditis elegans*. Genome 31:690–697

Caspers GJ, Leunissen JA, de Jong WW (1995) The expanding small heat-shock protein family, and structure predictions of the conserved "alpha-crystallin domain". J Mol Evol 40:238–248

Ding L, Candido EPM (2000a) HSP25, a small heat shock protein associated with dense bodies and M-lines of body wall muscle in *Caenorhabditis elegans*. J Biol Chem 275:9510–9517

Ding L, Candido EPM (2000b) HSP43, a small heat-shock protein localized to specific cells of the vulva and spermatheca in the nematode *Caenorhabditis elegans*. Biochem J 349:409–412

Ding L, Candido EPM (2000c) Association of several small heat shock proteins with reproductive tissues in the nematode *Caenorhabditis elegans*. Biochem J 351:13–17

Dixon DK, Jones D, Candido EPM (1990) The differentially expressed 16-kD heat shock genes of *Caenorhabditis elegans* exhibit differential changes in chromatin structure during heat shock. DNA Cell Biol 9:177–191

Duggan A, Ma C, Chalfie M (1998) Regulation of touch receptor differentiation by the *Caenorhabditis elegans mec-3* and *unc-86* genes. Development 125:4107–4119

Fire A (1999) RNA-triggered gene silencing. Trends Genet 15(9):358–363

Francis R, Waterston RH (1991) Muscle cell attachment in *Caenorhabditis elegans*. J Cell Biol 114:465–479

Fraser AG, James C, Evan GI, Hengartner MO (1999) *Caenorhabditis elegans* inhibitor of apoptosis protein (IAP) homologue BIR-1 plays a conserved role in cytokinesis. Curr Biol 9:292–301

Harris J, Honigberg L, Robinson N, Kenyon C (1996) Neuronal cell migration in *C. elegans*: regulation of *Hox* gene expression and cell position. Development 122:3117-3131

Herman MA, Vassilieva LL, Horvitz HR, Shaw JE, Herman RK (1995) The *C. elegans* gene *lin-44*, which controls the polarity of certain asymmetric cell divisions, encodes a Wnt protein and acts cell nonautonomously. Cell 83:101-110

Herman T, Horvitz HR (1999) Three proteins involved in *Caenorhabditis elegans* vulval invagination are similar to components of a glycosylation pathway. Proc Natl Acad Sci USA 96: 974-979

Hockertz MK, Clark-Lewis I, Candido EPM (1991) Studies of the small heat shock proteins of *Caenorhabditis elegans* using anti-peptide antibodies. FEBS Lett 280:375-378

Hong Y, Roy R, Ambros V (1998) Developmental regulation of a cyclin-dependent kinase inhibitor controls postembryonic cell cycle progression in *Caenorhabditis elegans*. Development 125: 3585-3597

Jones D, Candido EPM (1999) Feeding is inhibited by sublethal concentrations of toxicants and by heat stress in the nematode *Caenorhabditis elegans*: relationship to the cellular stress response. J Exp Zool 284:147-157

Jones D, Russnak RH, Kay RJ, Candido EPM (1986) Structure, expression, and evolution of a heat shock gene locus in *Caenorhabditis elegans* that is flanked by repetitive elements. J Biol Chem 261:12006-12015

Jones D, Dixon DK, Graham RW, Candido EPM (1989) Differential regulation of closely related members of the *hsp16* gene family in *Caenorhabditis elegans*. DNA 8:481-490

Jones D, Stringham EG, Babich SL, Candido EPM (1996) Transgenic strains of the nematode *C. elegans* in biomonitoring and toxicology: effects of captan and related compounds on the stress response. Toxicology 109:119-127

Kay RJ, Boissy RJ, Russnak RH, Candido EPM (1986) Efficient transcription of a *Caenorhabditis elegans* heat shock gene pair in mouse fibroblasts is dependent on multiple promoter elements which can function bidirectionally. Mol Cell Biol 6:3134-3143

Kim KK, Kim R, Kim SH (1998) Crystal structure of a small heat-shock protein. Nature 394: 595-599

Kim S, Wadsworth WG (2000) Positioning of longitudinal nerves in *C. elegans* by nidogen. Science 288:150-154

Kokke BP, Leroux MR, Candido EPM, Boelens WC, de Jong WW (1998) *Caenorhabditis elegans* small heat-shock proteins Hsp12.2 and Hsp12.3 form tetramers and have no chaperone-like activity. FEBS Lett 433:228-232

Lambert H, Charette SJ, Bernier AF, Guimond A, Landry J (1999) HSP27 multimerization mediated by phosphorylation-sensitive intermolecular interactions at the amino terminus. J Biol Chem 274:9378-9385

Leroux MR, Melki R, Gordon B, Batelier G, Candido EPM (1997a) Structure-function studies on small heat shock protein oligomeric assembly and interaction with unfolded polypeptides. J Biol Chem 272:24646-24656

Leroux MR, Ma BJ, Batelier G, Melki R, Candido EPM (1997b) Unique structural features of a novel class of small heat shock proteins. J Biol Chem 272:12847-12853

Lillibridge CD, Rudin W, Philipp MT (1996) *Dirofilaria immitis*: ultrastructural localization, molecular characterization, and analysis of the expression of p27, a small heat shock protein homolog of nematodes. Exp Parasitol 83:30-45

Linder B, Jin Z, Freedman JH, Rubin CS (1996) Molecular characterization of a novel, developmentally regulated small embryonic chaperone from *Caenorhabditis elegans*. J Biol Chem 271: 30158-30166

Mehra A, Gaudet J, Heck L, Kuwabara PE, Spence AM (1999) Negative regulation of male development in *Caenorhabditis elegans* by a protein-protein interaction between TRA-2A and FEM-3. Genes Dev 13:1453-1463

Merck KB, De Haard-Hoekman WA, Essink BBO, Bloemendal H, De Jong WW (1992) Expression and aggregation of recombinant alpha-A crystallin and its two domains. Biochim Biophys Acta 1130:267-276

Mutwakil MHAZ, Reader JP, Holdich DM, Smithurst PR, Candido EPM, Jones D, Stringham EG, de Pomerai DI (1997) Use of stress-inducible transgenic nematodes as biomarkers of heavy metal pollution in water samples from an English river system. Arch Environ Contam Toxicol 32:146-153

Page RD (1996) TreeView: an application to display phylogenetic trees on personal computers. Comput Appl Biosci 12:357-358

Pruss RM, Mirsky R, Raff MC, Thorpe R, Dowding AJ, Anderton BH (1981) All classes of intermediate filaments share a common antigenic determinant defined by a monoclonal antibody. Cell 27:419-428

Russnak RH, Candido EPM (1985) Locus encoding a family of small heat shock genes in *Caenorhabditis elegans*: two genes duplicated to form a 3.8-kilobase inverted repeat. Mol Cell Biol 5:1268-1278

Russnak RH, Jones D, Candido EPM (1983) Cloning and analysis of cDNA sequences coding for two 16 kilodalton heat shock proteins (hsps) in *Caenorhabditis elegans*: homology with the small hsps of *Drosophila*. Nucleic Acids Res 11:3187-3205

Satyal SH, Schmidt E, Kitagawa K, Sondheimer N, Lindquist S, Kramer JM, Morimoto RI (2000) Polyglutamine aggregates alter protein folding homeostasis in *Caenorhabditis elegans*. Proc Natl Acad Sci USA 97:5750-5755

Seydoux G, Mello CC, Pettitt J, Wood WB, Priess JR, Fire A (1996) Repression of gene expression in the embryonic germ lineage of *C. elegans*. Nature 382:713-716

Snutch TP, Baillie DL (1983) Alterations in the pattern of gene expression following heat shock in the nematode *Caenorhabditis elegans*. Can J Biochem Cell Biol 61:480-487

Stringham EG, Candido EPM (1993) Targeted single-cell induction of gene products in *Caenorhabditis elegans*: a new tool for developmental studies. J Exp Zool 266:227-233

Stringham EG, Candido EPM (1994) Transgenic *hsp16-lacZ* strains of *Caenorhabditis elegans* as biological monitors of environmental stress. Environ Toxicol Chem 13:1211-1220

Stringham EG, Dixon DK, Jones D, Candido EPM (1992) Temporal and spatial expression patterns of the small heat shock (hsp16) genes in transgenic *Caenorhabditis elegans*. Mol Biol Cell 3:221-233

Sym M, Robinson N, Kenyon C (1999) MIG-13 positions migrating cells along the anteroposterior body axis of *C. elegans*. Cell 98:25-36

The *C. elegans* Sequencing Consortium (1998) Genome sequence of the nematode *C. elegans*: a platform for investigating biology. Science 282:2012-2018

Thompson JD, Higgins DG, Gibson TJ (1994) CLUSTAL W: improving the sensitivity of progressive multiple sequence alignment through sequence weighting, position-specific gap penalties and weight matrix choice. Nucleic Acids Res 22:4673-4680

Van Luenen HG, Colloms SD, Plasterk RH (1993) Mobilization of quiet, endogenous Tc3 transposons of *Caenorhabditis elegans* by forced expression of Tc3 transposase. EMBO J 12:2513-2520

Vaux DL, Weissman IL, Kim SK (1992) Prevention of programmed cell death in *Caenorhabditis elegans* by human *bcl-2*. Science 258:1955-1957

Wadsworth WG, Riddle DL (1988) Acidic intracellular pH shift during *Caenorhabditis elegans* larval development. Proc Natl Acad Sci USA 85:8435-8438

Waterston RH (1988) Muscle. In: Wood WB (ed) The nematode *Caenorhabditis Elegans*. Cold Spring Harbor Laboratory, NY, pp 281-335

Way JC, Wang L, Run JQ, Wang A (1991) The *mec-3* gene contains cis-acting elements mediating positive and negative regulation in cells produced by asymmetric cell division in *Caenorhabditis elegans*. Genes Dev 5:2199-2211

Zhu J, Fukushige T, McGhee JD, Rothman JH (1998) Reprogramming of early embryonic blastomeres into endodermal progenitors by a *Caenorhabditis elegans* GATA factor. Genes Dev 12:3809-3814

Drosophila Small Heat Shock Proteins: Cell and Organelle-Specific Chaperones?

Sébastien Michaud, Geneviève Morrow, Julie Marchand, and Robert M. Tanguay[1]

1
Introduction

The cellular response to a heat shock treatment was originally observed in *Drosophila* by the appearance of specific puffs on polytene chromosomes (Ritossa 1962). These puffs are characterised by a high level of transcriptional activity. Concomitant with this physical manifestation is the strong induction of a restricted number of specific polypeptides thereby named Heat Shock Proteins (HSP). In *Drosophila melanogaster*, the major HSP were first identified through ^{35}S-labelling experiments on *Drosophila* tissue culture cells and salivary glands (Tissières et al. 1974) and have been commonly divided into three subfamilies based on their apparent molecular weight on SDS-PAGE. The Hsp83 and Hsp60 species are each sole members of their class, while many different genes encode for the highly conserved members of the Hsp70 subfamily. The small heat shock proteins (sHSP) group includes four polypeptides encoded by identified genes (*hsp22, hsp23, hsp26* and *hsp27*) which are all found within the same locus on chromosome 3 (67B). However, additional genes carrying open reading frames (ORF) which could potentially encode for proteins carrying the α-crystallin domain, hallmark domain of the sHSP family, have been readily identified both within (*hsp67a, hsp67b* and *hsp67c* – formerly known as *gene1, gene2* and *gene3*) and outside (*l(2)efl*) the 67B locus.

Classically defined by their heat-induced nature, many characteristics of sHsp in *Drosophila* and other model organisms now suggest that each of these proteins may provide distinct functional activities in vivo both during normal development and following exposure to stress. Since *Drosophila* sHsp have readily been the subjects of multiple previous reviews (Pauli and Tissières 1990; Arrigo and Tanguay 1991; Arrigo and Landry 1994; Michaud et al. 1997a; Joanisse et al. 1998b; Tanguay et al. 1999), we will focus here on key biological observations which are challenging the classical view of *Drosophila* sHsp and are now prompting us to model these proteins as being more than only stress-related chaperones.

[1] Laboratory of Cell and Developmental Genetics, Department of Medicine, Pavillon Marchand, Université Laval, Ste-Foy, Québec, G1K 7P4, Canada

Progress in Molecular and Subcellular Biology, Vol. 28
A.-P. Arrigo and W.E.G. Müller (Eds.)
© Springer-Verlag Berlin Heidelberg 2002

2
Members of the Small Heat Shock Proteins Family

Classical cloning approaches combined with the recent completion of the *Drosophila melanogaster* genomic sequence have revealed many putative ORF encoding for supplementary peptide members of the sHSP family. These genes can be found at four main chromosomal loci throughout the genome. The 67B cluster is by far the one containing the largest number of sHSP sequence-related genes; in addition to the genes encoding for the four main sHsp (Corces et al. 1980), it also includes three other developmentally and stress-regulated genes (*hsp67a*, *hsp67b* and *hsp67c*; Ayme and Tissières 1985) along with a novel putative ORF (CG4461) located between the *hsp22* and *hsp26* genes. Other putative ORF are each associated with a distinct chromosomal region: CG14207 (18D7), CG13133 (31A2), *l(2)efl* (59F4) (Kurzik-Dumke and Lohman 1995) and CG7409 (66A11–12).

Most of these *Drosophila* sHSP-related genes are constituted of a simple coding sequence devoid of intron. Only the coding sequence for the putative l(2)efl, Hsp67b and CG14207 peptides are interrupted by at least one intron. An alignment of identified and putative sHsp polypeptides shows that there are three conserved domains of homology within the sHsp species (Fig. 1). The main conserved stretch of amino acid is the α-crystallin domain (Ingolia and Craig 1982). Two other domains of lower homology can also be identified: an hydrophobic region in the amino terminal region (Southgate et al. 1983) and a 12-amino acids stretch located downstream of the α-crystallin domain in Hsp23, 26 and 27. Whether these subdomains fulfil specific functional aspects of sHSP biology remains to be determined.

3
Transcriptional Regulation

3.1
Organisation of Chromatin

Since sHSP genes transcription is strongly and rapidly activated following stress, the chromatin context of two *shsp* genes (*hsp26* and *hsp27*) has been studied in detail in order to unveil the functional role of surrounding DNA with regards to this physiological response. The first major observation is that these genes possess a preset state where factors of the transcriptional machinery such as TFIID are sitting on the DNA in the vicinity of the TATAA box, both before and after heat shock (Thomas and Elgin 1988). In addition, a paused RNA polymerase II (PolII) can also be readily found on these genes, a few nucleotides downstream of the transcription start site; this preinitiated state allows for a rapid induction of transcription through the elimination of the recruitment step of RNA PolII and its associated factors.

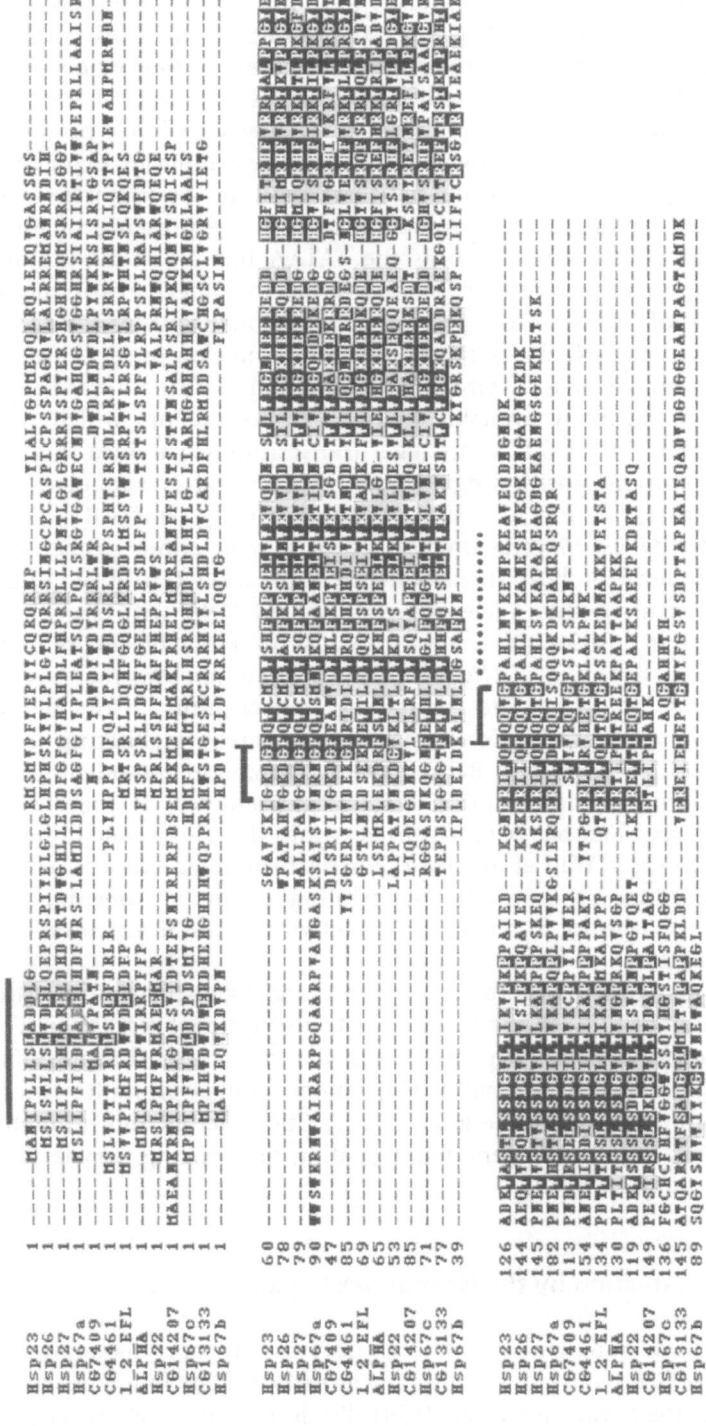

Fig. 1. Alignment of sHSP species in *Drosophila melanogaster*. A blast search with the human αB-crystallin on the whole *Drosophila* genome sequence was performed using the Berkeley *Drosophila* Blast server (http://www.fruitfly.org/blast/index.html). Twelve different open reading frames were retrieved. These were further aligned with the human αB-crystallin (designated*ALPHA*) using the *BCM Search Launcher: Multiple Sequence Alignments* program provided by the Human Genome Sequencing Center of Baylor College of Medicine (http://dot.imgen.bcm.tmc.edu:9331/multi-align/multi-align.html). Finally, *shading* of homologies for the aligned sequences was obtained using a BOXSHADE server (http://www.ch.embnet.org/software/BOX_form.html). Three homology regions are indicated above the alignment: the α-crystallin domain (*between open brackets*), the amino-terminal hydrophobic region (*plain line*) and a short region conserved between Hsp23, Hsp26 and Hsp27 (*dotted line*)

The GAGA factor, which binds specifically to (CT)n regions (Gilmour et al. 1989), was shown to play a crucial role in establishing the chromatin organisation of *shsp* promoters. It was demonstrated that these (CT)n regions were important for the formation of DNaseI hypersensitive (DH) sites (Glaser et al. 1990; Lu et al. 1992), and their deletion from an endogenous *hsp26* promoter resulted in a strong reduction (down to 23%) of heat-induced expression (Glaser et al. 1990). Both of these observations indicate the importance in defining accessible regions within promoters even in the presence of other functional regulatory elements (such as the TATAA box and DNA-binding sites for transcriptional factors). These studies led to a mechanistic model where the binding of GAGA to the specific (CT)n regions regulates the accessibility of surrounding DNA by properly positioning nucleosomes (Lu et al. 1993). This three-dimensional chromatin conformation in turn allows specific regulatory sites to be freely available to transcription factors which then ultimately act on the RNA PolII and trigger transcription. The idea that the function of nucleosome positioning is mainly structural is supported by the observation that deletion of DNA sequences usually wrapped around a nucleosome in the *hsp26* gene do not interfere with either proper DH site formation or level of heat shock-induced expression (Lu et al. 1995). However, replacement of this nucleosome-associated DNA with sequences altering nucleosome positioning results in a loss of inducible expression. Another functional aspect of DNA wrapping around nucleosomes serves to position distant regulatory sites bound by transcription factors in the vicinity of the poised transcriptional machinery, thereby facilitating transcriptional induction (Quivy and Becker 1996). Chromatin remodelling resulting from binding of transcription factors to promoters is manifested by nucleosome sliding and has been reported within the *hsp27* gene to be an ATP-dependent process (Wall et al. 1995). Finally, histone acetylation, which facilitates the recruitment of given transcription factors by opening the chromatin structure, results in an increased activity of the *hsp26* promoter (Nightingale et al. 1998).

This overall finely tuned structure of promoters was globally shown at the genomic level to be of utmost importance for transcriptional regulation. For example, insertion of functional genes within the vicinity of chromosomal environments such as heterochromatin, which blocks proper promoter potentiation by inhibiting interaction between transcription factors and promoter cis-acting elements, resulted in gene misregulation or silencing (Cryderman et al. 1999).

3.2
Stress-Induced Activation by the Heat Shock Factor

Heat shock-induced transcription of *shsp* genes is a coordinate process (Shopland and Lis 1996) and can be observed as early as 300 s following stress (O'Brien and Lis 1993; Vazquez et al. 1993). While protein levels of Hsp22,

Hsp23, Hsp26 and Hsp27 are markedly increased after heat shock, there is a clear difference with regards to the level of transcription for each gene and for the stability of the resulting mRNA (Vitek and Berger 1984). This induction has so far been shown to solely depend on the activation of a unique transcription factor, the Heat Shock Transcription Factor (HSF). Although HSF is constantly expressed, its activation is marked by a transition from a monomeric to trimeric form (Westwood et al. 1991). Once activated, HSF binds with high affinity to specific regulatory sequences named Heat Shock Elements (HSE) which are located in the promoter regions of *shsp* genes (Hsp22: Klemenz and Gehring 1986; Hsp23: Mestril et al. 1985; Hsp26: Cohen and Meselson 1985; Hsp27: Riddihough and Pelham 1986) to activate transcription. While multiple HSE may be present in the promoter of *shsp* genes, only a subset of these may be functionally relevant for heat-induced transcription. For instance, only three of the seven HSE identified on the *hsp26* promoter are necessary and sufficient for heat induction (Thomas and Elgin 1988; Glaser et al. 1990).

Heat-induced transcription of the *hsp67a* and *hsp67b* genes displays unusual behaviour as it is restricted to a specific developmental stage in *Drosophila melanogaster* or requires prior ecdysone treatment in cultured cells (Pauli et al. 1988; Vazquez 1991). Whether this conditional activation is dependent on ecdysone-induced protein synthesis, or is a mere reflection of chromatin rearrangement modulated by the hormone treatment, at these loci remains to be tested.

A recent report on the polymorphism of specific repeated sequences (CATA) found in the *hsp23* promoter has shown that genetic variation of these repeats is independent of selection for heat tolerance (Frydenberg et al. 1999), therefore suggesting that these sequences have no impact on Hsp23 expression or that Hsp23 is not involved in heat tolerance mechanisms *in vivo*.

3.3
Developmental Transcription of *shsp* Genes

3.3.1
Regulation Cascade Induced by Ecdysone

Consistent with their developmental expression profile, an implication of ecdysteroids in *shsp* transcriptional induction has been documented. This hormone was known to induce sHsp expression in both S3 cells (Ireland and Berger 1982) and salivary glands (Ireland et al. 1982) and to directly alter chromatin structure around *shsp* genes as revealed by modulation of DH sites (Kelly and Cartwright 1989). Interestingly, the same β-ecdysone hormone was reported to act as a teratogen in primary embryonic cells, but resulted in the activation of Hsp22 and Hsp23 only (Buzin and Bournias-Vardiabasis 1984). Such a differential response between primary embryonic cells and more fully differentiated cells of the salivary glands or established cell line cultures (S3 or

Kc) may reflect a requirement for intracellular factor(s) which would act in potentiating cells to properly respond to ß-ecdysone stimulation.

The activated ecdysone receptor/Ultraspiracle (EcR/USP) dimer binds to the ecdysone response element (EcRE) found in the promoter region of multiple genes, including *shsp*. Many of these EcRE may be necessary for full developmental expression of *shsp* genes; for example, three EcRE in *hsp23* promoter were shown to be involved in ecdysone response of this gene (Mestril et al. 1986; Dubrovsky et al. 1996). The DNA-binding capacity of EcR to EcRE is highly activated by hormone treatment (Luo et al. 1991). Other active EcRE have been mapped to the promoter of *hsp22* (Klemenz and Gehring 1986) and *hsp27* (Hoffman and Corces 1986; Hoffman et al. 1987; Riddihough and Pelham 1987). It was also shown that EcRE could act, in the absence of hormone, as repressors of promoter activity thereby suggesting that inactive EcR could still bind to these regions and inhibit transcription from a basal promoter (Dobens et al. 1991). This repressor effect is, however, dependent on the proximity of these EcRE to the TATAA box, implying that the silencing action could be mediated through physical steric inhibition of proper transcription factor/RNA PolII interaction caused by the presence of EcR sitting on the DNA. Recent data obtained *in vivo* through clonal analysis of cells mutant for USP have confirmed the importance of the repressor effect of non-activated EcR/USP dimers in the regulation of secondary ecdysone-responsive genes (Schubiger and Truman 2000).

Genes such as *hsp27* and *hsp23*, whose products accumulate in large amounts after exposure to ecdysterone (Berger 1984), have been shown to be differentially regulated by this hormone; while *hsp27* was originally classified as a primary response gene (i.e. directly induced by the ecdysone receptor activated by the ligand), *hsp23* was defined as a secondary response gene, dependent on primary target gene expression and consequently on *de novo* protein expression (Amin et al. 1991). An identified mediator of both *hsp23* and *hsp27* response to ecdysone is the Broad-Complex (BR-C) as the deletion of this locus, which encodes for multiple genetic functions, results in a 95–99% inhibition of production for both sHsp in salivary glands (Dubrovsky et al. 1994). This strong downregulation also correlates with the loss of a major DH site in the promoter of *hsp23* (Dubrovsky et al. 1996). Which of the BR-C isoforms accounts for each domain of expression of sHSP is still unknown. However, it is clear that variable levels of the different isoforms of BR-C affect different tissues. Furthermore, the complex and specific patterns of expression of Hsp23 cannot simply be linked to this sole transcription factor as the expression level of this sHsp within the larval brain is not abrogated in BR-C mutants, implying that other factors must impact on *hsp23* promoter to regulate its developmental expression (Dubrovsky et al. 1996).

3.3.2
Tissue- and Cell-Specific Enhancers

Transcriptional induction of sHSP within given adult organs has also been shown to be highly regulated by multiple regulatory elements. However, none of the precise factors which bind to identified crucial enhancers for sHSP expression during either spermatogenesis (Hsp26: Glaser and Lis 1990) or oogenesis (Hsp26: Cohen and Meselson 1985; Frank et al. 1992; Hsp27: Hoffman et al. 1987) have so far been identified.

4
Intracellular Localisation - Analysis of Targeting Signals

Despite their extensive level of conservation at the amino acid level and similar biochemical behaviour such as the formation of oligomers under native state, each sHsp clearly exhibits distinct intracellular localisation both in *Drosophila* cells (see below) and when transfected into mammalian cultured cells (Tanguay et al. 1999). Recently, Hsp22 has been shown to localise within the mitochondrial matrix. Amino acids 1–28 of Hsp22 contain a functional mitochondrial targeting signal sufficient, when fused to the GFP reporter protein, to carry it into mitochondria of mammalian cells (Morrow et al. 2000). This stretch of residues shares many features of the usual matrix targeting sequence found in other proteins (Reviewed in Neupert 1997); it contains five positively charged residues, two unessential negatively charged residues and many hydroxylated ones. Five amino acids in this targeting sequence are essential for importation of Hsp22 in mitochondria (Arg2, Trp8, Arg9, Met10 and Ala11), four of which are enclosed in a theoretical targeting signal predicted by PSORT (Nakai and Kanehisa 1992).

Both Hsp23 (Duband et al. 1986) and Hsp26 (Tanguay et al. 1999) are found in the cytoplasm of cultured cell lines and salivary glands (Arrigo and Ahmed-Zadeh 1981) under normal conditions. However, their repartition within the cytoplasm suggests that they may associate with different structures of this intracellular compartment. Following stress, Hsp23 has been shown to accumulate in both the cytoplasm and the nucleolus of Kc cells and in the nuclei of salivary glands (Duband et al. 1986).

Hsp27 is a nuclear protein both after stress and following ecdysone stimulation in cultured cell lines. Furthermore, stress conditions also result in a dramatic decrease in its solubility (Beaulieu et al. 1989). Many observations suggest that the nature and thereby intracellular locale of this protein can be different when expressed in a non-stress context, as Hsp27 is also detected in a perinuclear distribution during specific developmental instances, such as in ovarian follicle cells (Marin and Tanguay 1996). These observations may be clues indicating that modulation of the intracellular localisation and biochemical properties could serve to ultimately dictate its biological activity. To

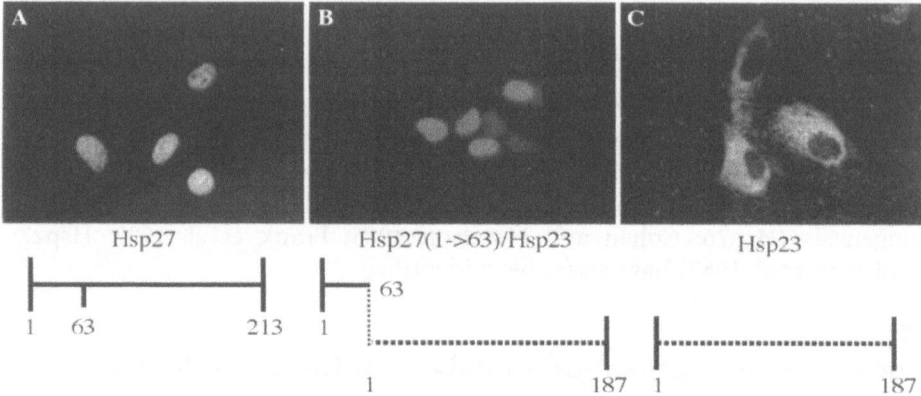

Fig. 2. The first 63 amino acids of Hsp27 contain a functional nuclear localisation signal. HeLa cells were transfected either with **A** pRc/CMV-Hsp27, **B** a pRc/CMV vector encoding a chimera protein made of the first 63 amino acids of Hsp27 and the whole Hsp23 polypeptide or with **C** pRc/CMV-Hsp23. Following expression for 24 h, immunofluorescence was performed with monoclonal antibodies directed against Hsp27 (**A**) or Hsp23 (**B** and **C**)

address the functional relevance of Hsp27 localisation, we first set out to experimentally define its nuclear localisation signal (NLS). Localisation of a chimeric protein made up of the first 63 amino acids of Hsp27 fused at the amino terminal region of the whole Hsp23 polypeptide demonstrates that all the information for proper nuclear targeting is contained within that stretch (Fig. 2). Ongoing studies in our laboratory are aimed at more precisely defining the NLS of Hsp27 (Marchand et al., unpubl.).

5
Biochemical Properties and Post-Translational Modifications

Despite the fact that the biological functions of sHsp during normal development or after stress remain unknown, biochemical observations have provided clear evidence that they were capable of forming oligomers *in vivo* (Arrigo et al. 1985; de Sa et al. 1989; Morrow et al. 2000). Whether the cytosolic Hsp23 and Hsp26 form mixed oligomers with each other or distinct homo-oligomers is unclear at this time. Density sucrose gradients analysis shows that these sHsp sediment as large oligomers in *Drosophila* cells or in mammalian cells after transfection and that Hps23 and Hsp26 peaks are separate from each other in such gradients (Tanguay, unpubl.). The sHsp have also been reported to be associated with both hnRNA, in heat shocked *Drosophila* tissue culture cells (Kloetzel and Bautz 1983), and with different cytoplasmic particles. The 19S particle isolated from post-polysomal supernatants of *Drosophila* embryo and cultured cells contains Hsp23 (Schuldt and Kloetzel 1985), while a 16S particle containing at least Hsp27 was shown to rearrange into large aggregates upon heat shock (Haass and Kloetzel 1990). Although oligomerisation of sHsp in mammals has been readily shown to be important for function (see Arrigo and Paul, this Vol.), further biochemical studies within tightly controlled experimental systems will be required to fully assess the functional significance of this complex and dynamic associative behaviour of *Drosophila* sHsp.

The sole post-translational modification reported so far for Hsp26 and Hsp27 is phosphorylation. This modification has been shown to occur in response to heat shock or ecdysterone treatment of *Drosophila* Kc cells (Rollet and Best-Belpomme 1986). It is worth pointing out that two isoforms of Hsp23 and four of Hsp27 have been identified through isoelectrofocusing (Marin et al. 1996b). However, one of the Hsp23 forms and one of the four isoforms of Hsp27 seem to be insensitive to phosphatase treatment, implying that other post-translational modifications may occur on sHsp. Interestingly, both of these sHsp display a stage-specific profile of phosphorylation state during normal *Drosophila* development (Marin et al. 1996b); while heads and testes contain all isoforms for both Hsp23 and Hsp27, only a subset of two isoforms of Hsp27 and a single Hsp23 specie could be detected throughout ovarian and embryonic development. Whether this differential regulation at the post-translational level reflects a functional implication in regulating sHsp functions remains to be assessed.

6
Stress-Induced Expression of sHsp

Heat treatment of *Drosophila* cultured cells or embryos not only results in massive accumulation of HSP, but it also modifies basic functions within the cells. A defined characteristic of the heat shock response is the rapid production of novel mRNA which is translated to HSP coupled with the inhibition of normal protein production, both at the transcriptional and translational level (Hultmark et al. 1986).

Although expression of heat shock proteins has been classically observed in response to a heat stress, multiple stresses of a different nature (chemical, physical or molecular) inducing a similar response have been identified. For instance, intracellular oxidative stress induced by H_2O_2 treatment (Courgeon et al. 1988) or a transition from anaerobiosis to normoxia (Ropp et al. 1983), has been shown to induce the heat shock response, although with differences in intensity and in the species of HSP induced. A differential non-coordinate response of the *shsp* genes generated by specific inducers has also been reported. Some metal ions such as nickel and zinc as well as other teratogens have been shown to selectively induce Hsp22 and Hsp23 but not Hsp26 and Hsp27 (Buzin and Bournias-Vardiabasis 1984; Bournias-Vardiabasis et al. 1990) while exposure to other compounds such as arsenite triggers the expression of high molecular weight HSP without a detectable production of sHsp (Vincent and Tanguay 1982).

6.1
Cell-Specific Response of sHsp

Multiple observations *in vivo* suggest that the general heat shock response may be totally or partially abrogated in defined situations. Such inhibition was first observed in young pre-gastrula embryo (less than 3 h), where no sHsp mRNA were induced by heat shock treatment (Dura 1981; Zimmerman et al. 1983).

Other observations within specific tissues of *Drosophila melanogaster* also indicate that additional factors beside the HSF may serve to modulate the heat shock response. For example, a cell-specific response following heat treatment was uncovered in the ommatidial units of *Drosophila* eyes. This peculiar phenomenon is depicted by a restricted heat-induced expression of Hsp23 in the cone cell lineage while other cell lineages (photoreceptors and pigment cells), although competent to respond to stress through the induction of Hsp26 and Hsp27, remain unable to engender Hsp23 expression (Marin et al. 1996a). These observations therefore suggest that the peculiar expression of Hsp23 in ommatidia after heat shock is attributable to a cell-specific inhibition mechanism. A similar situation was observed in testes where distinct cell lineages (cysts, primary spermatocytes) expressing Hsp23 or Hsp27 do not respond to heat induction as indicated by a stable level of these proteins while Hsp22 is highly induced (Michaud et al. 1997b). Theoretically, such silencing phenomenon may be linked to non-permissive chromatin structure produced by the binding of cell-specific transcription factors to the endogenous promoters resulting in inhibition of HSF binding to the HSE. It will be interesting to verify if this cell-specific inhibition serves a functional role *in vivo* or if it merely represents a consequence of the transcriptional regulation program acquired through cell differentiation.

6.2
Functions of sHsp Under Stress Conditions

Even if the functions and mechanisms of action are well understood for many members of the larger Hsp60, 70 and 90 families, the *in vivo* functions of sHSP remain unclear. It was demonstrated both in *Drosophila* tissue culture cells and pupae that sHsp expression was correlated with the acquisition of thermotolerance, as measured by increased survival following stress (Berger and Woodward 1983). Similarly, expression of *Drosophila* Hsp27 in heterologous mammalian cell lines has also been shown to confer thermotolerance (Rollet et al. 1992) and protection against TNF-α, H_2O_2, menadione and staurosporine-induced apoptosis (Mehlen et al. 1995; Mehlen et al. 1996b). The protection against TNF-α was further shown to be accompanied by an increase in intracellular glutathione (Mehlen et al. 1996a). Expression of *D. melanogaster* Hsp27 in this system therefore resulted in protective properties which were similar to those induced by overexpression of the endogenous mammalian Hsp27. However as these two proteins respectively reside in different intracellular compartments (cytoplasmic for mammalian Hsp27 versus nuclear for *Drosophila* Hsp27), they may exert their protective effects through different pathways. Therefore, it would be interesting to experimentally test if the protection conferred by co-expression of these two proteins is additive or not, thereby implying different or similar pathways of action.

In vitro mammalian Hsp27 has been shown to bind unfolded proteins and prevent their aggregation under stress conditions (reviewed in Fink 1999). We performed in vitro chaperone assays to verify if *Drosophila* sHsp could also carry out this function. From citrate synthase aggregation experiments, it stands out that the sHsp tested (Hsp22 and 23) can prevent heat-induced aggregation (Fig. 3), although with different efficiencies (Morrow et al. Unpubl.). These observations support a putative general chaperoning function for these proteins under stress conditions. Whether *Drosophila* sHsp can create a reservoir of unfolded protein in vivo and participate in protein refolding still remain to be tested. However as discussed below, examination of the pattern of expression of the different sHsp during development suggests that they play distinct functions in vivo or act as cell- or molecule-specific chaperones.

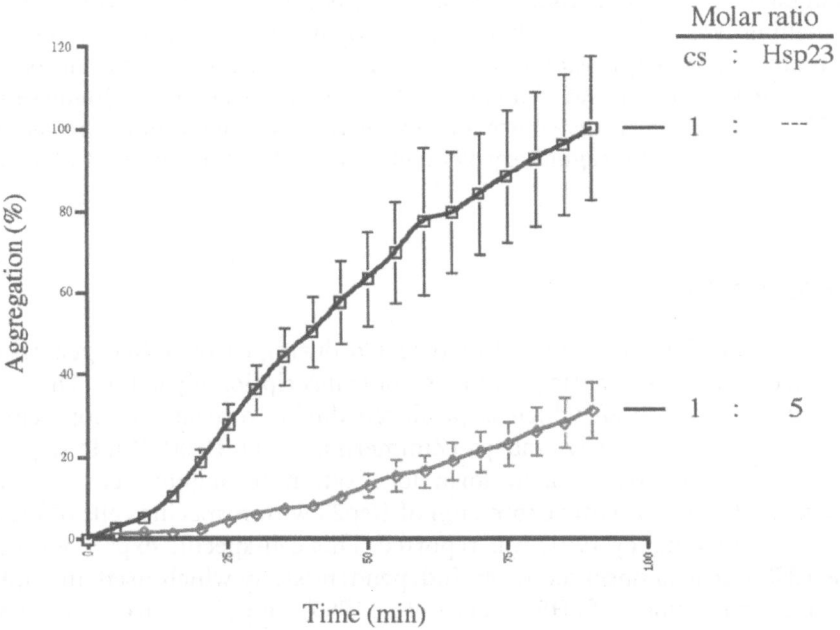

Fig. 3. Prevention of citrate synthase heat-induced aggregation by Hsp23. Citrate synthase (*cs* 0.1 µM) was incubated at 42 °C alone (*open square*) or in the presence of Hsp23 (0.5 µM; *open lozenge*). Aggregation was measured as an increase in absorbency at 320 nm

7
Developmental Expression of sHsp

In contrast to their coordinate induction following heat stress exposure, developmental expression of sHsp is uncoordinated both at the mRNA and protein level. The sHsp also display tissue- and stage-specific expression. Multiple studies have now unveiled valuable information regarding both the domain of expression throughout the fly life cycle and the transcriptional regulation mechanisms dictating this differential expression. Unfortunately, the biological significance of this highly-regulated expression still remains a mystery as in vivo data concerning the role of sHsp during fly development is still unavailable.

7.1
Stage, Tissue, and Cell Specificity

Most of the early developmental studies reported the expression of sHsp during the larval third instar stage (Cheney and Shearn 1983; Arrigo and Pauli 1988; Pauli and Tissières 1990; Vazquez 1991; Dubrovsky et al. 1994). This finding is consistent with the fact that transcription of the *shsp* genes is induced by the activated EcR/USP dimer in response to ecdysone at this stage, when abundant USP protein (Henrich et al. 1994) is concerted with strong ecdysone signalling. We will present here a non-exhaustive description of the major domain of sHsp expression during specific stages of the fly life cycle as previous reviews (Arrigo and Tanguay 1991; Michaud et al. 1997a; Joanisse et al. 1998b; Tanguay et al. 1999) provided extensive coverage of both the cell types expressing sHsp and regulatory elements involved in this differential developmental expression.

7.1.1
Embryogenesis

Many of the sHsp proteins can be detected during early embryogenesis even at stages when the zygotic genome is not transcriptionally active. These sHsp originate from mRNA which is produced during ovarian development and accumulated in the early embryo (Zimmerman et al. 1983). While Hsp27 has been shown to possess a uniform localisation throughout embryogenesis (Pauli et al. 1990), zygotic expression of Hsp23 within specific cells of the CNS (Arrigo and Tanguay 1991) was reported. This cell-specific expression within the CNS was also observed by an independent study which used an antibody recognising a subset of sHSP (Haass et al. 1990). Ongoing work in our laboratory using an antibody specific to Hsp23 has defined precise domains of expression for this protein throughout embryogenesis. For example, a strong pulse of Hsp23 expression can be observed in amnioserosa cells during dorsal closure and in a distinct segmentally-repeated group of lateral cells through-out late embryogenesis. These cells are tentatively identified as the oenocytes with regards to their ectodermal position which is juxtaposed to the neuronal cluster of the lateral chordotonal organs of the peripheral nervous system (Fig. 4). An ectodermal metameric expression of Hsp22 mRNA in stage 11 embryos has also been recently reported (Leemans et al. 2000). Since no corresponding

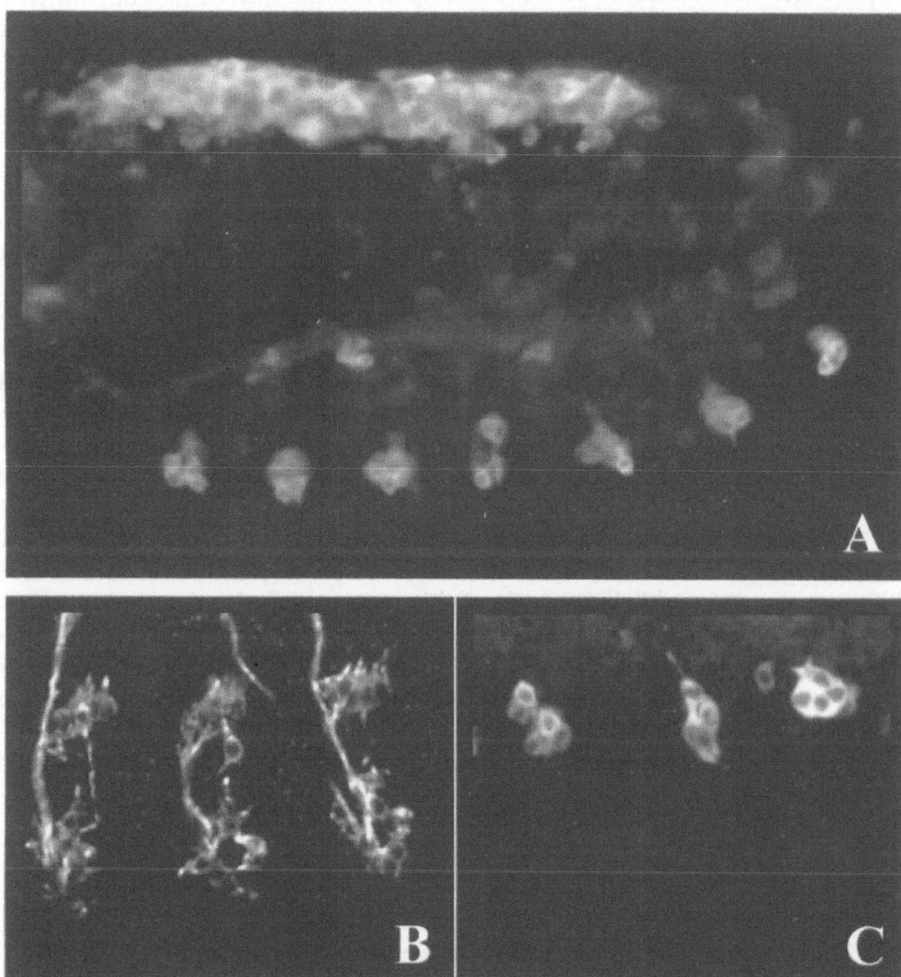

Fig. 4A–C. Hsp23 expression in stage 14 embryos. Whole mount immunohistochemistry on stage 14 embryos, using an antibody specific to Hsp23, reveals restricted expression to cells of the amnioserosa and lateral ectodermal clusters of cells restricted to abdominal segments (A). Those ectodermal cells are tentatively identified as the oenocytes (close-up in C) based on their juxta-posed localisation to neuronal clusters (visualised by the monoclonal 22C10 antibody in B) of lateral chordotonal organs of the peripheral nervous system

protein is detected during embryogenesis (Morrow, unpubl.), this either suggests that Hsp22 mRNA is translated at an undetectable level or that its translation is stress-dependent.

Profiles of transcription of *shsp* genes for which protein species remain to be identified also display distinct developmental regulation; *l(2)efl* (Kurzik-Dumke and Lohman 1995) and *hsp67c* mRNA (Pauli and Tonka 1987) are both found during embryogenesis until the beginning of pupation. Another gene without any detected protein product, *hsp67b*, displays peculiar behaviour as it produces two different poly-A transcripts: a small mRNA (560 bp) which is present during embryogenesis and a large one (780 bp) which is restricted to the male germ line (Pauli et al. 1988).

7.1.2
Germ Line

With the exception of Hsp22, all small heat shock polypeptides studied so far have been shown to be expressed both within the male and female germ lines.

Hsp23 expression in the male germ line was originally found to be restricted to somatic cells (cyst and epithelial), although an unclear association of this protein with elongated tails of spermatids was reported (Michaud et al. 1997b). Recent observations in our laboratory have confirmed that the *hsp23* promoter is indeed active in mature spermatids (Michaud and Tanguay, unpubl.). In the same organ, Hsp27 is expressed in both the cyst cells and primary spermatocytes (Michaud et al. 1997b), while Hsp26 was observed in the cytoplasm of spermatocytes (Marin et al. 1993). Both Hsp26 (Marin et al. 1993) and Hsp27 were found to be highly expressed in ovaries. Hsp27 was mainly found in the nuclei of nurse cells, where it dramatically shifted to a perinuclear localisation between stages 6 and 8. Expression of this protein was also detected in posterior follicle cells but only during stages 8–10 (Marin and Tanguay 1996). It is noteworthy to point out that most of the cells expressing Hsp27 (nurse cells and primary spermatocytes) have common intrinsic biological properties such as a high transcriptional rate and mRNA storage and export; however, no functional or biochemical evidence so far indicates that Hsp27 is in fact involved in these processes.

7.1.3
Ageing

Hsp22 is the prevalent transcriptionally-activated gene during normal ageing in flies as the corresponding mRNA level increases by factors of 2.5-fold in the thorax and up to 60-fold in the head (Wheeler et al. 1995; King and Tower 1999). An accumulation of Hsp23 mRNA is also observed in the thorax, albeit at a moderate extent (four- to eight-fold). No significant upregulation in either Hsp26 (Zou et al. 2000) or Hsp27 levels of mRNA has been detected during ageing (King and Tower 1999). Increasing Hsp22 and Hsp23 mRNA levels were also shown to correlate with increased life span and stress resistance in genetically selected *Drosophila melanogaster* lines (Kurapati et al. 2000).

7.2
Functions of shSp During Normal Development

Since each of the sHsp has a distinct expression pattern during fly development, it seems unlikely that these proteins would simply act as general chaperones in their in vivo context. Chaperone activity of sHsp on defined target proteins could however serve to modulate specific intracellular functions. A peculiar detail hindering the elucidation of sHSP function is the lack of null mutations for each of the *shsp* genes in *Drosophila*. While an isolated deletion polymorphism within the Hsp26 ORF was without any observable detrimental effect on homozygous flies (Sirotkin et al. 1986), a classical mutagenesis screen of the 67A-D region failed to identify lethal or visible mutations within any *shsp* genes (Leicht and Bonner 1988). These observations suggest that either the sHsp have redundant functions during development or that the gene cluster within the 67B region is somehow shielded from exogenous mutagenic compounds by unknown factors or properties. Nevertheless, a battery of P-elements insertions (Rorth 1996; Eissenberg and Elgin 1987; Fig. 5) have nowadays been isolated within the 67B locus. As the majority of these insertions are located in promoter regions, most do not result in null alleles at the protein level, but rather lead to misregulation of the targeted gene.

A specific insertion of a defective P-element in the regulatory region of *hsp27* (Hsp28[stl]; Eissenberg and Elgin 1987) results in abrogated developmental and heat-induced expression of the normal Hsp27 mRNA, although allowing a reduced level of normal Hsp27 mRNA to be produced in females. An additional strictly heat-inducible modified mRNA specie which contains at least part of the P-element as well as the Hsp27 ORF was also detected in this

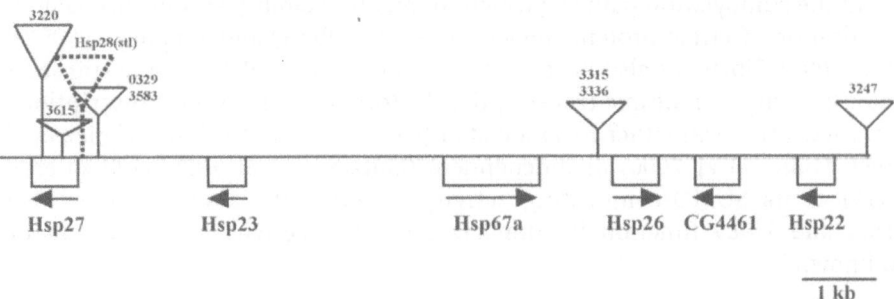

Fig. 5. Representation of the 67B locus. Schematic drawing of a specific region of the 67B locus displaying multiple sHsp open reading frames (*boxes*) along with their coding direction (*arrows*). Isolated P-element insertions (*inverted open triangles*) are identified by their strain number. The sole insertion discussed in the text (Hsp28[stl]) is represented by a *dotted line*

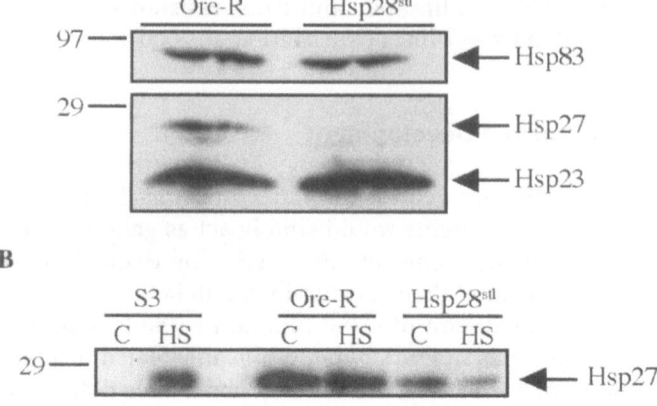

Fig. 6. Abrogation of Hsp27 heat-induced expression in Hsp28^{stl} flies. **A** Whole fly extracts from heat-shocked Ore-R and Hsp28^{stl} strains were submitted to SDS-PAGE and transferred to nitro-cellulose membrane for western analysis of different HSP species (Hsp23, Hsp27 and Hsp83) using specific monoclonal antibodies. The signal for Hsp83 is used as a loading control. **B** Protein extracts from normal and heat-shocked S3 cells and ovaries from Ore-R and Hsp28^{stl} strains were analysed as described in **A**

specific fly strain. This modified mRNA however displays a drastically reduced level of induction. As shown here, this RNA profile is directly transposable to the protein level of Hsp27 present in whole fly (Fig. 6A) or in ovarian tissues extracts (Fig. 6B).

7.2.1
Interaction with the SUMO-Conjugating Enzyme Ubc9

A two-hybrid approach has revealed that two members of the *Drosophila* sHSP (Hsp23 and Hsp27) associate *in vivo* with a specific member of the ubiquitin conjugating enzyme, Ubc9 (Joanisse et al. 1998a). This enzyme is a key effec-tor in the sumoylation pathway which modulates stability and/or intracellular localisation of target protein (reviewed in Kretz-Remy and Tanguay 1999; Yeh et al. 2000). Ubc9 has also been shown to act as a modulator of the transacti-vator capacity of multiple transcription factors. This effect on transcriptional transactivation can either be associated (Gostissa et al. 1999; Rodriguez et al. 1999; Müller et al. 2000) or independent (Saltzman et al. 1998; Poukka et al. 1999) of its SUMO-conjugating activity. Whether the association between sHsp and Ubc9 functionally impacts on either activity of Ubc9 remains unknown.

7.2.2
Modulation of Specific Biological Activity

An exciting observation on Hsp27 biology stems from a genetic screen which was based on a gain-of-function approach (Rorth et al. 1998). This experimental approach demonstrated that targeted overexpression of Hsp27 in ovarian border cells suppresses the defect engendered by a hypomorph muta-tion (*slbo¹*) at the *slow border cell* locus. The *slbo¹* mutation, which consists of a P-element insertion in the proximal region of the *slbo* promoter, hinders the developmentally regulated migration of border cells during oogenesis by downregulating the level of Slbo protein produced (Montell et al. 1992). It has been shown recently that the level of the Slbo protein needs to be tightly reg-ulated in border cells to allow their proper migration (Rorth et al. 2000). As the SUMO-conjugating activity of Ubc9 has been related to transcription factor activation or to regulation of protein turnover and activity, it is tempting to speculate on three alternative models which could account for the *slbo¹* rescue by Hsp27 (Fig. 7).

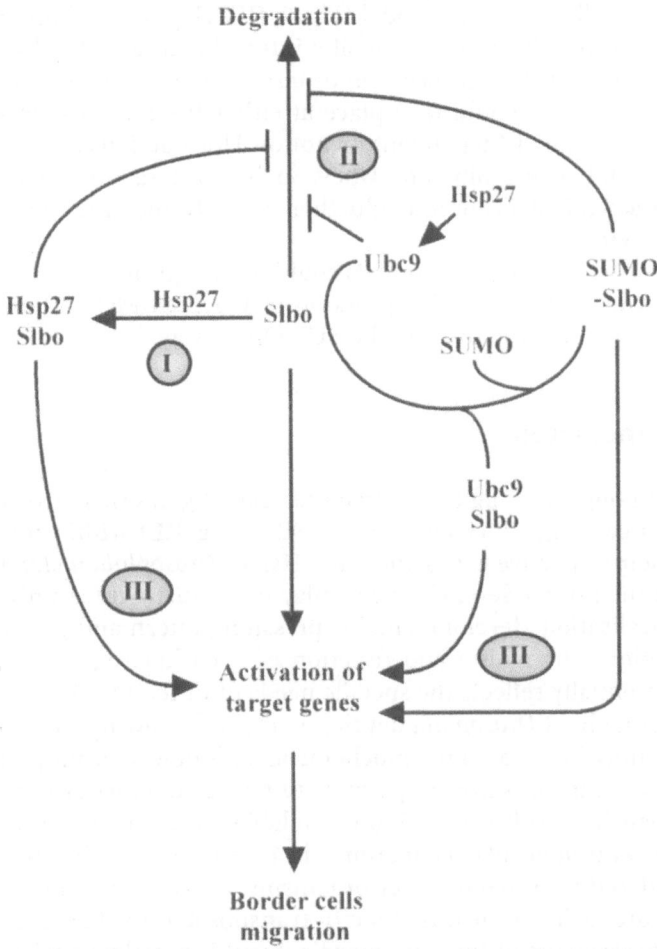

Fig. 7. Functional model of *slbo¹* rescue by Hsp27. Visual representation of 3 different modes of action (indicated by I, II and III; see Sect. 7.2.2 for detailed description) which could account for suppression of the *slbo¹* mutation in ovarian border cells by targeted overexpression of Hsp27

The first model (Fig. 7, I) would solely take into account the putative chaperone function of Hsp27. As migration depends on a specific level of Slbo, overexpression of a cellular chaperone could protect Slbo from targeted degradation and thereby increase its intracellular level or its half-life. This stabilisation could directly compensate for the original low level of expression, resulting in a level of target gene activation sufficient for border cells migration to occur.

A second possibility (Fig. 7, II) would implicate the modulation of the SUMO-conjugating activity of Ubc9 by the overexpression of Hsp27. Such alteration of Ubc9 activity could result in a decrease in the Slbo turnover rate mediated either by protecting Slbo itself by its conjugation to SUMO (as previously observed for Iκ-B; Desterro et al. 1998), or by affecting the activity of any other protein implicated in the Slbo degradation process.

Finally, in a third model (Fig. 7, III), Hsp27 could rescue the *slbo¹* mutation not by modification of the Slbo intracellular level, but by increasing its potential to activate transcription of target genes. The positive impact of Hsp27 on Slbo activity could take place at either the recruitment step to proper chromosomal loci (dependent or not on Ubc9 activity) or by mediating an interaction between Slbo and Ubc9. Such an interaction, in a SUMO-dependent or independent fashion, could then directly increase Slbo transcription factor activity.

These elementary models could be straightforwardly tested in vivo through border cells-directed expression of Ubc9, SUMO and any other components abrogating or activating the SUMO pathway.

8
Conclusion

Although sharing basic attributes like oligomeric native state and presence of similar regulatory elements (HSE and EcRE) within the promoter region of their respective genes, the four sHsp of *Drosophila melanogaster* studied so far at the protein level clearly display many distinct features such as intracellular localisation, developmental expression pattern and post-translational modification. The nature and function of such heterogeneity is still unknown but potentially reflects the specific needs of different cell types and compartments throughout *Drosophila* development and following stress. As early studies have mainly focused on the biochemical and molecular properties of sHsp over the past years, it is now imperative to take a step forward and attempt to identify their in vivo function. From available data, it is tempting to suggest a role of general molecular chaperones in response to environmental stresses as considered for sHsp of other organisms. However, it would be surprising if such a broad function was directly transposable to the developmental context as observations of the non-coordinate and spatially restricted pattern of expression of sHsp are now accumulating. It will also be of great interest to characterise the other putative sHsp ORF found with the help of the *Drosophila* Genome Project as these may provide additional clues related to sHSP biology.

References

Amin J, Mestril R, Voellmy R (1991) Genes for *Drosophila* small heat shock proteins are regulated differently by ecdysterone. Mol Cell Biol 11:5937–5944

Arrigo AP, Ahmed-Zadeh C (1981) Immunofluorescence localization of a small heat shock protein (Hsp23) in salivary gland cells of *Drosophila melanogaster*. Mol Gen Genet 184:73–79

Arrigo AP, Landry J (1994) Expression and function of the low-molecular-weight heat shock proteins. In: Morimoto RI, Tissières A, Georgopoulos C (eds) The biology of heat shock proteins and molecular chaperones. Cold Spring Harbor Laboratory Press, New York, pp 335–373

Arrigo AP, Pauli D (1988) Characterization of Hsp27 and three immunologically related polypeptides during *Drosophila* development. Exp Cell Res 175:169–183

Arrigo AP, Tanguay RM (1991) Expression of heat shock proteins during development in *Drosophila*. In: Hightower L, Nover L (eds) Heat shock and development. Springer, Berlin Heidelberg New York, pp 106–119

Arrigo A-P, Darlix J-L, Khandjian EW, Simon M, Spahr PF (1985) Characterization of the prosome from *Drosophila* and its similarity to the cytoplasmic structures formed by the low molecular weight heat-shock proteins. EMBO J 4:399–406

Ayme A, Tissières A (1985) Locus 67B of *Drosophila melanogaster* contains seven, not four, closely related heat shock proteins. EMBO J 4:2949–2954

Beaulieu JF, Arrigo AP, Tanguay RM (1989) Interaction of *Drosophila* 27,000 Mr heat-shock protein with the nucleus of heat-shocked and ecdysone-stimulated culture cells. J Cell Sci 92: 29–36

Berger EM (1984) The regulation and function of small heat-shock protein synthesis. Dev Genet 4:255–265

Berger EM, Woodward MP (1983) Small heat shock proteins in *Drosophila* may confer thermal tolerance. Exp Cell Res 147:437–442

Bournias-Vardiabasis N, Buzin C, Flores J (1990) Differential expression of heat shock protein in *Drosophila* embryonic cells following metal ion exposure. Exp Cell Res 189:177–182

Buzin CH, Bournias-Vardiabasis N (1984) Teratogens induce a subset of small heat shock proteins in *Drosophila* primary embryonic cell cultures. Proc Natl Acad Sci USA 81:4075–4079

Cheney CM, Shearn A (1983) Developmental regulation of *Drosophila* imaginal disc proteins: synthesis of a heat-shock protein under non-heat-shock conditions. Dev Biol 95:325–330

Cohen RS, Meselson M (1985) Separate regulatory elements for the heat-inducible and ovarian expression of the *Drosophila* hsp26 gene. Cell 43:737–746

Corces V, Holmgren R, Freund R, Morimoto R, Meselson M (1980) Four heat shock proteins of *Drosophila melanogaster* coded within a 12-kilobase region in chromosome subdivision 67B. Proc Natl Acad Sci USA 77:5390–5393

Courgeon A-M, Rollet E, Becker J, Maisonhaute C, Best-Belpomme M (1988) Hydrogen peroxide (H_2O_2) induces actin and some heat-shock proteins in *Drosophila* cells. Eur J Biochem 171: 163–170

Cryderman DE, Tang H, Bell C, Gilmour DS, Wallrath LL (1999) Heterochromatic silencing of *Drosophila* heat shock genes acts at the level of promoter potentiation. Nucleic Acids Res 27: 3364–3370

De Sa CM, Rollet E, de SA M-F, Tanguay RM, Best-Belpomme M, Scherrer K (1989) Prosomes and heat shock complexes in *Drosophila melanogaster* cells. Mol Cell Biol 9:2672–2681

Desterro JMP, Rodriguez MS, Hay RT (1998) SUMO-1 modification of IκBα inhibits NF-κB activation. Mol Cell 2:233–239

Dobens L, Rudolph K, Berger EM (1991) Ecdysterone regulatory elements functions as both transcriptional activators and repressors. Mol Cell Biol 11:1846–1853

Duband JL, Lettre F, Arrigo AP, Tanguay RM (1986) Expression and localization of Hsp23 in unstressed and heat-shocked *Drosophila* cultured cells. Can J Genet Cytol 28:1088–1092

Dubrovsky EB, Dretzen G, Bellard M (1994) The *Drosophila* broad-complex regulates developmental changes in transcription and chromatin structure of the 67B heat-shock gene cluster. J Mol Biol 241:353–362

Dubrovsky EB, Dretzen G, Berger EM (1996) The broad-complex gene is a tissue-specific modulator of the ecdysone response of the *Drosophila hsp23* gene. Mol Cell Biol 16:6542–6552

Dura J-M (1981) Stage dependent synthesis of heat shock induced proteins in early embryos of *Drosophila melanogaster*. Mol Gen Genet 184:381–385

Eissenberg JC, Elgin SCR (1987) *Hsp28^{stl}*: a P-element insertion mutation that alters the expression of a heat shock gene in *Drosophila melanogaster*. Genetics 115:333–340

Fink AL (1999) Chaperone-mediated protein folding. Physiol Rev 79:425–449

Frank LH, Cheung H-K, Cohen RS (1992) Identification and characterization of *Drosophila* female germ line transcriptional control elements. Development 114:481–491

Frydenberg J, Pierpaoli M, Loeschcke V (1999) *Drosophila melanogaster* is polymorphic for a specific repeated (CATA) sequence in the regulatory region of *hsp23*. Gene 236:243–250

Gilmour DS, Thomas GH, Elgin SCR (1989) *Drosophila* nuclear proteins bind to regions of alternating C and T residues in gene promoters. Science 245:1487–1490

Glaser RL, Lis JT (1990) Multiple, compensatory regulatory elements specify spermatocyte-specific expression of the *Drosophila melanogaster hsp26* gene. Mol Cell Biol 10:131–137

Glaser RL, Thomas GH, Siegfried E, Elgin SC, Lis JT (1990) Optimal heat-induced expression of the *Drosophila hsp26* gene requires a promoter sequence containing (CT)n.(GA)n repeats. J Mol Biol 211:751–761

Gostissa M, Hengstermann A, Fogal V, Sandy P, Schwarz SE, Scheffner M, Del Sal G (1999) Activation of p53 by conjugation to the ubiquitin-like protein SUMO-1. EMBO J 18:6462–6471

Haass C, Klein U, Kloetzel P-M (1990) Developmental expression of *Drosophila melanogaster* small heat-shock proteins. J Cell Sci 96:413–418

Haass C, Kloetzel P-M (1990) Molecular analysis of α-ecdysone induced 16 S complexes in *Drosophila* Schneider's S3 cells. Biochem Biophys Res Comm 168:314–319

Henrich VC, Szekely AA, Kim SJ, Brown NE, Antoniewski C, Hayden MA, Lepesant JA, Gilbert LI (1994) Expression and function of the *ultraspiracle* (*usp*) gene during development of *Drosophila melanogaster*. Dev Biol 165:38–52

Hoffman E, Corces V (1986) Sequences involved in temperature and ecdysterone-induced transcription are located in separate regions of a *Drosophila melanogaster* heat shock gene. Mol Cell Biol 6:663–673

Hoffman EP, Gerring SL, Corces VG (1987) The ovarian, ecdysterone, and heat-shock-response promoters of the *Drosophila melanogaster hsp27* gene react very differently to perturbations of DNA sequence. Mol Cell Biol 7:973–981

Hultmark D, Klemenz R, Gehring WJ (1986) Translational and transcriptional control elements in the untranslated leader of the heat-shock gene *hsp22*. Cell 44:429–438

Ingolia TD, Craig EA (1982) Four small *Drosophila* heat-shock proteins are related to each other and to mammalian α-crystallin. Proc Natl Acad Sci USA 79:2360–2364

Ireland RC, Berger EM (1982) Synthesis of low molecular weight heat shock peptides stimulated by ecdysterone in a cultured *Drosophila* cell line. Proc Natl Acad Sci USA 79:855–859

Ireland RC, Berger E, Sirotkin K, Yund MA, Osterbur D, Fristrom J (1982) Ecdysterone induces the transcription of four heat-shock genes in *Drosophila* S3 cells and imaginal discs. Dev Biol 93:498–507

Joanisse DR, Inaguma Y, Tanguay RM (1998a) Cloning and developmental expression of a nuclear ubiquitin-conjugating enzyme (DmUbc9) that interacts with small heat shock proteins in *Drosophila melanogaster*. Biochem Biophys Res Commun 244:102–109

Joanisse DR, Michaud S, Inaguma I, Tanguay RM (1998b) Small heat shock proteins of *Drosophila*: Developmental expression and functions. J Biosci 23:369–376

Kelly SE, Cartwright IL (1989) Perturbation of chromatin architecture on ecdysterone induction of *Drosophila melanogaster* small heat shock protein genes. Mol Cell Biol 9:332–335

King V, Tower J (1999) Aging-specific expression of *Drosophila* Hsp22. Dev Biol 207:107–118

Klemenz R, Gehring WJ (1986) Sequence requirement for expression of the *Drosophila melanogaster* heat shock protein *hsp22* gene during heat shock and normal development. Mol Cell Biol 6:2011–2019

Kloetzel P-M, Bautz EKF (1983) Heat-shock proteins are associated with hnRNA in *Drosophila melanogaster* tissue culture cells. EMBO J 2:705–710

Kretz-Remy C, Tanguay RM (1999) SUMO/sentrin: protein modifiers regulating important cellular functions. Biochem Cell Biol 77:299–309

Kurapati R, Brar Passananti H, Rose MR, Tower J (2000) Increased *hsp22* RNA levels in *Drosophila* lines genetically selected for increased longevity. J Gerontol 55A:B552–B559

Kurzik-Dumke U, Lohman E (1995) Sequence of the new *Drosophila melanogaster* small heat-shock-related gene, lethal (2) essential for life [l (2) efl], at locus 59F4,5. Gene 154:171–175

Leemans R, Egger B, Loop T, Kammermeier L, He H, Hartmann B, Certa U, Hirth F, Reichert H (2000) Quantitative transcript imaging in normal and heat-shocked *Drosophila* embryos by using high-density oligonucleotide arrays. Proc Natl Acad Sci USA 97:12138–12143

Leicht BG, Bonner JJ (1988) Genetic analysis of chromosomal region 67A-D of *Drosophila melanogaster*. Genetics 119:579–593

Lu Q, Wallrath LL, Allan BD, Glaser RL, Lis JT, Elgin SC (1992) Promoter sequence containing (CT)n.(GA)n repeats is critical for the formation of the DNase I hypersensitive sites in the *Drosophila hsp26* gene. J Mol Biol 225:985–998

Lu Q, Wallrath LL, Granok H, Elgin SCR (1993) (CT)n.(GA)n repeats and heat shock elements have distinct roles in chromatin structure and transcriptional activation of the *Drosophila hsp26* gene. Mol Cell Biol 13:2802–2814

Lu Q, Wallrath LL, Elgin SCR (1995) The role of a positioned nucleosome at the *Drosophila melanogaster hsp26* promoter. EMBO J 14:4738–4746

Luo Y, Amin J, Voellmy R (1991) Ecdysterone receptor is a sequence-specific transcription factor involved in the developmental regulation of heat shock genes. Mol Cell Biol 11:3660–3675

Marin R, Tanguay RM (1996) Stage-specific localization of the small heat shock protein Hsp27 during oogenesis in *Drosophila melanogaster*. Chromosoma 105:142–149

Marin R, Valet JP, Tanguay RM (1993) *hsp23* and *hsp26* exhibit distinct spatial and temporal patterns of constitutive expression in *Drosophila* adults. Dev Genet 14:69–77

Marin R, Demers M, Tanguay R (1996a) Cell-specific heat-shock induction of Hsp23 in the eye of *Drosophila melanogaster*. Cell Stress Chaperones 1:40–46

Marin R, Landry J, Tanguay RM (1996b) Tissue-specific post-translational modification of the small heat shock protein Hsp27 in *Drosophila*. Exp Cell Res 223:1–8

Mehlen P, Preville X, Chareyron P, Briolay J, Klemenz R, Arrigo AP (1995) Constitutive expression of human Hsp27, *Drosophila* Hsp27 or human αB-crystallin confers resistance to TNFα- and oxidative stress-induced cytotoxicity in stably transfected murine L929 fibroblasts. J Immunol 154:363–374

Mehlen P, Kretz-Remy C, Préville X, Arrigo AP (1996a) Human Hsp27, *Drosophila* Hsp27 and human αB-crystallin expression-mediated increase in glutathione is essential for the protective activity of these proteins against TNFα-induced cell death. EMBO J 15:2695–2706

Mehlen P, Schulze-Osthoff K, Arrigo AP (1996b) Small stress proteins as novel regulators of apoptosis. J Biol Chem 271:16510–16514

Mestril R, Rungger D, Schiller P, Voellmy R (1985) Identification of a sequence element in the promoter of the *Drosophila melanogaster hsp23* gene that is required for its heat activation. EMBO J 4:2971–2976

Mestril R, Schiller P, Amin J, Klapper H, Ananthan J, Voellmy R (1986) Heat shock and ecdysterone activation of the *Drosophila melanogaster hsp23* gene; a sequence element implied in developmental regulation. EMBO J 5:1667–1673

Michaud S, Marin R, Tanguay RM (1997a) Regulation of heat shock gene induction and expression during *Drosophila* development. Cell Mol Life Sci 53:104–113

Michaud S, Marin R, Westwood JT, Tanguay RM (1997b) Cell-specific expression and heat-shock induction of Hsps during spermatogenesis in *Drosophila melanogaster*. J Cell Sci 110: 1989–1997

Montell DJ, Rorth P, Spradling AC (1992) Slow border cells, a locus required for a developmentally regulated cell migration during oogenesis, encodes *Drosophila* C/EBP. Cell 71:51–62

Morrow G, Inaguma Y, Kato K, Tanguay RM (2000) The small heat-shock protein Hsp22 of *Drosophila melanogaster* is a mitochondrial protein displaying oligomeric organization. J Biol Chem 275:31204–31210

Müller S, Berger M, Lehembre F, Seeler J-S, Haupt Y, Dejean A (2000) c-Jun and p53 activity is modulated by SUMO-1 modification. J Biol Chem 275:13321–13329

Nakai K, Kanehisa M (1992) A knowledge base for predicting protein localisation sites in eukaryotic cells. Genomics 14:897–911

Neupert W (1997) Protein import into mitochondria. Annu Rev Biochem 66:863–917

Nightingale KP, Wellinger RE, Sogo JM, Becker PB (1998) Histone acetylation facilitates RNA polymerase II transcription of the *Drosophila hsp26* gene in chromatin. EMBO J 17:2865–2876

O'Brien T, Lis JT (1993) Rapid changes in *Drosophila* transcription after an instantaneous heat shock. Mol Cell Biol 13:3456–3463

Pauli D, Tissières A (1990) Developmental expression of the heat shock genes in *Drosophila melanogaster*. In: Morimoto R, Tissières A, Georgopoulos C (eds) Stress proteins in biology and medicine. Cold Spring Harbour Laboratory Press, New York, pp 361–378

Pauli D, Tonka CH (1987) A *Drosophila* heat shock gene from locus 67B is expressed during embryogenesis and pupation. J Mol Biol 198:235–240

Pauli D, Tonka CH, Ayme-Southgate A (1988) An unusual split *Drosophila* heat shock gene expressed during embryogenesis, pupation and in testes. J Mol Biol 200:47–53

Pauli D, Tonka CH, Tissieres A, Arrigo AP (1990) Tissue-specific expression of the heat shock protein Hsp27 during *Drosophila melanogaster* development. J Cell Biol 111:817–828

Poukka H, Aarnisalo P, Karvonen U, Palvimo JJ, Jänne OA (1999) Ubc9 interacts with the androgen receptor and activates receptor-dependent transcription. J Biol Chem 274:19441–19446

Quivy J-P, Becker PB (1996) The architecture of the heat-inducible *Drosophila hsp27* promoter in nuclei. J Mol Biol 256:249–263

Riddihough G, Pelham HRB (1986) Activation of the *Drosophila hsp27* promoter by heat shock and by ecdysone involves independent and remote regulatory sequences. EMBO J 5:1653–1658

Riddihough G, Pelham HRB (1987) An ecdysone response element in the *Drosophila hsp27* promoter. EMBO J 6:3729–3734

Ritossa F (1962) A new puffing pattern induced by temperature shock and DNP in *Drosophila*. Experientia 18:571–573

Rodriguez MS, Desterro JMP, Lain S, Midgley CA, Lane DP, Hay RT (1999) SUMO-1 modification activates the transcriptional response of p53. EMBO J 18:6455–6461

Rollet E, Best-Belpomme M (1986) Hsp 26 and 27 are phosphorylated in response to heat shock and ecdysterone in *Drosophila melanogaster* cells. Biochem Biophys Res Commun 141: 426–433

Rollet E, Lavoie JN, Landry J, Tanguay RM (1992) Expression of *Drosophila*'s 27 kDa heat shock protein into rodent cells confers thermal resistance. Biochem Biophys Res Commun 185: 116–120

Ropp M, Courgeon A-M, Calvayrac R, Best-Belpomme M (1983) The possible role of the superoxyde ion in the induction of heat-shock and specific proteins in aerobic *Drosophila* cells during return to normoxia after a period of anaerobiosis. Can J Biochem Cell Biol 61:456–461

Rorth P (1996) A modular misexpression screen in *Drosophila* detecting tissue-specific phenotypes. Proc Natl Acad Sci USA 93:12418–12422

Rorth P, Szabo K, Bailey A, Laverty T, Rehm J, Rubin GM, Weigmann K, Milán M, Benes B, Ansorge W, Cohen SM (1998) Systematic gain-of-function genetics in *Drosophila*. Development 125: 1049–1057

Rorth P, Szabo K, Texido G (2000) The level of C/EBP protein is critical for cell migration during *Drosophila* oogenesis and is tightly controlled by regulated degradation. Mol Cell 6:23–30

Saltzman A, Searfoss G, Marcireau C, Stone M, Ressner R, Munro R, Franks C, D'Alonzo J, Tocque B, Jaye M, Ivaschenko Y (1998) hUBC9 associates with MEKKI and type I TNFα receptor and stimulates NFκB activity. FEBS Letters 425:431–435

Schubiger M, Truman JW (2000) The RXR ortholog USP suppresses early metamorphic processes in *Drosophila* in the absence of ecdysteroids. Development 127:1151–1159

Schuldt C, Kloetzel P-M (1985) Analysis of cytoplasmic 19 S ring-type particles in *Drosophila* which contain Hsp 23 at normal growth temperature. Dev Biol 110:65–74

Shopland LS, Lis JT (1996) HSF recruitment and loss at most *Drosophila* heat shock loci is coordinated and depends on proximal promoter sequence. Chromosoma 105:158–171

Sirotkin K, Bartley N, Perry III WL, Briggs D, Grell EH, Morganelli C, Berger EM, Bonner JJ, Leicht B (1986) Deletion polymorphism in a *Drosophila melanogaster* heat shock gene. Mol Gen Genet 204:266–272

Southgate R, Ayme A, Voellmy R (1983) Nucleotide sequence analysis of the *Drosophila* small heat shock gene cluster at locus 67B. J Mol Biol 165:35–57

Tanguay RM, Joanisse DR, Inaguma Y, Michaud S (1999) Small heat shock proteins: in search of functions in vivo. In: Storey KB (ed) Environmental stress and gene regulation. BIOS Scientific Publishers Ltd, Oxford, pp 125–138

Thomas GH, Elgin SCR (1988) Protein/DNA architecture of the DNase I hypersensitive region of the *Drosophila hsp26* promoter. EMBO J 7:2191–2201

Tissières A, Mitchell HK, Tracy UM (1974) Protein synthesis in salivary glands of *Drosophila melanogaster*: relation to chromosome puffs. J Mol Biol 84:389–398

Vazquez J (1991) Response to heat shock of *gene 1*, a *Drosophila melanogaster* small heat shock gene, is developmentally regulated. Mol Gen Genet 226:393–400

Vazquez J, Pauli D, Tissières A (1993) Transcriptional regulation in *Drosophila* during heat shock: a nuclear run-on analysis. Chromosoma 102:233–248

Vincent M, Tanguay RM (1982) Different intracellular distributions of heat-shock and arsenite-induced proteins in *Drosophila* Kc cells. J Mol Biol 162:365–378

Vitek MP, Berger EM (1984) Steroid and high-temperature induction of the small heat-shock protein genes in *Drosophila*. J Mol Biol 178:173–189

Wall G, Varga-Weisz PD, Sandaltzopoulos R, Becker PB (1995) Chromatin remodeling by GAGA factor and heat shock factor at the hypersensitive *Drosophila hsp26* promoter in vitro. EMBO J 14:1727–1736

Westwood JT, Clos J, Wu C (1991) Stress-induced oligomerization and chromosomal relocalization of heat-shock factor. Nature 353:822–827

Wheeler JC, Bieschke ET, Tower J (1995) Muscle-specific expression of *Drosophila* Hsp70 in response to aging and oxidative stress. Proc Natl Acad Sci USA 92:10408–10412

Yeh ETH, Gong L, Kamitani T (2000) Ubiquitin-like proteins: new wines in new bottles. Gene 248:1–14

Zimmerman JL, Petri W, Meselson M (1983) Accumulation of a specific subset of *D. melanogaster* heat shock mRNAs in normal development without heat shock. Cell 32:1161–1170

Zou S, Meadows S, Sharp L, Jan LY, Nung Jan Y (2000) Genome-wide study of aging and oxydative stress response in *Drosophila melanogaster*. Proc Natl Acad Sci USA 97:13726–13731

The Developmental Expression of Small HSP

Sean M. Davidson, Marie-Thérèse Loones, Olivier Duverger,
and Michel Morange[1]

1
Introduction

This chapter is based on the hypothesis that chaperones and heat shock proteins have specific developmental functions. In other words, at different developmental and differentiation stages, the function of these proteins could be crucially required, as they are the "limiting step" in this process. These functions might be the normal, chaperone function of these proteins, or more specific ones. It is now widely realized that even the most ubiquitous chaperones, such as HSP/HSC70 have both general functions – in the case of HSP/HSC70, these being the refolding of denatured proteins or stabilization of nascent ones – and more specific functions, such as the activity of HSP/HSC70 in uncoating clathrin cages. At least two of the functions thus far demonstrated for small HSP (sHSP) – protection against apoptosis and modulation of the cytoskeletal structure – are amply sufficient to explain a developmental role of sHSP.

Over the past few years, the number of known genes coding for mammalian sHSP has increased to eight, and they all seem to play important roles during development. To harmonize this contribution with the other chapters in this book, and to clarify the presentation of data from diverse sources, we will adopt the following strategies:

1. This chapter will focus on development in mammals since other organisms are covered more completely in other chapters. We will only refer to the expression of sHSP during development of non-mammalian organisms at the end of this chapter, to confirm some of our conclusions, to outline some new developmental functions which emerged from these studies, and to underline the interest of these animal models and of the molecular technologies which can be applied to them to reveal sHSP functions.
2. We will describe successively the pattern of expression of sHSP during mammalian development and the possible changes in their phosphorylation state, what is known about their regulation (for which the only data available concern HSP25 and α-crystallins) and finally attempt to give a

[1] Unité de Génétique Moléculaire, Ecole normale supérieure, 46 rue d'Ulm, 75230 Paris Cedex 05, France

Progress in Molecular and Subcellular Biology, Vol. 28
A.-P. Arrigo and W.E.G. Müller (Eds.)
© Springer-Verlag Berlin Heidelberg 2002

physiological meaning to these data by discussing them with respect to what is known of the functions of sHSP.

3. The data are organized from the point of view of a developmental biologist, considering the expression of sHSP in the three different cell lineages. We hope that this organization will help to reveal interesting parallels. However, despite their abundance, these observations remain punctual and there will be huge gaps in our developmental description.

4. A large place is reserved in this review for the expression of sHSP in the heart and muscles. The expression of most sHSP is very high in these tissues. We have integrated unpublished data from our own lab concerning the expression of sHSP in these tissues.

Unless otherwise specified, data refer to results from analyses of mouse embryos. We frequently use the term HSP25 to refer to both the human HSP27 and its murine ortholog.

2
Small HSP in Extraembryonic Development

HSP25 was found to be particularly abundant in the apical projections of the endometrial epithelium at the time of implantation, suggesting that it may be involved in the attachment of the blastocyst to the uterine wall (Ciocca et al. 1983, 1996). In extra embryonic layers, the expression of HSP25 is detected by immunohistochemistry in structures which participate in feto-maternal exchanges during pregnancy: in differentiating giant trophoblastic cells from day 11 to day 16 of gestation in mouse, then in endothelial cells of fetal blood vessels (Ciocca et al 1996), and during the first two trimesters of pregnancy in humans (Shah et al. 1998).

HSP25 is present at a low level in the preimplantation embryo and accumulates in a differentiation-dependent manner in embryonic carcinoma and stem cells in culture (Stahl et al. 1992; Davidson and Morange 2000).

3
Small HSP in Ectodermal Lineages

3.1
Neural Crest Cell Derivatives

HSP25 is present in migrating rhombomeric neural crest cells (NCC) and in the cephalic and the spinal ganglia at the initial phase of their formation in the mouse (Loones et al. 2000). The first endothelial cells originating from the NCC which migrate into the neural tube where they develop into brain capillaries are strongly HSP25 positive (Gernold et al. 1993; Loones et al. 2000). HSP25 is present in neurons and neurites of the differentiating trigeminal or spinal ganglia, its expression varying between different neurons (Gernold et al. 1993; Loones et al. 2000), and during development. For example, HSP25 appears only postnatally in the spinal sensory dorsal root ganglia in rat (Costigan et al. 1998).

3.2
Neurectoderm

HSP25 expression follows the formation of the first tracts, and occurs in many differentiating fields, where neurons leave the proliferating zone to undergo differentiation and morphogenesis. HSP25 is distributed differentially in the various layers of the developing neocortex and hippocampus. At E12.5, HSP25 is expressed in radial glial cells and is also higher in the first differentiating neurons of the peripheral layer. From E15.5, HSP25 is strongly expressed in the cytoplasm of hippocampal and cortical postmigratory neurons of the marginal zone (Loones et al. 2000).

HSP25 is also differentially expressed in the cerebellum during its formation: Purkinje cells are not organized in their characteristic single layer until P10. At E15.5, although the molecular layer and the Purkinje cells, which will express HSP25 (Gernold et al. 1993), are not yet distinguishable, a layer internal to the external granular layer is clearly immunoreactive for HSP25 (Loones, unpublished data).

In the adult mouse (but not in rat), HSP25 expression reveals a uniquely complex cerebellar organization with expression of HSP25 limited to a small subset of Purkinje cells distributed in parasagittal bands which alternate with unstained bands (C.L. Armstrong et al. 2000; Plumier et al. 1997).

The highest level of expression of HSP25 in the brain is found in specific nuclei. In hypoglossal nucleus at E12.5, nascent neurons are weakly HSP25 immunoreactive. From E15.5, the stage corresponding to the onset of muscular contraction, when the XIIth cranial nerve innervates the tongue muscles, the huge hypoglossal motoneurons become heavily labeled (Fig 1C). The coincidence of these two events call to mind reciprocal interactions between the motoneurons and their target. HSP25 is also abundant in other nuclei in the ventral area of the pontic region and in the nucleus in the differentiating field of the tegmentum. As in ganglia, not all neurons of the tegmentum nucleus are labeled, and in the positive neurons HSP25 is abundant in axons and dendrites.

αB-crystallin is detected at a very low level during embryonic development of the rat cortex, cerebellum and brainstem (Kato et al. 1991). The level increases between the 5th and 9th weeks post partum, when the adult level is reached: neurons in the spinal cord, brainstem and hippocampus are weakly immunoreactive, while glial cells (mainly oligodendrocytes) are intensely reactive as are mitral cells of the olfactory bulb and Schwann cells in the peripheral nerves. In vitro analyses of differentiating oligodendrocytes (Neri et al. 1997) suggest that αB-crystallin may permit changes in cytoskeletal organiza-

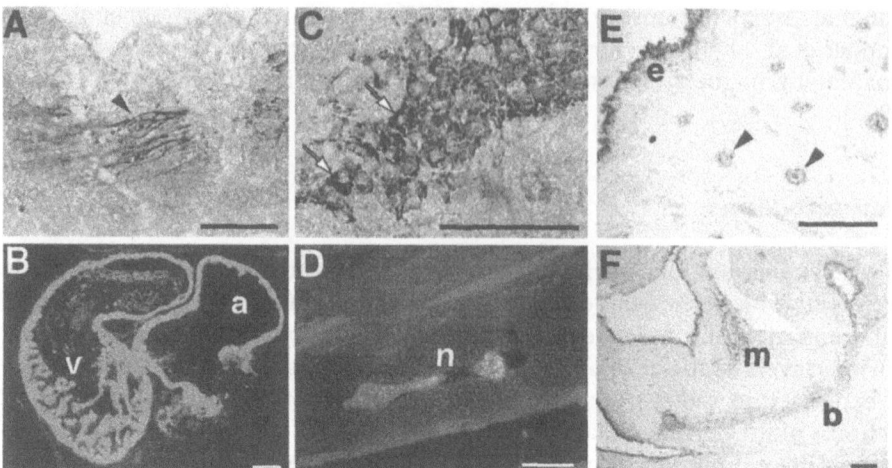

Fig. 1. HSP25 immunolocalization using immunoperoxidase (A, C, E, F) or immunofluoresence (B, D) on parasagittal sections of mouse embryos. **A** E9.5: myoblasts migrating out of the somitic myotomes (*arrowheads*) express HSP25. **B** E10.5: HSP25 strongly accumulates in the myocardium of the developing heart (*a* atrium; *v* ventricle). **C, D, E, F** E15.5; **C** in the hypoglossal nucleus, HSP25 is strongly detected in the cytoplasm of its large neurons (*arrows*). **D** HSP25 is strongly expressed in the cells of the regressing notochord (*n*). **E** Vibrissae (*arrowheads*) and epidermis (*e*) express HSP25. **F** HSP25 is expressed in bladder epithelium (*b*), skeletal muscle (*m*) and in epidermis. *Bar* = 100 μm

tion during flattening of the oligodendrocytes processes, an event which precedes myelin sheath compaction.

HSP25 colocalizes with αB-crystallin in the eye lens (Gernold et al 1993). HSP25 is synthesized in the developing eye from E10.5 in a different area from αB-crystallin (Loones, unpubl. observ.). Later, HSP25 is synthesized in the lens as it is forming, and here it co-localizes with αB-crystallin, as it does in the pigmentary epithelium.

3.3
Expression in the Skin

HSP27 is first detected in human skin at week 14 of gestation (Jantschitsch et al. 1998). At this stage, the epidermis consists of a basal cell layer, several layers of intermediate cells and a periderm. HSP27 staining is confined to the periderm, and increases with the distance of the keratinocytes from the basal layer, in parallel with the extent of keratinization (Viac et al. 1989; Jantschitsch et al. 1998). The expression pattern of HSP27 can be recapitulated using an in vitro system of keratinocyte differentiation. HSP27 is expressed at a low level in an undifferentiated human keratinocytic cell line which appears as non-cornified cuboidal cells, whereas a two-fold increase is observed when the cells are induced to differentiate into more spindle-shaped, cornified cells (Kindas-Mugge and Trautinger 1994).

4
Small HSP in Mesodermal Lineages

Interestingly, sHSP (except αA-crystallin) are most highly expressed during the differentiation of tissues deriving from the mesoderm. Paraxial mesoderm generates the sclerotome (which ultimately becomes the chondrocytes of the vertebrae and ribs), and the dermomyotome (which gives rise to the dermis of the back and most skeletal muscle). Lateral mesoderm generates the heart and circulatory system. The intermediate mesoderm generates the kidneys and the gonads. Each of these tissues expresses HSP25 at certain stages during development. The mouse embryonic notochord, which is derived from the chordamesoderm, also expresses HSP25 and αB-crystallin (Fig 1D; Gernold et al. 1993).

4.1
Small HSP in Muscle Development

Outside the eye, the tissue containing the highest levels of sHSP is muscle. These proteins are typically present at particularly high levels during muscle formation, starting from the earliest stages of myocyte differentiation (Fig 1A), although within these tissues certain family members appear to show distinct patterns of expression or localization which may indicate specialization of function. Since most sHSP are expressed in the heart, a major challenge lies in distinguishing the overlapping and unique roles which undoubtedly exist for these proteins.

4.1.1
The Expression Pattern of sHSP in the Heart

In situ hybridisation and transgenic reporter studies indicate that throughout mammalian development the αB-crystallin protein expression pattern recapitulates that of the mRNA (Dubin et al. 1991; Gopal-Srivastava et al. 1995; Haynes et al. 1996; Benjamin et al. 1997). Thus, regulation is likely to be primarily at a transcriptional level. The first αB-crystallin transcripts can be detected in the mouse embryo at E8.5, when they are restricted to the cardiomyocytes of the primitive heart tube (Benjamin et al. 1997). With the formation of the heart chambers from E9.5, αB-crystallin and HSP25 are uniformly expressed in the cardiomyocytes of the atrium and ventricle (Fig 1B; Haynes et al. 1996; Benjamin et al. 1997). As heart development proceeds, αB-crystallin and HSP25 increase to a maximum at E10.5 (Benjamin et al. 1997; Loones, unpubl. results).

The expression of αB-crystallin and HSP25 appears to be restricted to the myocyte lineage; they are not detected in the endothelial cells of the heart which contribute to the aortic outflow tract, atrio-ventricular canal, endocardial cushions and developing aortic valves. The cells of the cardiac conduction system, which set and maintain the rhythmic contractions of the heart, are formed by cells recruited from within the cardiomyogenic lineage (Cheng et al. 1999). In the conduction system, the nodes which generate the pace-making signal express lower levels of αB-crystallin than surrounding cardiomyocytes, while in the bundle fibers which conduct the signals, αB-crystallin levels are higher (Leach et al. 1994). Intriguingly, the expression pattern of αB-crystallin in the heart is coincident with that of desmin, a protein with which it is proposed to functionally interact (Bennardini et al. 1992; Leach et al. 1994).

After birth, the regulation of HSP25 and αB-crystallin in the heart apparently becomes independent, since HSP25 levels decrease in rat hearts while αB-crystallin remains constant (Lutsch et al. 1997), and in adult mouse and rat hearts the levels of HSP25 and αB-crystallin are significantly different (Klemenz et al. 1993).

The small HSP MKBP/HSPB2, was cloned by virtue of its association with myotonic dystrophy protein kinase (DMPK) from human muscle, and a northern blot indicates it is weakly ubiquitous but highly expressed in cardiac and skeletal muscle (Suzuki et al. 1998). Before birth, MKBP is present in the rat heart at significant levels, but its concentration has approximately doubled in the neonate heart, possibly due to oxidative stress (Shama et al. 1999). On the other hand, MKBP is expressed during skeletal differentiation in vitro (i.e. without oxidative stress; Sugiyama et al. 2000). After this maximal expression in neonatal rat and human hearts, it decreases to 10% of initial levels by 13 weeks in the rat, and similarly in human hearts (Shama et al. 1999).

A cDNA encoding an sHSP was isolated from a human heart library and called HSPL27 (Lam et al. 1996), but was subsequently found to be a chimeric artifact of cDNA library construction. It has been re-cloned from a human expressed sequence tag (EST) and is now referred to as HSPB3 (Boelens et al. 1998a). This sHSP is a relatively distant member of the family, and is highly expressed in the heart (Lam et al. 1996; Boelens et al. 1998a; Sugiyama et al. 2000).

A computational approach to identify ESTs which are selectively expressed in the heart resulted in the identification of a new sHSP called cvHSP (Krief et al. 1999). As expected, using Northern blot or RT-PCR, the mRNA is found to be highly present in human heart (including fetal heart).

HSP20 was purified from rat and human muscle after co-elution with αB-crystallin and HSP27, but is also expressed in the heart (Kato et al. 1994). Although αA-crystallin is primarily restricted to the ocular lens, RT-PCR has been used to detect low quantities of αA-crystallin mRNA in the heart (Srinivasan et al. 1992).

The latest addition to the family of mammalian sHSP, HSP22 (Genbank accession number AF250139), is also expressed in the heart (M.J. Welsh, pers. comm.).

4.1.2
The Expression Pattern of shSP in Skeletal and Smooth Muscle

By in situ hybridization of mouse embryos, αB-crystallin mRNA can be detected at E11.5 in muscle ventral to the developing vertebral column (Benjamin et al. 1997). At E12.5 it is detected in pre-muscle masses of the forelimb buds, at E13.5 in the muscle bundles of the pharynx, tongue and mandible (Benjamin et al. 1997), and at E15.5-E16.5 in limb muscles, intercostal muscles, and diaphragm (Benjamin et al. 1997). A similar expression pattern was found for an αB-crystallin promoter transgenically expressed in mice, although this technique additionally enabled detection of transcriptional activity as early as E9.5 in the myotome of the somites (Haynes et al. 1996). αB-crystallin is predominantly expressed in type I (slow-twitch-oxidative) muscle fibers and type 2A (fast-twitch oxidative-glycolytic) rather than type 2B (fast-twitch glycolytic) muscle fibers of skeletal muscles (Atomi et al. 1991b; Haynes et al. 1996; T. Iwaki et al. 1990; Kato et al. 1992). In the few days after birth, αB-crystallin and HSP27 increase dramatically in the hindlimb muscle of the rat, probably due to increased use of these muscles, since transection of the sciatic nerve decreases the level of these sHSPs (Inaguma et al. 1993).

HSP25 protein can be detected in myoblasts migrating from the somatic myotomes of the trunk at E9.5, and then in developing back muscles and the pharynx at E13 in the mouse (Gernold et al. 1993). From E15 until birth it is strongly expressed in all muscles, including those surrounding the intestine, though the heart displays the strongest labeling (Gernold et al. 1993). HSP25 is detected in the smooth muscle of large vessels and blood capillaries of adult human ventricles, and of rat ventricles from the postnatal age of 30 days, but these cells remain negative for αB-crystallin (Lutsch et al. 1997).

HSP20 is preferentially expressed in skeletal muscle, particularly in slow-twitch muscles (Inaguma et al. 1996), but at lower levels than αB-crystallin (Kato et al. 1994). In skeletal muscle, MKBP is localized at Z lines and the neuromuscular junction (Suzuki et al. 1998).

HSP22, cvHSP, HSPB3 and MKBP are detected in extracts of skeletal muscle and possibly smooth muscle, though to a lesser extent than in the heart (M.J. Welsh, pers. comm.; Suzuki et al. 1998; Krief et al. 1999; Sugiyama et al. 2000).

The levels of MKBP/HSPB2, HSPB3 and αB-crystallin increase during in vitro differentiation of C2C12 myoblasts (Sugiyama et al. 2000). This study found the level of HSP27 mRNA remained constant throughout C2C12 differentiation, although by immunofluorescence we have detected high levels of HSP25 protein in differentiated C2C12 cells (unpublished results) which may indicate that there is post-transcriptional regulation in this system. A strong increase in HSP25 is observed during in vitro differentiation of an embryonic carcinoma cell line into skeletal muscle or cardiomyocytes (Davidson and Morange 2000).

4.1.3
The Subcellular Localization of sHSP in Muscle

The intracellular localization of sHSP changes from cytoplasmic to fibrillar during muscle development, reflecting the formation of the myofibrils. Immediately after birth, HSP25 and αB-crystallin localization is nearly identical in rat cardiomyocytes, being distributed equally between the cytoplasm and myofibrils in the ventricle (Lutsch et al. 1997). Eleven days later, these proteins are mainly myofibrillar, being localized in the I band and M-line (Lutsch et al. 1997). In myofibrils isolated from skeletal muscle and in cultured cardiomyocytes the insoluble fraction of αB-crystallin is localized to the Z bands (i.e. the region within the I band where the actin filaments are anchored, and where desmin is located), by immunohistochemistry and electron microscopy (Atomi et al. 1991b; Bennardini et al. 1992). αB-crystallin has also been localized to the Z lines and intercalated discs of human adult and fetal hearts (Leach et al. 1994).

Curiously, although MKBP is colocalized with desmin at the intercalated discs and Z lines of the rat heart, the postnatal increase in MKBP is predominantly cytosolic (Shama et al. 1999). No filamentous localization of MKBP could be detected in an in vitro skeletal muscle differentiation system (Sugiyama et al. 2000).

HSP20 is localized in the cytoplasm of unstressed neonatal rat cardiomyocytes in vitro and is primarily soluble (van de Klundert and de Jong 1999). Although it has not yet been examined, it may be localized differently to other sHSP during muscle differentiation since, in contrast to αB-crystallin or HSP25, barely any association with sarcomeres is detected in heat shocked cardiomyocytes (van de Klundert and de Jong 1999). On the other hand, it has been observed to interact with actin in carotid artery smooth muscle (Brophy et al. 1999b) and to redistribute to the insoluble fraction of heat shocked rat diaphragm muscle (Kato et al. 1994).

4.2
Small HSP in Other Mesodermal Lineages

4.2.1
Expression in Cartilage and Bones

Cartilage expresses HSP25, in both embryonic and adult mice, although more weakly than in heart and skeletal muscles (Loones et al. 1997). The different HSP are expressed in a dynamic spatio-temporal pattern along the mouse endochondral bone, which forms from a cartilagenous precursor (Loones and Morange 1998). The quantity of HSP25 increases during chondrocyte maturation from pre-chondrocytes to hypertrophic chondrocytes, in which it is associated with apoptosis (Tiffee et al. 2000). HSP25 is also expressed in osteoblasts lining newly formed bones (Tiffee et al. 2000).

Recent studies suggest that endothelin-1 (ET-1) stimulates HSP27 induction in cultured human osteoblast-like MC3T3-E1 cells and that p38 MAP kinase activation, mediated by protein kinase C (PKC), is involved in this induction of HSP27 (Kawamura et al. 1999). The authors further postulate that, in vivo, ET-1 produced and secreted from endothelial cells in the bone microenvironment (capillaries), stimulates the induction of HSP27 through specific receptors. However, the level of HSP27 also increases during differentiation of isolated osteoblasts in vitro (Shakoori et al. 1992).

4.2.2
Expression in the Derivatives of the Notochord

Throughout mouse embryogenesis, HSP25 and αB-crystallin accumulate in the notochord (Gernold et al 1993). During vertebrae formation, notochord cells become confined to intervertebral spaces where they undergo an intervertebral disk fate. These remaining notochordal cells, called *nucleus pulposus* in human, are very strongly immunoreactive (Fig 1D).

4.2.3
Expression in the Kidney

The global level of HSP25 in the kidney is low. However, its distribution is not uniform: HSP25 is barely detectable in the cortex, whereas it is abundant in the inner medulla (Müller et al. 1996; Beck et al. 2000). This high level of expression could result from the extracellular tonicity and associated with the rearrangements of the cytoskeleton which occur in the cells. HSP25 is highly expressed in the mesangial cells and might participate in the contraction of these cells (Müller et al. 1999). Inhibition of p38 kinase reduces HSP25 phosphorylation and abolishes cell contractions: through its phosphorylation of HSP25, p38 might be involved in the response to angiotensin II (Müller et al. 1999). It has been observed that experimental nephrotic syndrome is associated with an increase in HSP25 in the glomerular epithelial cells (Smoyer et al. 1996), underlining once again the role of HSP25 in the normal or pathological physiology of the kidney.

4.2.4
Expression in the Adrenal Glands

HSP25 is undetectable in the adrenal gland during embryogenesis, but it is primarily localized in the cortex of the gland in the adult mouse, mostly in fascicular and reticular zones whose cells are responsive to adrenocorticotropic hormone (ACTH; Wakayama and Iseki 1998).

4.2.5
Expression in the Gonads

In both the testis and ovary, the expression of HSP25 is low in somatic components. In testis, HSP25 is expressed postnatally with the first wave of spermatogenesis and its expression is maintained during the reproductive period. HSP25 is limited to early meiotic spermatocytes I (leptotene, zygotene and early pachytene; Wakayama and Iseki 1999). A possible role in the pairing of homologous chromosomes and/or in formation of synaptonemal complex has been suggested. αB-crystallin has also been detected in spermatocytes of rat testis (Kato et al. 1991).

4.2.6
Expression in the Blood Cells

The expression of sHSP in blood cells has not been comprehensively studied, but HSP27 is expressed in platelets (Mendelsohn et al. 1991), is more strongly expressed in immature human T cells compared to mature T cells (Hanash et al. 1993), and increases transiently in promyelocytic HL-60 cells differentiating in vitro (Spector et al. 1993) and in activated B cells undergoing growth arrest immediately before terminal differentiation (Spector et al. 1992).

5
Small HSP in Endodermal Lineages

5.1
Expression in the Upper Digestive Tract

A strong immunoreactivity for HSP25 is first detected at E9.5 in the pharyngial epithelium around the pharyngial pouches. The reactivity progresses dorsally and caudally, to the whole pharynx and to the oesophagus (Loones, unpubl. data). In the adult, HSP25 transcripts are found in the stratified squamous epithelium and in the skeletal muscular layer (Wakayama and Iseki 1998).

5.2
Expression in the Stomach and Intestine

Only the smooth muscle layer (of mesodermal origin) expresses HSP25 whereas the epithelium cells are devoid of HSP25 during embryogenesis (Gernold et al. 1993 and our observations). In the adult, Wakayama and Iseki (1998) detect HSP25 transcripts in the stratified squamous epithelium of the fore-stomach, which resembles oesophagus epithelium.

5.3
Expression in the Bladder

HSP25 strongly accumulates in the bladder epithelium from E15.5 in mouse (Fig 1F). In the adult both the epithelial cells and the smooth muscle cells show HSP25 transcripts (Wakayama and Iseki 1998). αB-crystallin colocalizes with HSP25 in the embryo (Gernold et al. 1993), but is expressed to a lower extent in the adult (Klemenz et al. 1993).

6
Phosphorylation of sHSP During Development

Many different types of stresses are known to induce phosphorylation of HSP25 at three (human) or two (mouse) serine residues (see K. Kato, Chap. 7, and M. Gaestel, Chap. 8, this Vol.), a change which appears to correlate with the dissociation of HSP25 from a large oligomer into tetramers (Rogalla et al. 1999). Removal of Leukemia Inhibitory Factor (LIF) from cultures of mouse embryonic stem cells causes differentiation accompanied by a rapid dephosphorylation of HSP27 and a simultaneous, though transient increase in its oligomeric size (Mehlen et al. 1997). However, this correlation may not hold in all differentiation systems, since during in vitro differentiation of HL-60 cells, HSP27 undergoes a transient phosphorylation which precedes an increase in the apparent molecular weight of HSP27 complexes (Chaufour et al. 1996). A possible explanation for these differences is that the oligomeric complexes being detected may consist of tetramers of HSP25 complexed with other proteins rather than consisting solely of HSP25 (Rogalla et al. 1999).

In the case of muscle and heart, stress-induced relocalization of αB-crystallin and HSP25 to the insoluble sarcomeres is believed to be phosphory-lation-dependent (Barbato et al. 1996; S.C. Armstrong et al. 2000; Knoepp et al. 2000; Sakamoto et al. 2000). In contrast, although the phosphorylation of HSP25 during development has not been extensively studied, it does not appear to correlate well with its intracellular localization. For example, the extent of HSP25 phosphorylation in the ventricle reduces between E20 and the new-born rat, just as sarcomeric association is increasing (Lutsch et al. 1997). In adult mouse heart, HSP25 is predominantly, but not exclusively, non-phosphorylated (Klemenz et al. 1993). Another study found that HSP25 was not extensively phosphorylated during muscle development although 30% is insoluble in control adult rat hearts (Sakamoto et al. 2000). By two-dimensional gel electrophoresis, 43% of HSP27 was found to be phosphorylated in normal adult rat hearts, but this form was found equally distributed between the soluble and insoluble fractions (Yoshida et al. 1999). The phosphorylation of HSP25 does not appear to be essential, at least for in vitro differentiation, since it can be inhibited without affecting differentiation of an embryonic carcinoma into cardiomyocytes (Davidson and Morange 2000).

The postnatal developmental phosphorylation of αB-crystallin in the rat lens has been examined using phospho-specific antibodies, and although phosphorylation was barely detectable at birth, phosphorylation at Ser-45 increased to a maximum at 8 weeks of age (Ito et al. 1999). The ERK1/ERK2 MAPK pathway is believed to be responsible for phosphorylation at this site (Kato et al. 1998) while phosphorylation at other sites known to be targets of MAPKAPK-2, was observed in cardiac and skeletal muscle (Ito et al. 1999). Another study found phosphorylation of αA-crystallin and αB-crystallin to increase in the human lens between the fetal stage and 3 years of age (Ma et al. 1998). HSP25 is also phosphorylated during lens cell differentiation (Chiesa et al. 1997).

Phosphorylated isoforms of HSP20 were detected on 2D western blots of the purified protein from skeletal muscle (Kato et al. 1994). HSP20 is phosphorylated after cyclic nucleotide-dependent relaxation of vascular smooth muscle (Beall et al. 1997), and its phosphorylation in vascular smooth muscle in response to cyclic nucleotide-dependent pathways appears to partially dissociate macromolecular aggregates of HSP20 (Brophy et al. 1999a).

Only a single species of MKBP is detected on 2D gels (Yoshida et al. 1999) and HSPB3 does not appear to have potential phosphorylation sites, at least for MAPKAPK2 (Boelens et al. 1998a).

HSP22 can be phosphorylated by PKC and probably other kinases, but not MAPKAPkinase2/3. HSP22 interacts strongly with phosphorylated HSP25, when the latter protein is phosphorylated; it was in fact identified through this interaction (M. Welsh, pers. comm.).

7
Developmental Regulation of Small HSP Gene Expression

The regulation of mammalian sHSP is considered to be primarily at the level of transcription. The promoters of the sHSP have not been extensively studied, with the exception of those of the α-crystallins. The well-known sequence for the heat shock element (HSE) has been detected in only those sHSP which are inducible by heat, i.e. HSP25 (Hickey et al. 1986; Frohli et al. 1993; Gaestel et al. 1993), and αB-crystallin (Klemenz et al. 1991), and these have been shown to be functional. However, it seems likely that binding of heat shock transcription factors (HSFs) to the HSEs is not involved in developmental expression, at least in most cases. Developmental expression is not altered in the HSF1 knockout (Xiao et al. 1999) and no HSF2 protein (Davidson, unpublished results) or HSF2 promoter activity (V. Mezger, M. Morange, personal communication) can be detected in the embryonic mouse heart, in which there is a high level of expression of sHSP during development.

7.1
Regulation of αA-Crystallin Gene Transcription

Lens-specific expression of a reporter gene under the control of the promoter of the mouse αA-crystallin gene (*cryaA*) requires only the region (−364/+45) of the promoter relative to the start site of transcription, though expression can also be regulated by a negatively-acting region at (−1556/−1165) (Sax et al. 1994). The proximal region of the promoter includes various positive and negative elements. For example upstream transcription factor (USF), a member of the basic helix loop helix zipper transcription factor family, binds to (−7/+5) (Sax et al. 1997), Pax-6 binds to (−48/−33) (Cvekl et al. 1995) and a CREB/ATF family member binds to a site at (−111/−93) (Cvekl et al. 1995). An unknown protein appears to bind to site (−25/−12) and interacts with TBP binding at the TATA box (−31/−26) (Sax et al. 1995). α-Crystallin binding protein 1 (α-cryBP1) is a large (2688 amino acid) ubiquitous protein which interacts specifically with a functionally important element at (−66/−57) of the murine *cryaA* promoter (Nakamura et al. 1990). Antisense inhibition of α-cryBP1 expression reduces expression under the control of the *cryaA* enhancer (Brady et al. 1995).

7.2
Regulation of αB-Crystallin Gene Transcription

Transcription of *cryaB* in most tissues is directed by a transcription site which is located downstream of a TATA box, but alternative transcription start sites between −285 and −266 are used in rat brain (A. Iwaki et al. 1990) and a transcription start at −474 is used in mouse brain and lung cells (Frederikse et al. 1994). Heat-induced transcription of murine *cryaB* is probably under the control of a HSE at position (−39/−53) and possibly a shorter HSE at (−376/−385) (Klemenz et al. 1991). Also, a glucocorticoid receptor sensitive element appears to be present between position −465 and −389 (Scheier et al. 1996).

Lens-specific activity of the *cryaB* promoter has been demonstrated to require only the region (−164/+44) of the promoter, which comprises two sites: LSR1 (−147/−118) and LSR2 (−78 and −46) (Gopal-Srivastava and Piatigorsky 1994). These can both be activated by Pax-6 and retinoic acid receptors in co-transfection experiments, but presumably there are also other factors involved which are lens-specific. During development, an upstream enhancer at position (−429/−259) of the *cryaB* promoter is required for all non-lens promoter activity of a transgenic reporter-gene in mice (Gopal-Srivastava et al. 1995, 2000). This enhancer contains at least 5 cis-elements called αBE1, αBE4, αBE2, αBE3, and MRF. The role of these elements in cardiac regulation of αB-crystallin transcription has been studied using cardiac cell lines in tissue culture, in cultured neonatal rat cardiomyocytes, and by transgenic expression of reporter genes in mice.

Virtually nothing is known about the factor binding αBE1 (−407/−397), except that this site is occupied in extracts from murine skeletal muscle, cardiomyocytes, lung, lens and even L929 cells which barely express HSP25 (Gopal-Srivastava and Piatigorsky 1993; Gopal-Srivastava et al. 1995; Haynes et al. 1995).

αBE4 (−386/−377) is used only in myocardial cells. It contains an overlapping HSE and a reverse CArG box [5′-GG(A/T)$_6$CC-3′] to which an SRF-like factor binds (Gopal-Srivastava et al. 1995). It seems likely that other factors interact at this site since SRF in myotube nuclear extracts does not appear to bind the αBE4 site (Gopal-Srivastava and Piatigorsky 1993). The αBE2 site (−360/−327) has an AP-2-like binding sequence (Gopal-Srivastava and Piatigorsky 1993) and footprinting experiments indicate it is occupied, but the factor binding it has not been identified. The αBE3 site (−317/−306) is used in cells of all tissues tested except for lung (Haynes et al. 1995).

The MRF binding site (−300/−288) is used selectively in skeletal muscle and myocardial cells (Gopal-Srivastava and Piatigorsky 1993; Gopal-Srivastava et al. 1995).The *cryaB* MRF site contains a single E-box [CANNTG] which, in skeletal muscle, has been proposed to bind to a MyoD family member and to interact with at least one additional element (Gopal-Srivastava and Piatigorsky 1993). In skeletal muscle Myf-5 may be the only myogenic factor expressed early enough to regulate *cryaB* in mouse myotomes (Haynes et al. 1996). The MRF site is also essential for promoter activity in transfected primary cardiomyocytes but rather than binding a MyoD member, it appears to bind a member of the USF family (Gopal-Srivastava et al. 1995).

Curiously, the *cryaB* gene is head-to-head with the HSPB2/MKBP gene at 11q22-3 in the human, separated by a distance of less than 1 kb. A similar arrangement is present in the mouse, but not the duck (Iwaki et al. 1997). The possibility obviously arises that they share promoter elements. There is a GC box but no obvious TATA box immediately upstream of HSPB2. There are two conserved HSE, although they may only be involved in the induction of *cryaB* transcription since MKBP transcription is not heat inducible (Suzuki et al. 1998; Sugiyama et al. 2000). There are two E boxes present at −140 and −166 of MKBP, which may be involved in muscle-specific transcription.

7.3
Regulation of HSP25 Gene Transcription

The murine *hsp25* gene contains two HSE at (−199/−186) and (−179/−166) involved in stress induction via binding to HSF1 (Gaestel et al. 1993). There is also an HSE in the first intron of mammalian *hsp27* genes, which may mediate negative regulation of *hsp27* transcription (Cooper et al. 2000). Other putative promoter control elements include two GC signals at (−113/−108) and (−103/−98), a CAAT signal (−78/−71) and a TATA signal (−27/−21).

The first 200 bp of the *hsp27* promoter are responsible for the majority of the promoter activity in human breast cancer cells (Oesterreich et al. 1996). A protein, called HET (HSP27-ERE-TATA-binding protein) had been identified as binding within this region, to an imperfect estrogen response element (ERE) (−87/−83), but this appears to be indirect, via an interaction between HET and the estrogen receptor, previously known to bind to the ERE (Oesterreich et al. 2000). Intriguingly, HET has been identified as an hnRNP protein which relocates to nuclear granules after heat shock, where HSF1 is also located (Weighardt et al. 1999). The significance of these findings is not yet clear.

The murine *hsp25* promoter contains a 9/10 match to the mouse MEF2 consensus binding site [5′-CT($^A/_t$)($^a/_t$)AAATAG-3′] (Andres et al. 1995) at position (−51/−42) relative to the transcription start site. In skeletal muscle, MEF2 factors interact with members of the MyoD family of basic helix-loop-helix transcription factors and activate muscle-specific genes (Molkentin et al. 1995) so it is conceivable that MEF2 is involved in *hsp25* expression in muscle. Although no cardiac transcription factors are known to interact with MEF2, candidate molecules include GATA4, SRF and the cardiac basic helix-loop-helix proteins dHAND and eHAND (discussed in Black and Olson 1998). On the other hand, the human *hsp27* gene contains a possibly equivalent site (CCAT-TAATAG) at (−81/−72) which is mid-way between the MEF2 and the SRF consensus binding site of [CC($^A/_T$)$_6$GG]. MEF2 and SRF are both members of the MADS-box family of transcription factors, which share a 57 amino-acid N-terminal motif important for mediating DNA binding and protein dimerization (reviewed in Black and Olson 1998). It would be interesting to examine whether the factor with "antigenic similarities to SRF" binding to αBE4 in the *cryaB* promoter (Gopal-Srivastava et al. 1995) is in fact a member of the MEF2 family with a general role in regulating muscle expression of sHSP.

Developmental expression of *Drosophila* sHSP is regulated mainly by steroid hormones binding to a beta-ecdysone response element in the promoter (Thomas and Lengyel 1986). Mammalian HSP27 has also been shown to be regulated by hormones in certain tissues such as the breast and the endometrium of the uterus (reviewed in Ciocca et al. 1993).

8
The Role of Small HSP in Development and Differentiation

HSP are widely known for their role during stress, which is to bind unfolded proteins, prevent them from aggregating and chaperone their refolding (or degradation). Gradually it has emerged that certain HSP play vital roles in the normal metabolism of the cell, aiding protein synthesis, folding and transport. However, the highly regionalized expression pattern of HSP (and particularly sHSP) during the development of various organisms including *Drosophila*, mammals and plants suggests that they are playing critical roles at specific differentiation steps and that the different members of the sHSP family may have overlapping or partially redundant roles. In myocytes, as in certain other non-stressed cells, HSP27, αB-crystallin, HSP20 and apparently HSP22 form heteroligomeric complexes (Boelens et al. 1998b; Kato et al. 1992, 1994; Zantema et al. 1992), but MKBP, HSPB2 and HSPB3 are present in a distinct, and smaller oligomeric complex (Suzuki et al. 1998; Sugiyama et al. 2000). The universal properties of sHSP are believed to be conferred by their defining feature of a conserved "crystallin" domain found in the C termini.

8.1
The Chaperone Function

Like most of the HSP, sHSP are capable of chaperoning proteins, by preventing their aggregation and potentiating their refolding by other HSP (see J. Buchner, Chap. 3, this Vol.). Mice lacking the αA-crystallin gene accumulate cytoplasmic inclusion bodies (which contain αB-crystallin and HSP25) in lens fiber cells (Brady et al. 1997) and develop cataracts. Thus, sHSP appear to have an important chaperone activity during development, though the range of this activity may be restricted. For example, the upregulation of MKBP (and not other sHSP) in the skeletal muscle of myotonic dystrophy patients, the interaction observed between MKBP and DMPK, and the colocalization at the neuromuscular junction suggest a chaperone function of MKBP which is restricted to this kinase (Suzuki et al. 1998).

8.2
The Direct or Indirect Actin-Binding Property

The chaperone activity of most sHSP seems directed towards structural and contractile proteins such as actin and the intermediate filaments. In fact, one of the earliest proposed roles of HSP25 was as an inhibitor of actin polymerization, based on an in vitro inhibitory activity of non-phosphorylated HSP25 oligomers (Miron et al. 1991; Benndorf et al. 1994) and this has since been confirmed in tissue culture (Schneider et al. 1998). HSP27, αA-crystallin and αB-crystallin may furthermore have a protective role in stabilizing the actin network (Guay et al. 1997), an effect which is phosphorylation-dependent (Wang and Spector 1996).

HSP27 and HSP20 have been proposed to mediate contraction of smooth muscle cells (Bitar et al. 1991; Brophy et al. 1999b). HSP27 has been found to co-immunoprecipitate not only with actin, but with myosin, tropomyosin and caldesmon in extracts from smooth muscle cells (Ibitayo et al. 1999). Phosphorylated HSP20 has also been shown to co-immunoprecipitate with nonpolymerized actin in extracts from smooth muscle. Additionally, cvHSP has been found to bind to α-filamin, and might conceivably alter a regulatory function of filamin on actin polymerization (Krief et al. 1999). On the other hand, we have shown that HSP25 is not essential for the formation of functional contractile actinomyosin filaments during in vitro differentiation of an embryonic carcinoma into cardiomyocytes (Davidson and Morange 2000).

In addition to its role in muscular contraction, actin is involved in cellular migration, and HSP27 has also been implicated in this role, both in smooth muscle cells (Hedges et al. 1999), and in arterial or umbilical vein endothelial cells (Rousseau et al. 1997; Piotrowicz et al. 1998). Compellingly, HSP25 emerged from a genetic screen in *Drosophila* for suppressors of a cell migration mutant (Rorth et al. 1998), indicating a possible role for sHSP in cell migration during development.

8.3
Small HSP and Intermediate Filaments

Much attention has been focussed recently on the role of sHSP with respect to the intermediate filament network (see R. Quinlan, Chap. 12, this Vol.). As mentioned, αB-crystallin and HSP25 colocalize in muscle with actin and desmin, apparently via direct interactions. Strong evidence that sHSP and desmin interact in a manner critical for cardiac function has recently been provided by the demonstration that a point mutation of αB-crystallin is responsible for a familial desmin myopathy (Vicart et al. 1998). The relatively late-age appearance of this phenotype may be due to redundancy between αB-crystallin and other sHSP. Alternatively, the role of sHSP in maintaining the desmin network may not be essential for cardiogenesis in vivo, since desmin itself is not essential for cardiac or skeletal muscle development in mice (Li et al. 1996; Milner et al. 1996) and skeletal differentiation can proceed in vitro even when desmin is made to form aggregates (Schultheiss et al. 1991). The function of desmin rather seems to be to confer enhanced resistance to physical stress, particularly as the demand on muscles and heart increases after birth (Li et al. 1997), and therefore sHSP may be instrumental in establishing and/or chaperoning this network under such circumstances.

8.4
Small HSP and Apoptosis

Many HSP can protect cells from various inducers of apoptosis (see P. Arrigo, Chap. 10, this Vol.). sHSP appear to be unusual in that their overexpression can increase resistance to apoptosis induced by either caspase-8 independent or dependent (i.e., fas ligand) pathways (Mehlen et al. 1996). This distinguishes the sHSP from HSP70 which exerts its anti-apoptotic role at the level of the apoptosome (Beere et al. 2000; Saleh et al. 2000). Thus, one role for sHSP in development may be to confer resistance to apoptosis. The transient increase in HSP27 expression which occurs during in vitro differentiation of embryonic stem cells or rat olfactory neurons counteracts apoptosis (Mehlen et al. 1997, 1999). Apoptosis is involved in normal heart development, primarily in non-myocytes (Poelmann et al. 2000), which, though it may be coincidental, are also those cells which do not express HSP25. Furthermore, mature hearts are protected against various inducers of apoptosis in vivo by virally-delivered HSP25

(Brar et al. 1999). However, it is obvious that not all cells which lack HSP25 die during development, therefore one must ask why certain cells need to fortify their resistance to apoptosis.

8.5
Small HSP and the Response to Stress

An alternative proposal is that, rather than sHSP being pre-emptively expressed before developmental apoptotic challenges (see above), or simultaneously as required for intermediate filament rearrangements (Sect. 7.3), developmental expression occurs in response to normal physiological stress. For instance, the expression of HSP25 in tissues such as kidney and bladder may be due to exposure to toxic environmental or metabolic products (Tanguay et al. 1993) or simply to changes in extracellular conditions, such as tonicity (Beck et al. 2000). The heart may increase expression of HSP25 in response to the oxidative stress experienced at birth (Shama et al. 1999), thus protecting the cells from reactive oxygen species, as sHSP have been shown to be capable of in vitro (Preville et al. 1999). Alternatively, the fact that HSP25 is induced in failing and hypertrophic adult hearts (Knowlton et al. 1998; see D. Latchman, Chap. 14, this Vol.) suggests that neonatal HSP25 expression may be due to increased load on the heart. sHSP are induced in skeletal muscle in response to the increased strain of muscular contractions which occur in the neonate, an effect which is prevented by denervation (Atomi et al. 1991a; Inaguma et al. 1993, 1996; Neufer and Benjamin 1996). However, it seems clear that the developmental expression of sHSP is not solely in response to extrinsic stresses, since it occurs equally in tissue culture differentiation systems (e.g. Davidson and Morange 2000; Sugiyama et al. 2000). Furthermore, as mentioned, developmental sHSP expression does not seem to involve the stress-related HSF.

8.6
Small HSP and Protein Degradation

One role of HSP such as HSP70 seems to be to chaperone proteins towards protein degradation pathways (Bercovich et al. 1997), and there is some evidence that sHSP may also be involved in proteolysis. sHSP have been detected in a number of neurodegenerative diseases colocalized with intracellular aggregates of ubiquitylated intermediate filaments, for example in the Lewy bodies of Parkinson's disease (Lowe et al. 1992). A type of intracellular protein aggregate called an "aggresome" has recently been described (Johnston et al. 1998), which are non-membrane bound accumulations of unfolded proteins which also include various HSP including HSP27 (Anton et al. 1999). The presence of the proteasomal machinery indicates that protein degradation is also occurring in these aggresomes. A subunit of the proteasome was found to be associated with the sHSP oligomer (de Sa et al. 1989), and an interaction has been detected between a *Drosophila* sHSP and DmUbc9 (Joanisse et al. 1998), one of the enzymes that conjugates ubiquitin to proteins and "marks" them for degradation.

Though it is currently only at speculation level, it is possible that developmental sHSP accumulation is related to the huge change in protein profile (i.e. degradation of old proteins and synthesis of new proteins) which must occur when a cell differentiates.

9
Small HSP in the Development of Non-Mammals

sHSP also have a highly restricted expression pattern in organisms other than mammals (discussed in other chapters of this Volume). A recent examination of the completed genome of *C. elegans* identified 16 sHSP, and at least one of these, called HSP25, is localized to dense bodies (the worm equivalent of the Z-lines of vertebrate muscle) and M-lines in body wall muscle, the lining of the pharynx and the junctions between the cells of the spermathecal wall (Ding and Candido 2000). This protein, which was shown to have chaperone properties, was found to interact with vinculin and alpha-actinin, but not actin, suggesting a role related to focal adhesions. However, double-stranded RNA-i silencing of HSP25 in *C.elegans* did not affect development (Ding and Candido 2000), in contrast to the essential role demonstrated for another worm sHSP, SEC-1, which is expressed uniquely during early development (Linder et al. 1996).

Four sHSP (HSP30A, C, D and E) have been cloned from *Xenopus laevis*, but these are not expressed during development in the absence of stress (reviewed in Heikkila et al. 1997). During embryonic development, HSP30 mRNA can be detected constitutively and transiently only in the cement gland, a mucus-secreting structure at the anterior end of the embryo which permits attachment of the embryo to a solid support (Lang et al. 1999). More recently, a second group of five more basic sHSP was cloned from *Xenopus*, and some of these appear to be expressed constitutively (Ohan et al. 1998).

Two avian sHSP have been cloned, and one of the proteins exhibits constitutive expression restricted to certain tissues, particularly the eye, heart, and digestive organs at embryonic day 12 (Kawazoe et al. 1999a). Since all three members of the chicken HSF family are expressed during development, they have been suggested to be responsible for the developmental expression of chicken HSP (Kawazoe et al. 1999b).

The examples cited in this chapter illustrate both the complex co-ordinated regulation of sHSP expression during development, and the difficulty in establishing the critical function which is common to the sHSP expressed in different tissues. As with many of the HSP, although the basic function of sHSP may be as a "chaperone", the primary domain of action appears to vary with the range of target proteins present in different cell types, and this may result in different end effects on apoptosis, cell migration, or other cellular processes.

Acknowledgements. We are indebted to Michael Welsh for providing unpublished information and to Yunhua Chang and Noëlle Favet for their help.

References

Andres V, Cervera M, Mahdavi V (1995) Determination of the consensus binding site for MEF2 expressed in muscle and brain reveals tissue-specific sequence constraints. J Biol Chem 270: 23246–23249

Anton LC, Schubert U, Bacik I, Princiotta MF, Wearsch PA, Gibbs J, Day PM, Realini C, Rechsteiner MC, Bennink JR, Yewdell JW (1999) Intracellular localization of proteasomal degradation of a viral antigen. J Cell Biol 146:113–124

Armstrong CL, Krueger-Naug AM, Currie RW, Hawkes R (2000) Constitutive expression of the 25-kDa heat shock protein Hsp25 reveals novel parasagittal bands of Purkinje cells in the adult mouse cerebellar cortex. J Comp Neurol 416:383–397

Armstrong SC, Shivell CL, Ganote CE (2000) Differential translocation or phosphorylation of alpha B crystallin cannot be detected in ischemically preconditioned rabbit cardiomyocytes. J Mol Cell Cardiol 32:1301–1314

Atomi Y, Yamada S, Nishida T (1991a) Early changes of alpha B-crystallin mRNA in rat skeletal muscle to mechanical tension and denervation. Biochem Biophys Res Commun 181: 1323–1330

Atomi Y, Yamada S, Strohman R, Nonomura Y (1991b) Alpha B-crystallin in skeletal muscle: purification and localization. J Biochem (Tokyo) 110:812–822

Barbato R, Menabo R, Dainese P, Carafoli E, Schiaffino S, Di Lisa F (1996) Binding of cytosolic proteins to myofibrils in ischemic rat hearts. Circ Res 78:821–828

Beall AC, Kato K, Goldenring JR, Rasmussen H, Brophy CM (1997) Cyclic nucleotide-dependent vasorelaxation is associated with the phosphorylation of a small heat shock-related protein. J Biol Chem 272:11283–11287

Beck F-X, Neuhofer W, Müller E (2000) Molecular chaperones in the kidney: distribution, putative roles and regulation. Am J Physiol Renal Physiol 279:203–215

Beere HM, Wolf BB, Cain K, Mosser DD, Mahboubi A, Kuwana T, Tailor P, Morimoto RI, Cohen GM, Green DR (2000) Heat-shock protein 70 inhibits apoptosis by preventing recruitment of procaspase-9 to the Apaf-1 apoptosome. Nat Cell Biol 2:469–475

Benjamin IJ, Shelton J, Garry DJ, Richardson JA (1997) Temporospatial expression of the small HSP/alpha B-crystallin in cardiac and skeletal muscle during mouse development. Dev Dyn 208:75–84

Bennardini F, Wrzosek A, Chiesi M (1992) Alpha B-crystallin in cardiac tissue. Association with actin and desmin filaments. Circ Res 71:288–294

Benndorf R, Hayess K, Ryazantsev S, Wieske M, Behlke J, Lutsch G (1994) Phosphorylation and supramolecular organization of murine small heat shock protein HSP25 abolish its actin polymerization-inhibiting activity. J Biol Chem 269:20780–20784

Bercovich B, Stancovski I, Mayer A, Blumenfeld N, Laszlo A, Schwartz AL, Ciechanover A (1997) Ubiquitin-dependent degradation of certain protein substrates in vitro requires the molecular chaperone Hsc70. J Biol Chem 272:9002–9010

Bitar KN, Kaminski MS, Hailat N, Cease KB, Strahler JR (1991) Hsp27 is a mediator of sustained smooth muscle contraction in response to bombesin. Biochem Biophys Res Commun 181: 1192–1200

Black BL, Olson EN (1998) Transcriptional control of muscle development by myocyte enhancer factor-2 (MEF2) proteins. Annu Rev Cell Dev Biol 14:167–196

Boelens W, van Boekel M, de Jong W (1998a) HspB3, the most deviating of the six known human heat shock proteins. Biochim Biophys Acta 1388:513–516

Boelens WC, Croes Y, de Ruwe M, de Reu L, de Jong WW (1998b) Negative charges in the C-terminal domain stabilize the alphaB-crystallin complex. J Biol Chem 273:28085–28090

Brady JP, Kantorow M, Sax CM, Donovan DM, Piatigorsky J (1995) Murine transcription factor alpha A-crystallin binding protein I. Complete sequence, gene structure, expression, and functional inhibition via antisense RNA. J Biol Chem 270:1221–1229

Brady JP, Garland D, Duglas-Tabor Y, Robison WG Jr, Groome A, Wawrousek EF (1997) Targeted disruption of the mouse alpha A-crystallin gene induces cataract and cytoplasmic inclusion bodies containing the small heat shock protein alpha B-crystallin. Proc Natl Acad Sci USA 94:884–889

Brar BK, Stephanou A, Wagstaff MJ, Coffin RS, Marber MS, Engelmann G, Latchman DS (1999) Heat shock proteins delivered with a virus vector can protect cardiac cells against apoptosis as well as against thermal or hypoxic stress. J Mol Cell Cardiol 31:135–146

Brophy CM, Dickinson M, Woodrum D (1999a) Phosphorylation of the small heat shock-related protein, HSP20, in vascular smooth muscles is associated with changes in the macromolecular associations of HSP20. J Biol Chem 274:6324–6329

Brophy CM, Lamb S, Graham A (1999b) The small heat shock-related protein-20 is an actin-associated protein. J Vasc Surg 29:326–333

Chaufour S, Mehlen P, Arrigo AP (1996) Transient accumulation, phosphorylation and changes in the oligomerization of Hsp27 during retinoic acid-induced differentiation of HL-60 cells: possible role in the control of cellular growth and differentiation. Cell Stress Chaperones 1:225–235

Cheng G, Litchenberg WH, Cole GJ, Mikawa T, Thompson RP, Gourdie RG (1999) Development of the cardiac conduction system involves recruitment within a multipotent cardiomyogenic lineage. Development 126:5041–5049

Chiesa R, Noguera I, Sredy J (1997) Phosphorylation of HSP25 during lens cell differentiation. Exp Eye Res 65:223–229

Ciocca DR, Asch RH, Adams DJ, McGuire WL (1983) Evidence for modulation of a 24 K protein in human endometrium during the menstrual cycle. J Clin Endocrinol Metab 57:496–499

Ciocca DR, Oesterreich S, Chamness GC, McGuire WL, Fuqua SA (1993) Biological and clinical implications of heat shock protein 27,000 (Hsp27): a review. J Natl Cancer Inst 85:1558–1570

Ciocca DR, Stati AO, Fanelli MA, Gaestel M (1996) Expression of heat shock protein 25,000 in rat uterus during pregnancy and pseudopregnancy. Biol Reprod 54:1326–1335

Cooper LF, Uoshima K, Guo Z (2000) Transcriptional regulation involving the intronic heat shock element of the rat hsp27 gene. Biochim Biophys Acta 1490:348–354

Costigan M, Mannion RJ, Kendall G, Lewis SE, Campagna JA, Coggeshall RE, Meridith-Middleton J, Tate S, Woolf CJ (1998) Heat shock protein 27: developmental regulation and expression after peripheral nerve injury. J Neurosci 18:5891–5900

Cvekl A, Kashanchi F, Sax CM, Brady JN, Piatigorsky J (1995) Transcriptional regulation of the mouse alpha A-crystallin gene: activation dependent on a cyclic AMP-responsive element (DE1/CRE) and a Pax-6-binding site. Mol Cell Biol 15:653–660

Davidson SM, Morange M (2000) Hsp25 and the p38 MAPK pathway are involved in differentiation of cardiomyocytes. Dev Biol 218:146–160

De Sa CM, Rollet E, de Sa MF, Tanguay RM, Best-Belpomme M, Scherrer K (1989) Prosomes and heat shock complexes in Drosophila melanogaster cells. Mol Cell Biol 9:2672–2681

Ding L, Candido EP (2000) HSP25, a small heat shock protein associated with dense bodies and M-lines of body wall muscle in Caenorhabditis elegans. J Biol Chem 275:9510–9517

Dubin RA, Gopal-Srivastava R, Wawrousek EF, Piatigorsky J (1991) Expression of the murine alpha B-crystallin gene in lens and skeletal muscle: identification of a muscle-preferred enhancer. Mol Cell Biol 11:4340–4349

Frederikse PH, Dubin RA, Haynes JI II, Piatigorsky J (1994) Structure and alternate tissue-preferred transcription initiation of the mouse alpha B-crystallin/small heat shock protein gene. Nucleic Acids Res 22:5686–5694

Frohli E, Aoyama A, Klemenz R (1993) Cloning of the mouse hsp25 gene and an extremely conserved hsp25 pseudogene. Gene 128:273–277

Gaestel M, Gotthardt R, Muller T (1993) Structure and organisation of a murine gene encoding small heat-shock protein Hsp25. Gene 128:279–283

Gernold M, Knauf U, Gaestel M, Stahl J, Kloetzel PM (1993) Development and tissue-specific distribution of mouse small heat shock protein hsp25. Dev Genet 14:103–111

Gopal-Srivastava R, Piatigorsky J (1993) The murine alpha B-crystallin/small heat shock protein enhancer: identification of alpha BE-1, alpha BE-2, alpha BE-3, and MRF control elements. Mol Cell Biol 13:7144–7152

Gopal-Srivastava R, Piatigorsky J (1994) Identification of a lens-specific regulatory region (LSR) of the murine alpha B-crystallin gene. Nucleic Acids Res 22:1281–1286

Gopal-Srivastava R, Haynes JI II, Piatigorsky J (1995) Regulation of the murine alpha B-crystallin/small heat shock protein gene in cardiac muscle. Mol Cell Biol 15:7081–7090

Gopal-Srivastava R, Kays WT, Piatigorsky J (2000) Enhancer-independent promoter activity of the mouse alphaB-crystallin/small heat shock protein gene in the lens and cornea of transgenic mice. Mech Dev 92:125–134

Guay J, Lambert H, Gingras-Breton G, Lavoie JN, Huot J, Landry J (1997) Regulation of actin filament dynamics by p38 map kinase-mediated phosphorylation of heat shock protein 27. J Cell Sci 110:357–368

Hanash SM, Strahler JR, Chan Y, Kuick R, Teichroew D, Neel JV, Hailat N, Keim DR, Gratiot-Deans J, Ungar D et al. (1993) Data base analysis of protein expression patterns during T-cell ontogeny and activation. Proc Natl Acad Sci USA 90:3314–3318

Haynes JI II, Gopal-Srivastava R, Frederikse PH, Piatigorsky J (1995) Differential use of the regulatory elements of the alpha B-crystallin enhancer in cultured murine lung (MLg), lens (alpha TN4-1) and muscle (C2C12) cells. Gene 155:151–158

Haynes JI II, Duncan MK, Piatigorsky J (1996) Spatial and temporal activity of the alpha B-crystallin/small heat shock protein gene promoter in transgenic mice. Dev Dyn 207:75–88

Hedges JC, Dechert MA, Yamboliev IA, Martin JL, Hickey E, Weber LA, Gerthoffer WT (1999) A role for p38(MAPK)/HSP27 pathway in smooth muscle cell migration. J Biol Chem 274: 24211–24219

Heikkila JJ, Ohan N, Tam Y, Ali A (1997) Heat shock protein gene expression during Xenopus development. Cell Mol Life Sci 53:114–121

Hickey E, Brandon SE, Potter R, Stein G, Stein J, Weber LA (1986) Sequence and organization of genes encoding the human 27 kDa heat shock protein [published erratum appears in Nucleic Acids Res 1986 Oct 24;14(20):8230]. Nucleic Acids Res 14:4127–4145

Ibitayo AI, Sladick J, Tuteja S, Louis-Jacques O, Yamada H, Groblewski G, Welsh M, Bitar KN (1999) HSP27 in signal transduction and association with contractile proteins in smooth muscle cells. Am J Physiol 277:G445–G454

Inaguma Y, Goto S, Shinohara H, Hasegawa K, Ohshima K, Kato K (1993) Physiological and pathological changes in levels of the two small stress proteins, HSP27 and alpha B crystallin, in rat hindlimb muscles. J Biochem (Tokyo) 114:378–384

Inaguma Y, Hasegawa K, Kato K, Nishida Y (1996) cDNA cloning of a 20-kDa protein (p20) highly homologous to small heat shock proteins: developmental and physiological changes in rat hindlimb muscles. Gene 178:145–150

Ito H, Iida K, Kamei K, Iwamoto I, Inaguma Y, Kato K (1999) AlphaB-crystallin in the rat lens is phosphorylated at an early post-natal age. FEBS Lett 446:269–272

Iwaki A, Iwaki T, Goldman JE, Liem RK (1990) Multiple mRNAs of rat brain alpha-crystallin B chain result from alternative transcriptional initiation. J Biol Chem 265:22197–22203

Iwaki A, Nagano T, Nakagawa M, Iwaki T, Fukumaki Y (1997) Identification and characterization of the gene encoding a new member of the alpha-crystallin/small hsp family, closely linked to the alphaB- crystallin gene in a head-to-head manner. Genomics 45:386–394

Iwaki T, Kume-Iwaki A, Goldman JE (1990) Cellular distribution of alpha B-crystallin in non-lenticular tissues. J Histochem Cytochem 38:31–39

Jantschitsch C, Kindas-Mugge I, Metze D, Amann G, Micksche M, Trautinger F (1998) Expression of the small heat shock protein HSP 27 in developing human skin. Br J Dermatol 139:247–253

Joanisse DR, Inaguma Y, Tanguay RM (1998) Cloning and developmental expression of a nuclear ubiquitin-conjugating enzyme (DmUbc9) that interacts with small heat shock proteins in *Drosophila melanogaster*. Biochem Biophys Res Commun 244:102–109

Johnston JA, Ward CL, Kopito RR (1998) Aggresomes: a cellular response to misfolded proteins. J Cell Biol 143:1883–1898

Kato K, Shinohara H, Kurobe N, Inaguma Y, Shimizu K, Ohshima K (1991) Tissue distribution and developmental profiles of immunoreactive alpha B crystallin in the rat determined with a sensitive immunoassay system. Biochim Biophys Acta 1074:201–208

Kato K, Shinohara H, Goto S, Inaguma Y, Morishita R, Asano T (1992) Copurification of small heat shock protein with alpha B crystallin from human skeletal muscle. J Biol Chem 267: 7718–7725

Kato K, Goto S, Inaguma Y, Hasegawa K, Morishita R, Asano T (1994) Purification and characterization of a 20-kDa protein that is highly homologous to alpha B crystallin. J Biol Chem 269:15302–15309

Kato K, Ito H, Kamei K, Inaguma Y, Iwamoto I, Saga S (1998) Phosphorylation of alphaB-crystallin in mitotic cells and identification of enzymatic activities responsible for phosphorylation. J Biol Chem 273:28346–28354

Kawamura H, Otsuka T, Matsuno H, Niwa M, Matsui N, Kato K, Uematsu T, Kozawa O (1999) Endothelin-1 stimulates heat shock protein 27 induction in osteoblasts: involvement of p38 MAP kinase. Am J Physiol 277:E1046–E1054

Kawazoe Y, Tanabe M, Nakai A (1999a) Ubiquitous and cell-specific members of the avian small heat shock protein family. FEBS Lett 455:271–275

Kawazoe Y, Tanabe M, Sasai N, Nagata K, Nakai A (1999b) HSF3 is a major heat shock responsive factor during chicken embryonic development. Eur J Biochem 265:688–697

Kindas-Mugge I, Trautinger F (1994) Increased expression of the M(r) 27,000 heat shock protein (hsp27) in in vitro differentiated normal human keratinocytes. Cell Growth Differ 5:777–781

Klemenz R, Frohli E, Steiger RH, Schafer R, Aoyama A (1991) Alpha B-crystallin is a small heat shock protein. Proc Natl Acad Sci USA 88:3652–3656

Klemenz R, Andres AC, Frohli E, Schafer R, Aoyama A (1993) Expression of the murine small heat shock proteins hsp 25 and alpha B crystallin in the absence of stress. J Cell Biol 120:639–645

Knoepp L, Beall A, Woodrum D, Mondy JS, Shaver E, Dickinson M, Brophy CM (2000) Cellular stress inhibits vascular smooth muscle relaxation. J Vasc Surg 31:343–353

Knowlton AA, Kapadia S, Torre-Amione G, Durand JB, Bies R, Young J, Mann DL (1998) Differential expression of heat shock proteins in normal and failing human hearts. J Mol Cell Cardiol 30:811–818

Krief S, Faivre JF, Robert P, Le Douarin B, Brument-Larignon N, Lefrere I, Bouzyk MM, Anderson KM, Greller LD, Tobin FL, Souchet M, Bril A (1999) Identification and characterization of cvHsp. A novel human small stress protein selectively expressed in cardiovascular and insulin-sensitive tissues. J Biol Chem 274:36592–36600

Lam WY, Wing Tsui SK, Law PT, Luk SC, Fung KP, Lee CY, Waye MM (1996) Isolation and characterization of a human heart cDNA encoding a new member of the small heat shock protein family–HSPL27. Biochim Biophys Acta 1314:120–124

Lang L, Miskovic D, Fernando P, Heikkila JJ (1999) Spatial pattern of constitutive and heat shock-induced expression of the small heat shock protein gene family, Hsp30, in Xenopus laevis tailbud embryos. Dev Genet 25:365–374

Leach IH, Tsang ML, Church RJ, Lowe J (1994) Alpha-B crystallin in the normal human myocardium and cardiac conducting system. J Pathol 173:255–260

Li Z, Colucci-Guyon E, Pincon-Raymond M, Mericskay M, Pournin S, Paulin D, Babinet C (1996) Cardiovascular lesions and skeletal myopathy in mice lacking desmin. Dev Biol 175:362–366

Li Z, Mericskay M, Agbulut O, Butler-Browne G, Carlsson L, Thornell LE, Babinet C, Paulin D (1997) Desmin is essential for the tensile strength and integrity of myofibrils but not for myogenic commitment, differentiation, and fusion of skeletal muscle. J Cell Biol 139:129–144

Linder B, Jin Z, Freedman JH, Rubin CS (1996) Molecular characterization of a novel, developmentally regulated small embryonic chaperone from Caenorhabditis elegans. J Biol Chem 271: 30158–30166

Loones M-T, Morange M (1998) Hsp and chaperone distribution during endochondral bone development in mouse embryo. Cell Stress Chaperones 3:237–244

Loones M-T, Rallu M, Mezger V, Morange M (1997) HSP gene expression and HSF2 in mouse development. Cell Mol Life Sci 53:179–190

Loones M-T, Chang YH, Morange M (2000) The distribution of heat shock proteins in the nervous system of the unstressed mouse embryo suggests a role in neuronal and non-neuronal differentiation. Cell Stress Chaperones 5:291–305

Lowe J, McDermott H, Pike I, Spendlove I, Landon M, Mayer RJ (1992) Alpha B crystallin expression in non-lenticular tissues and selective presence in ubiquitinated inclusion bodies in human disease. J Pathol 166:61–68

Lutsch G, Vetter R, Offhauss U, Wieske M, Grone HJ, Klemenz R, Schimke I, Stahl J, Benndorf R (1997) Abundance and location of the small heat shock proteins HSP25 and alphaB-crystallin in rat and human heart. Circulation 96:3466–3476

Ma Z, Hanson SR, Lampi KJ, David LL, Smith DL, Smith JB (1998) Age-related changes in human lens crystallins identified by HPLC and mass spectrometry. Exp Eye Res 67:21–30

Mehlen P, Schulze-Osthoff K, Arrigo AP (1996) Small stress proteins as novel regulators of apoptosis. Heat shock protein 27 blocks Fas/APO-1- and staurosporine-induced cell death. J Biol Chem 271:16510–16514

Mehlen P, Mehlen A, Godet J, Arrigo AP (1997) hsp27 as a switch between differentiation and apoptosis in murine embryonic stem cells. J Biol Chem 272:31657–31665

Mehlen P, Coronas V, Ljubic-Thibal V, Ducasse C, Granger L, Jourdan F, Arrigo AP (1999) Small stress protein Hsp27 accumulation during dopamine-mediated differentiation of rat olfactory neurons counteracts apoptosis. Cell Death Differ 6:227–233

Mendelsohn ME, Zhu Y, O'Neill S (1991) The 29-kDa proteins phosphorylated in thrombin-activated human platelets are forms of the estrogen receptor-related 27-kDa heat shock protein. Proc Natl Acad Sci USA 88:11212–11216

Milner DJ, Weitzer G, Tran D, Bradley A, Capetanaki Y (1996) Disruption of muscle architecture and myocardial degeneration in mice lacking desmin. J Cell Biol 134:1255–1270

Miron T, Vancompernolle K, Vandekerckhove J, Wilchek M, Geiger B (1991) A 25-kD inhibitor of actin polymerization is a low molecular mass heat shock protein. J Cell Biol 114:255–261

Molkentin JD, Black BL, Martin JF, Olson EN (1995) Cooperative activation of muscle gene expression by MEF2 and myogenic bHLH proteins. Cell 83:1125–1136

Müller E, Neuhofer W, Ohno A, Rucker S, Thurau K, Beck FX (1996) Heat shock proteins HSP25, HSP60, HSP72, HSP73 in isoosmotic cortex and hyperosmotic medulla of rat kidney. Pfluegers Arch 431:608–617

Müller E, Burger-Kentischer A, Neuhofer W, Fraek M-L, März J, Thurau K, Beck F-X (1999) Possible involvement of heat shock protein 25 in the angiotensin II-induced glomerular mesangial cell contraction via p38 MAP kinase. J Cell Physiol 181:462–469

Nakamura T, Donovan DM, Hamada K, Sax CM, Norman B, Flanagan JR, Ozato K, Westphal H, Piatigorsky J (1990) Regulation of the mouse alpha A-crystallin gene: isolation of a cDNA encoding a protein that binds to a cis sequence motif shared with the major histocompatibility complex class I gene and other genes. Mol Cell Biol 10:3700–3708

Neri CL, Duchala CS, Macklin WB (1997) Expression of molecular chaperones and vesicle transport proteins in differentiating oligodendrocytes. J Neurosci Res 50:769–780

Neufer PD, Benjamin IJ (1996) Differential expression of B-crystallin and Hsp27 in skeletal muscle during continuous contractile activity. Relationship to myogenic regulatory factors. J Biol Chem 271:24089–24095

Oesterreich S, Hickey E, Weber LA, Fuqua SA (1996) Basal regulatory promoter elements of the hsp27 gene in human breast cancer cells. Biochem Biophys Res Commun 222:155–163

Oesterreich S, Zhang Q, Hopp T, Fuqua SA, Michaelis M, Zhao HH, Davie JR, Osborne CK, Lee AV (2000) Tamoxifen-bound estrogen receptor (ER) strongly interacts with the nuclear matrix protein HET/SAF-B, a novel inhibitor of ER-mediated transactivation. Mol Endocrinol 14:369–381

Ohan NW, Tam Y, Fernando P, Heikkila JJ (1998) Characterization of a novel group of basic small heat shock proteins in Xenopus laevis A6 kidney epithelial cells. Biochem Cell Biol 76:665–671

Piotrowicz RS, Hickey E, Levin EG (1998) Heat shock protein 27 kDa expression and phosphorylation regulates endothelial cell migration. FASEB J 12:1481–1490

Plumier JC, Hopkins DA, Robertson HA, Currie RW (1997) Constitutive expression of the 27-kDa heat shock protein (Hsp27) in sensory and motor neurons of the rat nervous system. J Comp Neurol 384:409–428

Poelmann RE, Molin D, Wisse LJ, Gittenberger-de Groot AC (2000) Apoptosis in cardiac development. Cell Tissue Res 301:43–52

Preville X, Salvemini F, Giraud S, Chaufour S, Paul C, Stepien G, Ursini MV, Arrigo AP (1999) Mammalian small stress proteins protect against oxidative stress through their ability to increase glucose-6-phosphate dehydrogenase activity and by maintaining optimal cellular detoxifying machinery. Exp Cell Res 247:61–78

Rogalla T, Ehrnsperger M, Preville X, Kotlyarov A, Lutsch G, Ducasse C, Paul C, Wieske M, Arrigo AP, Buchner J, Gaestel M (1999) Regulation of Hsp27 oligomerization, chaperone function, and protective activity against oxidative stress/tumor necrosis factor alpha by phosphorylation. J Biol Chem 274:18947–18956

Rorth P, Szabo K, Bailey A, Laverty T, Rehm J, Rubin GM, Weigmann K, Milan M, Benes V, Ansorge W, Cohen SM (1998) Systematic gain-of-function genetics in Drosophila. Development 125: 1049–1057

Rousseau S, Houle F, Landry J, Huot J (1997) p38 MAP kinase activation by vascular endothelial growth factor mediates actin reorganization and cell migration in human endothelial cells. Oncogene 15:2169–2177

Sakamoto K, Urushidani T, Nagao T (2000) Translocation of HSP27 to sarcomere induced by ischemic preconditioning in isolated rat hearts. Biochem Biophys Res Commun 269:137–142

Saleh A, Srinivasula SM, Balkir L, Robbins PD, Alnemri ES (2000) Negative regulation of the Apaf-1 apoptosome by Hsp70. Nat Cell Biol 2:476–483

Sax CM, Cvekl A, Kantorow M, Sommer B, Chepelinsky AB, Piatigorsky J (1994) Identification of negative-acting and protein-binding elements in the mouse alpha A-crystallin –1556/–1165 region. Gene 144:163–169

Sax CM, Cvekl A, Kantorow M, Gopal-Srivastava R, Ilagan JG, Ambulos NP Jr, Piatigorsky J (1995) Lens-specific activity of the mouse alpha A-crystallin promoter in the absence of a TATA box: functional and protein binding analysis of the mouse alpha A-crystallin PE1 region. Nucleic Acids Res 23:442–451

Sax CM, Cvekl A, Piatigorsky J (1997) Transcriptional regulation of the mouse alpha A-crystallin gene: binding of USF to the –7/+5 region. Gene 185:209–216

Scheier B, Foletti A, Stark G, Aoyama A, Dobbeling U, Rusconi S, Klemenz R (1996) Glucocorticoids regulate the expression of the stressprotein alpha B-crystallin. Mol Cell Endocrinol 123:187–198

Schneider GB, Hamano H, Cooper LF (1998) In vivo evaluation of hsp27 as an inhibitor of actin polymerization: hsp27 limits actin stress fiber and focal adhesion formation after heat shock. J Cell Physiol 177:575–584

Schultheiss T, Lin Z, Ishikawa H, Zamir I, Stoeckert CJ, Holtzer H (1991) Desmin/vimentin intermediate filaments are dispensable for many aspects of myogenesis. J Cell Biol 114:953–966

Shah M, Stanek J, Handwerger S (1998) Differential localization of heat shock proteins 90, 70, 60 and 27 in human decidua and placenta during pregnancy. Histochem J 30:509–518

Shakoori AR, Oberdorf AM, Owen TA, Weber LA, Hickey E, Stein JL, Lian JB, Stein GS (1992) Expression of heat shock genes during differentiation of mammalian osteoblasts and promyelocytic leukemia cells. J Cell Biochem 48:277–287

Shama KM, Suzuki A, Harada K, Fujitani N, Kimura H, Ohno S, Yoshida K (1999) Transient upregulation of myotonic dystrophy protein kinase-binding protein, MKBP, and HSP27 in the neonatal myocardium. Cell Struct Funct 24:1–4

Smoyer WE, Gupta A, Mundel P, Ballew JD, Welsh MJ (1996) Altered expression of glomerular heat shock protein 27 in experimental nephrotic syndrome. J Clin Invest 97:2697–2704

Spector NL, Samson W, Ryan C, Gribben J, Urba W, Welch WJ, Nadler LM (1992) Growth arrest of human B lymphocytes is accompanied by induction of the low molecular weight mammalian heat shock protein (Hsp28). J Immunol 148:1668–1673

Spector NL, Ryan C, Samson W, Levine H, Nadler LM, Arrigo AP (1993) Heat shock protein is a unique marker of growth arrest during macrophage differentiation of HL-60 cells. J Cell Physiol 156:619–625

Srinivasan AN, Nagineni CN, Bhat SP (1992) alpha A-crystallin is expressed in non-ocular tissues. J Biol Chem 267:23337–23341

Stahl J, Wobus AM, Ihrig S, Lutsch G, Bielka H (1992) The small heat shock protein hsp25 is accumulated in P19 embryonal carcinoma cells and embryonic stem cells of line BLC6 during differentiation. Differentiation 51:33–37

Sugiyama Y, Suzuki A, Kishikawa M, Akutsu R, Hirose T, Waye MM, Tsui SK, Yoshida S, Ohno S (2000) Muscle develops a specific form of small heat shock protein complex composed of MKBP/HSPB2 and HSPB3 during myogenic differentiation. J Biol Chem 275:1095–1104

Suzuki A, Sugiyama Y, Hayashi Y, Nyu-i N, Yoshida M, Nonaka I, Ishiura S, Arahata K, Ohno S (1998) MKBP, a novel member of the small heat shock protein family, binds and activates the myotonic dystrophy protein kinase. J Cell Biol 140:1113–1124

Tanguay RM, Wu Y, Khandjian EW (1993) Tissue-specific expression of heat shock proteins of the mouse in the absence of stress. Dev Genet 14:112–118

Thomas SR, Lengyel JA (1986) Ecdysteroid-regulated heat-shock gene expression during *Drosophila melanogaster* development. Dev Biol 115:434–438

Tiffee JC, Griffin JP, Cooper LF (2000) Immunolocalization of stress proteins and extracellular matrix proteins in the rat tibia. Tissue Cell 32:141–147

Van de Klundert FA, de Jong WW (1999) The small heat shock proteins Hsp20 and alphaB-crystallin in cultured cardiac myocytes: differences in cellular localization and solubilization after heat stress. Eur J Cell Biol 78:567–572

Viac J, Su H, Reano A, Kanitakis J, Chardonnet Y, Thivolet J (1989) Distribution of an estrogen receptor-related protein (P29) in normal skin and in cultured human keratinocytes. J Dermatol 16:98–102

Vicart P, Caron A, Guicheney P, Li Z, Prevost MC, Faure A, Chateau D, Chapon F, Tome F, Dupret JM, Paulin D, Fardeau M (1998) A missense mutation in the alphaB-crystallin chaperone gene causes a desmin-related myopathy. Nat Genet 20:92–95

Wakayama T, Iseki S (1998) Expression and cellular localization of the mRNA for the 25-kDa heat-shock protein in the mouse. Cell Biol Int 22:295–304

Wakayama T, Iseki S (1999) Specific expression of the mRNA for 25 kDA heat-shock protein in the spermatocytes of mouse seminiferous tubules. Anat Embryol (Berl) 199:419–425

Wang K, Spector A (1996) alpha-crystallin stabilizes actin filaments and prevents cytochalasin-induced depolymerization in a phosphorylation-dependent manner. Eur J Biochem 242:56–66

Weighardt F, Cobianchi F, Cartegni L, Chiodi I, Villa A, Riva S, Biamonti G (1999) A novel hnRNP protein (HAP/SAF-B) enters a subset of hnRNP complexes and relocates in nuclear granules in response to heat shock. J Cell Sci 112:1465–1476

Xiao X, Zuo X, Davis AA, McMillan DR, Curry BB, Richardson JA, Benjamin IJ (1999) HSF1 is required for extra-embryonic development, postnatal growth and protection during inflammatory responses in mice. EMBO J 18:5943–5952

Yoshida K, Aki T, Harada K, Shama KM, Kamoda Y, Suzuki A, Ohno S (1999) Translocation of HSP27 and MKBP in ischemic heart. Cell Struct Funct 24:181–185

Zantema A, Verlaan-De Vries M, Maasdam D, Bol S, van der Eb A (1992) Heat shock protein 27 and alpha B-crystallin can form a complex, which dissociates by heat shock. J Biol Chem 267:12936–12941

Expression and Phosphorylation of Mammalian Small Heat Shock Proteins

Kanefusa Kato, Hidenori Ito, and Yutaka Inaguma[1]

1
Introduction

Mammalian small Hsps (shHsps) comprise 7 members found so far, αA- and αB-crystallin, Hsp25/27, Hsp20 (Kato et al. 1994a), MKBP (Suzuki et al. 1998), HspL27 (Lam et al. 1996), and cvHSP (Krief et al. 1999), all of which contain the α-crystallin domain in the carboxy-terminal half of the molecule.

Among these shHsps, only αB-crystallin and Hsp27 are stress-inducible. Although the regulation of expression of Hsp27 and αB-crystallin is not yet fully elucidated, it is believed to be mediated similarly to that of the Hsp70 gene by the binding of a heat shock factor (HSF) to the regulatory heat shock element (HSE) (Morimoto et al. 1990) found in the promoter region of their genes (Hickey et al. 1986; Klemenz et al. 1991; Gaestel et al. 1993). However, activation of HSE-binding activity of HSF does not always result in activation of genes for Hsps (Jurivich et al. 1992; Cotto et al. 1996). In addition to regulation at the level of activation of HSF, stress-induced synthesis of heat shock or stress proteins (Hsps) is also modulated by other transcription factors, such as a cyclic AMP responsive element binding protein (Choi et al. 1991). Steroid hormones, including estrogen, are known to be potent inducers of Hsp27 in mammalian cells (Edwards et al. 1980) as originally observed in *Drosophila* (Ireland and Berger 1982; Ireland et al. 1982). The induction of Hsp27 in chicken embryo cells (Edington and Hightower 1990) and of Hsp25 in Ehrlich ascites tumor cells by cisplatin can furthermore be enhanced without changes in transcriptional activity (Edington and Hightower 1990). The expression of Hsp27 and αB-crystallin seems to be regulated by several independent unknown systems.

Phosphorylation of serine residues is a common feature of small Hsps. Phosphorylated forms of αA- and αB-crystallins are known to be present in the ocular lens (Voorter et al. 1986; Chiesa et al. 1987) and Hsp27 can be phosphorylated at serine residues in response to stress (Welch 1985). αB- crystallin is also phosphorylated under stress conditions (Ito et al. 1997b) while Hsp20 in vascular smooth muscles is phosphorylated during cyclic nucleotide-

[1] Department of Biochemistry, Institute for Developmental Research, Aichi Human Service Center, 713-8 Kamiya, Kasugai, Aichi 480-0392, Japan

Progress in Molecular and Subcellular Biology, Vol. 28
A.-P. Arrigo and W.E.G. Müller (Eds.)
© Springer-Verlag Berlin Heidelberg 2002

dependent vasorelaxation (Beall et al. 1997). In this chapter we describe the modification of the expression of Hsp27 and αB-crystallin with various factors as well as the phosphorylation of Hsp27, Hsp20, and αB-crystallin.

2
Modification of the Expression of Hsp27 and αB-Crystallin

When cells are exposed to heat or chemical insult, expression of genes for Hsps is enhanced and the proteins accumulate in cells reaching a peak after about 10–16 h. Such cells exhibit tolerance to additional stress. In the case of heat shock for 15 min, responses of Hsps in cultured cells are detected at 43 °C or higher temperatures, but not at 42 °C (Fig. 1A,B). In contrast, when rats are subjected to a whole body heat shock at 42 °C for 15 min, expression of Hsp27 and αB-crystallin in the liver is markedly enhanced (Fig. 1C). Thus the response of

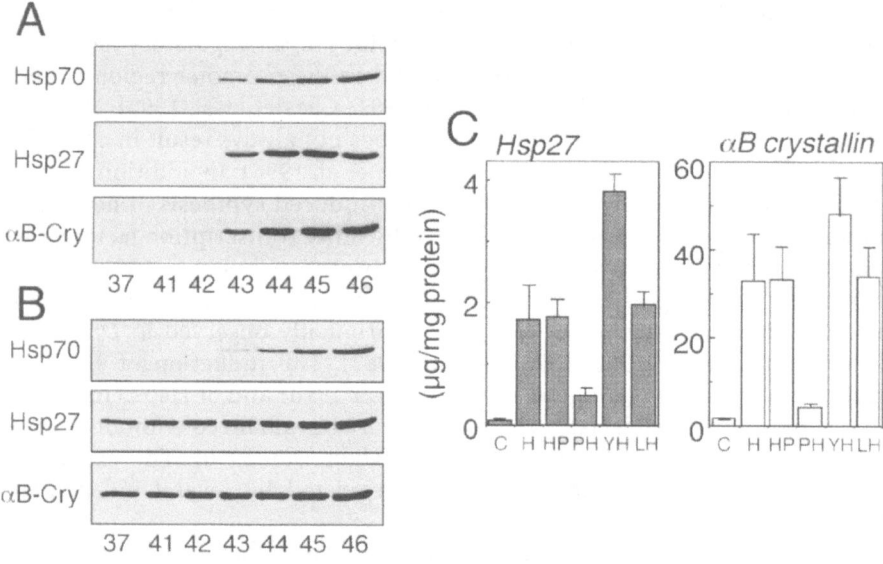

Temperature of heating for 15 min (°C)

Fig. 1. Induction of heat shock proteins by heat shock in cultured cells (**A** and **B**) or in rat liver. A C6 rat glioma cells or B BRL rat liver cells were subjected to heat shock for 15 min at the indicated temperature. After 16 h of culture at 37 °C, cells were harvested and each cell extract containing 20 µg of protein was subjected to SDS-PAGE and subsequent Western blot analysis with antibodies against Hsp70, Hsp27 and αB-crystallin. C Rats were given prazosin (*PH*), yohimbine (*YH*), or propranolol (*LH*) (0.8 mg/kg body weight, i.p.) before they were subjected to heat stress at 42 °C for 15 min (*H*), Rats in another group were injected with prazosin after heat treatment (*HP*). Untreated rats were used as controls (*C*). Rats were killed 16 h after the treatment, and levels of Hsp27 and αB-crystallin in the liver were determined. Each *bar* shows the mean ± SE of the results from 5 rats. (Inaguma et al. 1995)

Hsps in some tissues to stress loaded on to a whole living body is much more sensitive than that in cultured cells, suggesting that endogenous factors modulate the stress-induced expression. In fact, the induction of Hsp27 and αB-crystallin in rat liver after whole body heat shock (42 °C for 15 min) is almost completely suppressed by a prior injection (but not after heat shock) of prazosin, an α1-adrenergic antagonist (Fig. 1C; Inaguma et al. 1995).

It has been reported that the expression of Hsp70 in adrenal glands (Blake et al. 1991) and vascular smooth muscle (Udelsman et al. 1993) is enhanced in rats subjected to restraint for 1 h at normal temperature. Since such induction is not observed in hypophysectomized rats, stress responses in vivo appear to be modifiable by the endocrine system (Holbrook and Udelsman 1994).

Jurivich et al.(1994) reported that arachidonic acid can enhance the expression of Hsp70 by lowering the threshold of response in cultured cells. Prostaglandin J or A can also enhance the expression of Hsp70 in growing cells (Ohno et al. 1988; Santoro et al. 1989) and Milarski and Morimoto (1986) found an increase in the S phase during cell cycle progression. The heat-induced expression of Hsp70 is reported to be modulated via protein kinase A (PKA) and protein kinase C (PKC) (Choi et al. 1991; Kim et al. 1993; Pizurki and Polla 1994) and polyhydroxyl alcohols, including glycerol, sorbitol and mannitol, suppress heat-induced expression of Hsps (Edington et al. 1989; Kato et al. 1993). However, modulation of stress responses by various factors is often observed only in a specified cell line with different effects in others, reflecting a variation in cell functions.

Here, we summarize the responses of small Hsps, Hsp27 and αB-crystallin, to stress and their modification by various factors in rat C6 glioma cells.

2.1
Stimulation of Stress-Induced Expression of Hsp27 and αB-Crystallin by Modulators of the Arachidonic Acid Cascade

Normal C6 rat glioma cells express low levels of Hsp27 and αB-crystallin. However, expression of the two proteins is induced by exposure to sodium arsenite or heat, and in both cases, as with Hsp70, this is markedly stimulated by the presence of indomethacin, an inhibitor of cyclooxygenase (Smith 1989; Shimizu and Wolfe 1990), or nordihydroguaiaretic acid, an inhibitor of lipoxygenase (Domin et al. 1994), as detected by specific immunoassays (Fig. 2) or Western blot and Northern blot analyses (Ito et al. 1996). The presence of melittin or mastoparan, activators of phospholipase A2, during the stress period also stimulates the induction of the two small Hsps and Hsp70 (Ito et al. 1996; Kato et al. 1995). Exposure to single inhibitors alone does not induce expression. The stimulatory effects of these chemicals on the arsenite-induced expression of Hsp27 and αB-crystallin is predominant when the concentration of arsenite (magnitude of stress) is low (Fig. 2), suggesting that each lowers the threshold of stress induction. Induction of Hsp27 and αB-crystallin in adrenal glands of heat-stressed rats is also enhanced by a prior injection of aspirin,

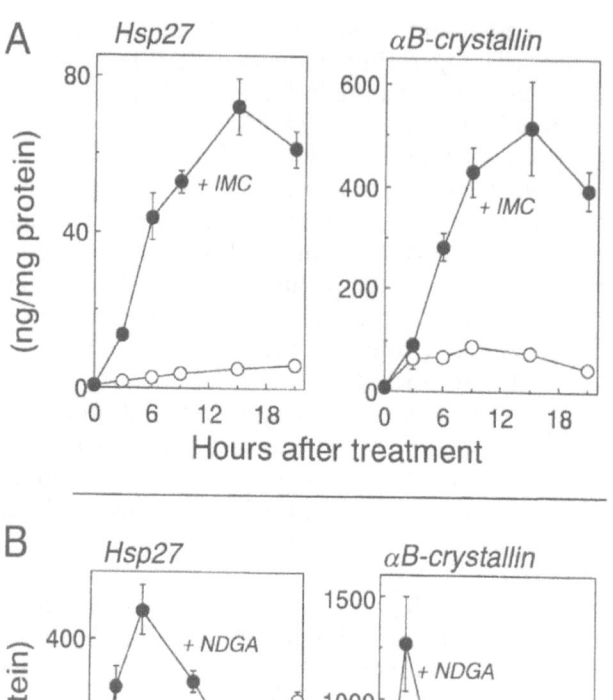

Fig. 2. A Stimulation by indomethacin or **B** nordihydroguaiaretic acid of the arsenite-induced accumulation of Hsp27 and αB-crystallin. **A** C6 cells were exposed to 50 μM arsenite (*open circles*) or 50 μM arsenite plus 100 μM indomethacin (*IMC, closed circles*) for 1 h at 37 °C and then cultured in normal medium for the indicated periods of time. **B** Cells were exposed to arsenite at various concentrations in the presence (*closed circles*) or absence (*open circles*) of 100 μM nordihydroguaiaretic acid (*NDGA*) at 37 °C for 1 h. After 16 h of culture in normal medium, cells were harvested for the quantitation of Hsp27 and αB-crystallin. Each point shows the mean ± SD of results from four dishes. (Ito et al. 1996)

another cyclooxygenase inhibitor (Ito et al. 1996). These results unequivocally indicate that the stress responses of Hsp27 and αB-crystallin in cells are enhanced by chemicals that inhibit the breakdown or stimulate the production of arachidonic acid. However, arachidonic acid added to the medium does not induce the expression nor does it stimulate the arsenite- or heat-induced responses of Hsp27 and αB-crystallin in C6 cells. The reasons for these different results obtained with modulators and arachidonic acid are unclear.

Curcumin, a major component of turmeric and a known anti-oxidant, anti-inflammatory and anti-carcinogenic agent, is a potent stimulator of the stress-induced expression of Hsp27, αB-crystallin and Hsp70 in cultured cells and in the liver and adrenal glands of heat-stressed rats (Kato et al. 1998b). Since curcumin is also a potent inhibitor of lipoxygenase and cyclooxygenase (Huang et al. 1991; Srivastava et al. 1995), it is suggested that the mechanism of the stimulation by curcumin is similar to that of salicylate, indomethacin and nordihydroguaiaretic acid.

The mechanism(s) of the stimulation of expression of Hsp27 and αB-crystallin by modulators of the arachidonic acid cascade is not fully elucidated. However, the following (Jurivich et al. 1992; Cotto et al. 1996) are suggestive in terms of the mechanism: Sodium salicylate and indomethacin activate HSF binding to HSE, but do not induce Hsp transcription; transcriptionally inert drug-induced HSF is not inducibly serine-phosphorylated; the intermediate state of HSF can be converted to the transcriptionally active state by subsequent exposure to heat stress. It is not yet known how the HSF is activated by the drug (arachidonic acid).

2.2
Modulation of the Expression of Hsp27 and αB-Crystallin by Protein Kinases

The finding that heat-induced expression of Hsp70 was inhibited by staurosporine, an inhibitor of PKC, indicates a role for PKC signaling in stress responses (Kim et al. 1993). Vasopressin is also known to stimulate the induction of Hsp27 and αB-crystallin in aortic smooth muscle via PKC activation (Kaida et al. 1999) and prostaglandin F2α induces the expression of Hsp27 via PKC and p44/42 MAP kinase activation (Kozawa et al. 1999c). C6 glioma cells exposed to phorbol 12-myristate 13 acetate (PMA), a phorbol ester and an activator of PKC, do not exhibit enhanced expression of Hsp27 and αB-crystallin. However, with exposure to arsenite or heat in the presence of PMA or okadaic acid, expression of Hsp27, αB-crystallin, and Hsp70 is markedly enhanced (Ito et al. 1995). The stimulatory effect of PMA is further increased in the presence of okadaic acid, but it is strongly inhibited by staurosporine. Since the arsenite-induced release of arachidonic acid from cells is also stimulated in the presence of PMA and/or okadaic acid, and the stimulatory effects of PMA and okadaic acid on the arsenite-induced accumulation of αB-crystallin and Hsp27 are strongly suppressed by quinacrine, an inhibitor of phospholipase A2, the effects may be caused, in part, by increased metabolic activity in the arachidonic acid cascade.

The induction of Hsp70 in vascular smooth muscle of intact rats by restraint stress is mediated by α1-adrenergic receptors (Udelsman et al. 1993; Holbrook and Udelsman 1994). Similarly, the induction of Hsp27 and αB-crystallin in adrenal glands and liver of heat-stressed rats is suppressed by an α1-adrenergic antagonist (Inaguma et al. 1995). These results suggest the involvement of cyclic AMP and PKA in the regulation of expression, not only of Hsp70 but also of Hsp27 and αB-crystallin.

The stress-induced expression of Hsp27 and αB-crystallin in C6 glioma cells is also modulated by cyclic AMP/PKA (Kato et al. 1996a). Induction of the expression of Hsp27 in C6 glioma cells by exposure to arsenite or heat is markedly enhanced in the presence of isoproterenol, forskolin, cholera toxin, or dibutyryl cyclic AMP, all of which can elevate intracellular levels of cyclic AMP. However, induction of Hsp70 by stress is barely stimulated by isoproterenol. In contrast, induction of αB-crystallin by arsenite or heat stress is suppressed by the above agents. The mechanisms by which cyclic AMP/PKA regulate expression of Hsp27 and αB-crystallin remain to be clarified.

Exposure of C6 cells to a low concentration (0.1–3 µg/ml) of anisomycin, an activator of p38 MAP kinase (Meier et al. 1996), for a few hours after heat stress (43 °C for 30 min) causes accumulation of Hsp27 and αB-crystallin and their mRNAs, but not Hsp70 (Kato et al. 1999). The results of reporter assays, using an αB-crystallin promoter fused to a luciferase reporter gene, suggest that the increase in levels of mRNA is due to de novo transcription. However, the DNA binding activity of HSF is barely affected by subsequent exposure to anisomycin. Stimulatory effects of anisomycin are also observed in cells exposed to arsenite. A specific inhibitor of p38 MAP kinase, SB202190 (Lee et al. 1994), suppresses the stimulation by anisomycin of the heat stress induced expression of Hsp27 and αB-crystallin. The osmotic induction, but not the thermal induction, of Hsp70 in canine renal epithelial cells is also dependent on the activation of p38 MAP kinase (Sheikh-Hamad et al. 1998).

In osteoblasts (Kozawa et al. 1999a) and aortic smooth muscle cells (Kozawa et al. 1999b) the expression of Hsp27 is induced by sphingosine 1-phosphate and this is suppressed by SB203580 (Lee et al. 1994), another inhibitor of p38 MAP kinase.

αB-crystallin in rat astrocytes is induced by hypertonic stress in the absence of activation of HSF without any change in expression of Hsp27 (Head et al. 1996). Synthesis and accumulation of αB-crystallin (but not Hsp27) in C6 glioma cells are induced by colchicine, vinblastine, colcemid, or nocodazole, which promote the disassembly of microtubules (Kato et al. 1996b). Induction of αB-crystallin by these drugs is suppressed by taxol, a microtubule-stabilizing anti-mitotic agent, and is sensitive to staurosporine, a protein kinase inhibitor (Kato et al. 1996b).

Reports suggesting the involvement of several protein kinases in the expression of Hsp27 and αB-crystallin have accumulated. However, the mechanisms in individual cases remain to be clarified.

2.3
Modulation of Expression of Hsp27 and αB-Crystallin by the Cellular Redox State

It has been suggested that loss of redox control with a shift to an oxidative state due to loss of protein thiols or glutathione might trigger expression of Hsps. In fact, thiol reactive reagents (p-chloromercuribenzoate and iodoacetamide) can induce Hsps in human or murine melanoma cells (Caltabiano et al. 1986) and reactive electrophiles induce Hsp70 in an epithelial cell line, LLC-PKI (Chen et al. 1992). Iodoacetamide (Liu et al. 1996) and diamide (Freeman et al. 1995) can activate HSF, resulting in induction of Hsp70. In contrast, a high concentration (2 mM) of dithiothreitol was found to inhibit the heat shock response in HeLa cells and mouse C2C12 myoblasts (Huang et al. 1994). Mehlen et al. (1996) reported that overexpression of Hsp27 in mouse L929 cells resulted in a decrease in reactive oxygen species and an increase in the levels of glutathione.

Exposure of C6 cells to diethyl maleate, a compound that binds free thiol groups (Ku and Billings 1986), results in the stimulation of HSE-binding activity of HSF. However, the activated HSF is not phosphorylated and is transcriptionally inert, because it causes neither accumulation of Hsp27, αB-crystallin and Hsp70 nor expression of mRNAs for these proteins. However, when cells are subjected to arsenite or heat stress in the presence of diethyl maleate, accumulation of the three Hsps and their mRNAs is markedly enhanced (Ito et al. 1998). The period of HSE-binding activity of HSF stimulated by arsenite is extended by addition of diethyl maleate. The duration of the induced phosphorylation state of HSF1 is also prolonged by diethyl maleate (Ito et al. 1998). A specific and irreversible inhibitor of γ-glutamyl-cystein synthetase, buthionine sulfoximine (Griffith 1982), affects the stress responses similar to those observed with diethyl maleate (Ito et al. 1998).

It has been shown that binding of HSF to HSE does not always result in activation of the transcription of genes for Hsps, For example, an alkaline shift in pH of the culture (Petronini et al. 1995), oxidative injury (Bruce et al. 1993), and anti-inflammatory drugs, such as salicylate and indomethacin (Jurivich et al. 1992; Cotto et al. 1996), activate HSF binding to HSE, but do not induce Hsp transcription. Diethyl maleate may induce an intermediate state of HSF that is bound to HSE but is transcriptionally inert, similar to that reported for salicylate and indomethacin (Cotto et al. 1996). Such an intermediate state of HSF might be expected to become transcriptionally active in the presence of stress, and together with the directly activated HSF synergistically contribute to transcription of Hsp genes.

It was once believed that reducing reagents act as suppressors of stress responses (Huang et al. 1994). However, the arsenite-induced expression of Hsp27, αB-crystallin and Hsp70 is markedly enhanced in C6 cells that have been exposed to arsenite (100 μM for 1 h) in the presence of a low concentration (0.03–0.1 mM) of dithiothreitol, 2-mercaptoethanol or dithioerythritol

(Kato et al. 1997). Enhanced expression of mRNAs for the three Hsps, as well as prolonged activation of HSF, are also observed in cells treated in the presence of 0.05 mM dithiothreitol. Arsenite-inducible expression of the three proteins is completely suppressed when dithiothreitol is present at concentrations above 1 mM during the stress period. Exposure of cells first to arsenite for 1 h and then to dithiothreitol results in a very effective suppression of the arsenite-inducible responses, and the responses are inhibited even by a low concentration of dithiothreitol. These results suggest that the signal transduction pathway for the arsenite-induced expression of Hsps involves at least two redox-sensitive steps: (1) a process that is stimulated by mild reducing power during the stress period; and (2) a process that is followed by the activation of HSF and is sensitive to suppression by reducing agents.

It has been demonstrated that the presence of a certain concentration of thiols could amplify the effects of reactive oxygen radicals by producing thiol radicals, which themselves remove hydrogen atoms from biomolecules, further perpetuating the damage (Kim et al. 1985; Hamazaki et al. 1989; Yim et al. 1994). As for the trigger of the arsenite-induced production of Hsps, it has been suggested that arsenite may shift the redox control in cells to the oxidative state by reacting with protein thiols (Klemperer and Pickart 1989; Beckmann et al. 1992). Therefore, the enhancement of the arsenite-induced responses by simultaneous exposure to low concentrations of dithiothreitol could be due to a redox cycling effect on free radical production.

Prostaglandins A and J are reported to induce the synthesis of Hsp70 and Hsp90 (Ohno et al. 1988; Santoro et al. 1989), as well as the activation of HSF (Amici et al. 1992; Holbrook et al. 1992), in proliferating cells. Exposure of C6 glioma cells to µM levels of prostaglandin A or J for 1 h at 37 °C stimulates HSE-binding activity of HSF, but barely induces any synthesis of Hsp27, αB-crystallin, or Hsp70. However, it is found that various prostaglandins do stimulate the stress-induced expression of Hsps, when they are added to the medium of confluent culture C6 cells during exposure to stress (Ito et al. 1997a). In cells exposed to arsenite in the presence of a prostaglandin, increased expression of mRNAs for Hsp27, αB-crystallin, and Hsp70 precedes enhanced accumulation of the respective proteins. HSE-binding activity of HSF, activated by the exposure to arsenite, is sustained for an extended period in cells treated with arsenite in the presence of a prostaglandin. Prostaglandins may thus induce an intermediate state of HSF that is bound to HSE but is transcriptionally inert, as described above. In fact, Western blot analysis of HSF1 reveals that activation by prostaglandins alone does not result in a phosphorylation-dependent mobility shift. The intermediate state of HSF becomes transcriptionally active in the presence of stress as described above.

It is not known how prostaglandins activate HSF. Thiol-reducing reagents inhibit the stress responses perhaps by preventing the oxidation process that is considered to be a trigger for the induction by stress of Hsps (Jacquier-Sarlin et al. 1995). Exogenously added prostaglandins have been reported to accumulate in cell nuclei as a consequence of their binding to nuclear proteins

(Narumiya and Fukushima 1986; Narumiya et al. 1986). It has also been reported that prostaglandins react with thiol groups of some nuclear proteins (Fukushima 1992), possibly resulting in a stimulation of stress responses. On the other hand, it has been found that some prostaglandins, when added exogenously, effect release of arachidonic acid from cells (Tokuda et al. 1992), which reflects breakdown of membrane phospholipids. The stimulatory effects of prostaglandins on stress-induced synthesis of Hsp27, αB-crystallin, and Hsp70 observed in C6 cells might be mediated, at least in part, by enhancement of the activity of the arachidonic acid cascade.

3
Phosphorylation of Small Hsps

Among the seven known members of the α-crystallin small Hsps family, the presence of phosphorylated forms has been demonstrated for αA-crystallin, αB-crystallin, Hsp27, and Hsp20. All of their phosphorylation sites are serine residues and some of the protein kinases responsible in each case have been identified. However, the physiological significance of phosphorylation of these Hsps is not fully understood. Reports for αA-crystallin are limited but phosphorylation of its Ser-122 in the lens (Voorter et al. 1986) is thought to occur via a cAMP-dependent pathway (Spector et al. 1985) or by an autokinase activity of the protein itself (Kantorow and Piatigorsky 1994; Kantorow et al. 1995). Below, we have concentrated attention on the phosphorylation of Hsp27, Hsp20, and αB-crystallin.

3.1
Phosphorylation of Hsp27

It has been shown that Hsp27 undergoes rapid post-translational phosphorylation, within minutes of exposure to heat (Arrigo and Welch 1987; Landry et al. 1991) or arsenite (Welch 1985), that results in the subsequent accumulation of Hsp27. In addition, a similarly enhanced phosphorylation of Hsp27 can be elicited by treatment of cells with phorbol ester, tumor necrosis factor, interleukin 1, or okadaic acid, which do not stimulate the synthesis of Hsp27 (Welch 1985; Arrigo 1990; Saklatvala et al. 1991; Guy et al. 1992; Landry et al. 1992). The phosphorylation of Hsp27 generates two or three more acidic isoforms (Arrigo and Welch 1987; Landry et al. 1992). It has been shown that human Hsp27 is phosphorylated at Ser-15, Ser-78, and Ser-82 (Landry et al. 1992; Stokoe et al. 1992) and mouse Hsp25 at Ser-15 and Ser-86 (Gaestel et al. 1991). The phosphorylation of each serine residue is catalyzed by MAPKAP kinase-2/-3, downstream of p38 MAP kinase, (Stokoe et al. 1992; Ludwig et al. 1996) and the delta isoform of protein kinase C (PKC) (Maizels et al. 1998). Selectivity of phosphorylation sites in Hsp27 or Hsp25 for these three enzymes has been reported to be apparently absent (Stokoe et al. 1992; Mclaughlin et al. 1996; Maizels et al. 1998). Dephosphorylation of Hsp27/25 in vivo seems to be catalyzed by protein phosphatase 2A, although protein phosphatase 2B also demonstrated activity in an in vitro system (Gaestel et al. 1992; Cairns et al. 1994).

3.1.1
Phosphorylation and Dissociation of Hsp27

Hsp27 in cells is present in two forms, an aggregated large form >500 kDa and a dissociated small form <100 kDa (Kato et al. 1994b), which are interconvertible on phosphorylation and dephosphorylation of serine residues. Phosphorylation of the aggregated form results in dissociation to small forms and dephosphorylation of the dissociated Hsp27 seems to cause aggregation (Kato et al. 1994b). Using various specific inhibitors for protein kinases, we have recently confirmed which enzymes participate in the dissociation (as a result of phosphorylation) of Hsp27 induced by various stimuli (Kato et al. 2001b). Dissociation of Hsp27 induced by exposure to arsenite (Fig. 3), CdCl$_2$, sorbitol, NaCl or anisomycin was completely suppressed by the presence of SB203580 (Lee et al. 1994) or PD169316 (Kummer et al. 1997), specific inhibitors of p38 MAP kinase, an enzyme upstream of MAPKAP kinase-2/-3, but not by inhibitors of PKC. However, phorbol ester (PMA)-induced dissociation of Hsp27 was completely suppressed by the presence of Go6983 (Gschwendt et al. 1996) or bisindolylmaleimide I (Toullec et al. 1991), specific inhibitors of PKC, but only partially by inhibitors of p38 MAP kinase (Fig. 3). The inhibitors of MAP kinase kinase, PD98059 (Alessi et al. 1995) and Uo126 (Favata et al. 1998), had no effect on the dissociation induced by the above chemicals.

Conversion of the non-phosphorylated, aggregated form of Hsp27 to the phosphorylated, dissociated form results in decreased tolerance to heat stress (Kato et al. 1994b). One non-phosphorylatable mutant of Hsp27 expressed in cells formed a large polymer like the wild type and was similarly protective against stress but blockage of its phosphorylation with SB203580 did not abolish protective activity against tumor necrosis factor α cytotoxicity (Mehlen et al. 1997; Preville et al. 1998). Another mutant Hsp27, in which the phosphorylatable serine residues were replaced by aspartic acid, did not form polymers and showed no protective effects (Rogalla et al. 1999). These results suggest that the cytoprotective activity of Hsp27 is associated with the large aggregate and phosphorylation of serine residues is not required for protection against cell assault. However, the physiological significance of phospho-

Fig. 3. Protein kinase inhibitors can suppress stress-induced dissociation of Hsp27. U251 MG human glioma cells were exposed to **A** 200 µM sodium arsenite (*As*) or **B** 1 µM *PMA* at 37 °C for 90 min with or without 10 µM each of SB203580 (*SB*), PD169316 (*PD16*), PD98059 (*PD98*), Uo126 (*Uo*), or 100 nM staurosporine (*STP*), 5 µM bisindolylmaleimide (*BIM*), or 5 µM Go6983 (*Go*). The cells were also exposed to each inhibitor alone. Each cell extract was subjected to sucrose density gradient (10–40%) centrifugation and the concentrations of Hsp27 in each fraction were determined with a specific immunoassay. (Kato et al. 2001b)

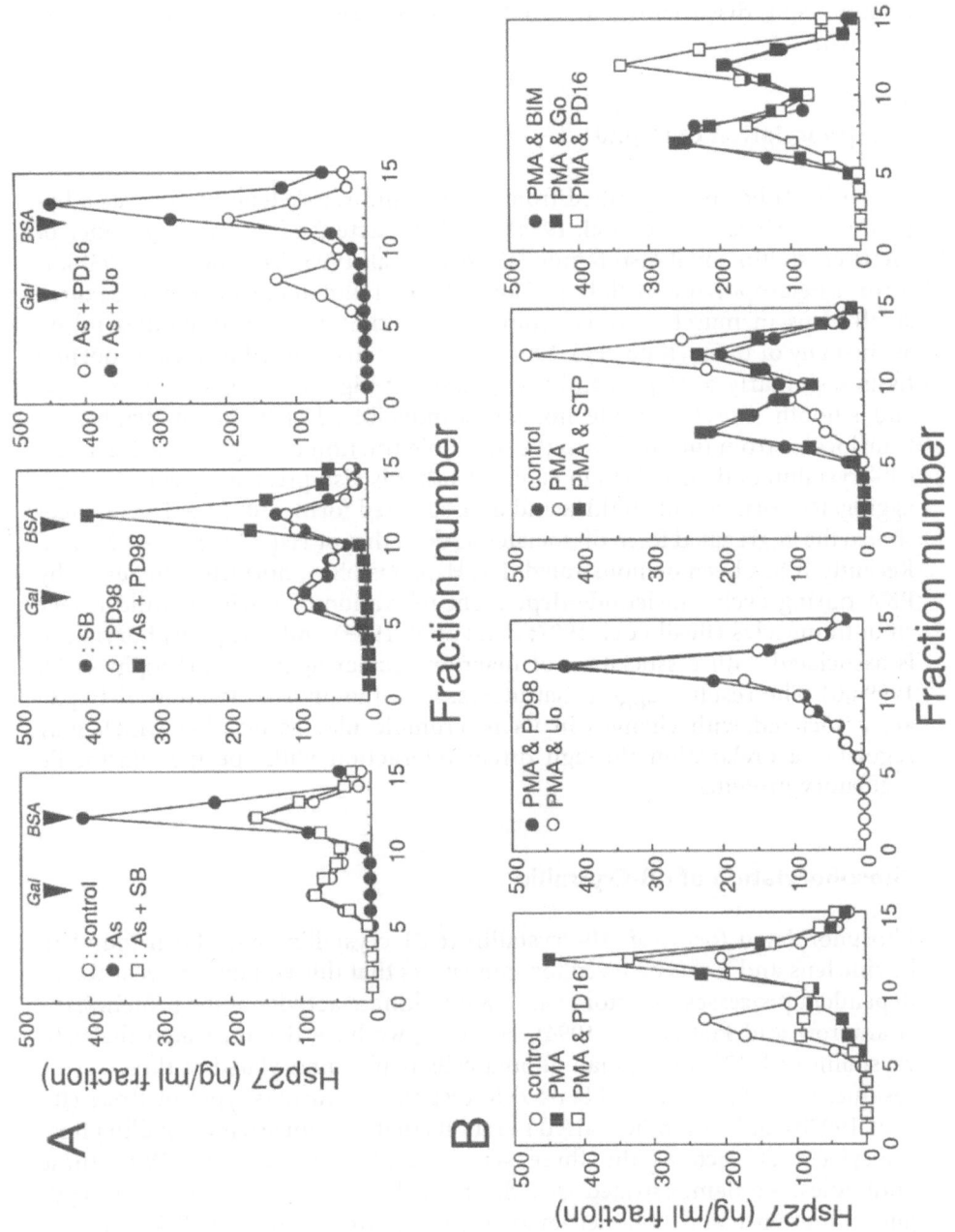

rylation and dissociation of Hsp27 in response to stress remains to be clarified.

3.2
Phosphorylation of Hsp20

Hsp20 (p20) has been purified from human and rat skeletal muscles as a by-product of Hsp27 (Kato et al. 1994a) and characterized as a new member of the α-crystallin small Hsp family (Caspers et al. 1995; De Jong et al. 1998). It forms a heteropolymer with Hsp27 and αB-crystallin in muscle and these three small Hsps in muscle extracts coimmunoprecipitate with antibodies raised against any of them (Kato et al. 1994a). Hsp20 is distributed in various normal tissues, similarly to Hsp27, and is expressed at high levels in skeletal, cardiac and smooth muscles. While not stress-inducible, Hsp20 in rat diaphragm translocates from the soluble to the insoluble fraction during heat shock as do αB-crystallin and Hsp27 (Kato et al. 1994a). It is also present in two forms, an aggregated form about 500 kDa and a dissociated form <100 kDa. During heat shock, the aggregated form dissociates as does that of Hsp27 in rat diaphragm. Recently, it has been demonstrated that Hsp20 is phosphorylated on Ser-16 by PKA during cyclic nucleotide-dependent relaxation of bovine carotid artery smooth muscles (Beall et al. 1997; Beall et al. 1999) and this phosphorylation is associated with dissociation of macromolecular aggregates (Brophy et al. 1999a,b). The results suggest that increases in the phosphorylation of Hsp20 are associated with changes in its macromolecular association which may regulate vasorelaxation through direct interaction with specific contractile regulatory proteins.

3.3
Phosphorylation of αB-Crystallin

Phosphorylated forms of αB-crystallin (αB1-crystallin) have been found in bovine lens and it was believed for some years that this was due to cyclic AMP-dependent processes (Spector et al. 1985) or kinase activity of the protein itself (Kantorow and Piatigorsky 1994). However, we have demonstrated that αB-crystallin in U373 MG human glioma cells is phosphorylated at three serine residues (Ser-19, Ser-45, and Ser-59) in response to various types of stress (Ito et al. 1997b) and antibodies raised in rabbits that recognize αB-crystallin phosphorylated at each of the three serine residues individually. With these antibodies, we demonstrated that the phosphorylation of each site is regulated differently and obtained evidence suggesting p44/42 MAP kinase and MAPKAP kinase-2 are responsible for phosphorylation of Ser-45 and Ser-59, respectively, in αB-crystallin in vivo (Kato et al. 1998a).

3.3.1
Phosphorylation of αB-Crystallin in Response to Stress
and Enzymatic Activities Responsible for Phosphorylation

Phosphorylation of αB-crystallin in U373 MG human glioma cells is stimulated by exposing the cells to various stimuli, which include heat, arsenite, phorbol 12-myristate 13-acetate (PMA), okadaic acid, H_2O_2, anisomycin, and high concentrations of sorbitol or NaCl but not agents elevating intracellular levels of cyclic AMP (Ito et al. 1997b). Cells exposed to PMA together with okadaic acid yield three bands of ^{32}P-labeled αB-crystallin on electrophoresis on isoelectric focusing gels, all of the phosphorylated residues being identified as serine. Structural analysis by mass spectrometry revealed that phosphorylation of αB-crystallin occurs at serines 19, 45, and 59. Although PMA and arsenite preferentially stimulate the phosphorylation of Ser-45 and Ser-59, respectively, all three serine residues are phosphorylated to some extent during exposure to each stress. However, the arsenite- and $CdCl_2$-induced phosphorylation is selectively suppressed by dithiothreitol and the PMA-induced phosphorylation is inhibited by staurosporine, Go 6983 or bisindolylmaleimide I, inhibitors of PKC. SB202190, an inhibitor of p38 MAP kinase, suppresses the phosphorylation induced by arsenite, anisomycin, H_2O_2, sorbitol, NaCl, and heat shock, but not that induced by PMA and okadaic acid. The PMA-induced phosphorylation is also selectively suppressed by an inhibitor of p44/42 MAP kinase kinase, PD98059. These results indicate that several kinase cascades might be associated with induction of phosphorylation of αB-crystallin. The agents that induce the phosphorylation of αB-crystallin described above are also effective stimulators of phosphorylation of Hsp27. Since Hsp27 is known to be phosphorylated by MAPKAP kinase-2/-3, which is under the control of p38 MAP kinase, it is suggested that p38 MAP kinase and p44/42 MAP kinase might be involved in the signal transduction cascade that leads to the phosphorylation of αB-crystallin.

In a cell-free system, we could not detect any phosphorylation of αB-crystallin. When bovine αB2-crystallin was incubated with extracts of U373 MG cells in the presence of $[P^{32}]$ATP and Mg^{2+}, hardly any radioactivity was incorporated into αB2-crystallin, even when extracts of cells exposed to PMA and okadaic acid were used (Fig. 4). However, when the amino-terminal 72 (N-72 K) peptide of αB2-crystallin that had been prepared by digestion of αB2-crystallin with lysyl endopeptidase, was employed as a substrate, the N-72 K peptide was phosphorylated significantly even with extracts of control cells, and markedly enhanced after exposure to PMA and okadaic acid (Fig. 4). These results suggest that the aggregation properties of the purified αB-crystallin might be different from those of the native form in cells and that protein kinases might be unable to gain access to sites of potential phosphorylation. Treatment of αB2-crystallin with lysyl endopeptidase results in the destruction of the α-crystallin domain and formation of the N-72K peptide, which contains each of the three phosphorylation sites in αB-crystallin accessible to

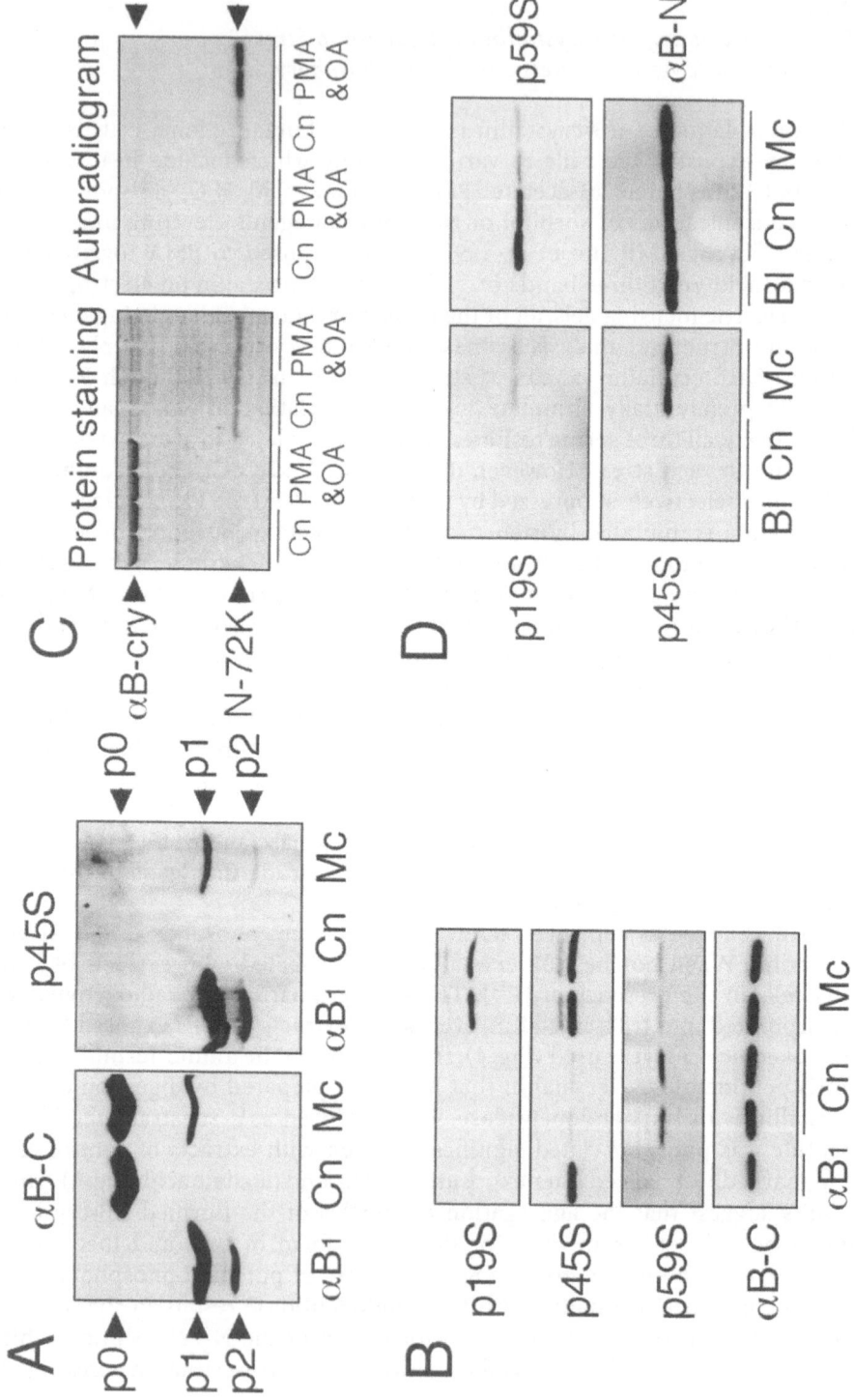

protein kinases. The protein kinases responsible for the phosphorylation of Ser-45 and Ser-59 in the N-72K peptide were partially purified from extracts of U373 MG human glioma cells that had been stimulated by exposure to H_2O_2 in the presence of calyculin A. The activities responsible for the phosphorylation of Ser-45 and Ser-59 eluted separately from a column of Superdex 200 at fractions corresponding to about 40 and 60 kDa, respectively, while the kinase for Ser-19 proved unstable. p44/42 MAP kinase and MAPKAP kinase-2 were concentrated in the Ser-45 kinase fraction and Ser-59 kinase fraction, respectively. Recombinant human p44 MAP kinase and MAPKAP kinase-2 purified from rabbit muscle selectively phosphorylate Ser-45 and Ser-59, respectively. While the Ser-45 kinase fraction and Ser-59 kinase fraction phosphorylate myelin basic protein and Hsp27, respectively. These results suggest that phosphorylation of Ser-45 and Ser-59 in αB-crystallin is catalyzed by p44/42 MAP kinase and MAPKAP kinase-2, respectively (Kato et al. 1998a).

3.3.2
Phosphorylation of αB-Crystallin in Mitotic Cells

Among the three antibodies that recognize each of the phosphorylated serine residues in αB-crystallin only anti-p45S proved applicable for immunocytochemistry. Therefore, we prepared new antibodies that recognize phosphorylated Ser-59 and Ser-19 in αB-crystallin with distinct amino acid sequences (unpubl.). For this purpose the human p59S peptide, FLRAPSWFDTG+cysteine, the bovine p59S peptide, FLRAPSWIDTG+cysteine, and the p19S peptide, FFPFHSPSRLFG+cystein, in which the serine residues were phosphorylated, were conjugated with hemocyanin and antibodies against human or bovine p59S or p19S were raised in rabbits, and purified with the respective peptide-coupled Sepharose as detailed earlier (Ito et al. 1997b).

◄
───────────────────────────────

Fig. 4. Phosphorylated αB-crystallin in mitotic cells (A, B) and protein kinase activities for the three serine residues (C, D). Extracts of control (*Cn*) or mitotic (*Mc*) U373 MG human glioma cells were subjected to IEF (A) or SDS-PAGE (B), together with bovine αB1-crystallin (*αB1*), and subsequent Western blot analysis with antibodies against the carboxyl-terminal peptide of αB-crystallin (*αB-C*) or against the p19S, p45S or p59S peptides. *p0*, unphosphorylated αB-crystallin; *p1* and *p2*, phosphorylated αB-crystallin. C Extracts of control cells (*Cn*) or extracts of cells that had been exposed for 60 min to 1 µM PMA plus 0.2 µM okadaic acid (*PMA and OA*) were incubated at 30 °C for 20 min with bovine αB2-crystallin or lysyl endopeptidase-digested αB2-crystallin in the presence of [^{32}P]ATP and Mg^{2+}. Aliquots of reaction mixture were subjected to tricine/SDS-PAGE, and the gel was stained with Coomassie Brilliant Blue and autoradiographed. D Extracts of control cells (*Cn*) and mitotic cells (*Mc*) were incubated at 30 °C for 20 min with lysyl endopeptidase-digested αB2-crystallin in the presence of ATP and Mg^{2+}, and aliquots of reaction mixture were subjected to tricine/SDS-PAGE and subsequent Western blot analysis with antibodies recognizing each of the phosphorylated serine residues (p19S, p45S, and p59S) or amino-terminal peptide of αB-crystallin (*αB-N*). *Bl*, incubated without a cell extract. (Kato et al. 1998a)

Immunofluorescence localization of αB-crystallin phosphorylated at each of the three serine residues in U373 MG human glioma cells upon exposure to PMA and okadaic acid revealed no characteristic differences from that in control cells. However, individual αB-crystallin phosphorylated forms were localized characteristically in mitotic cells, independent of their treatment with chemicals (Kato et al. 1998a).

Immunofluorescence localization of αB-crystallin phosphorylated at Ser-45 showed the cytoplasm of all mitotic cells to be intensely stained. This was the case for all cell lines examined, as well as sections of mouse embryos. SDS-PAGE and Western blot analysis of extracts of mitotic cells revealed that not only αB-crystallin phosphorylated at Ser-45 but also phosphorylation at Ser-19 was elevated (Fig. 4). Weak but significant staining of mitotic cells with antibodies against p19S was also observed (unpublished observations).

In contrast, we found the levels of αB-crystallin phosphorylated at Ser-59 to be considerably lower in mitotic than in control cells (Fig. 4; Kato et al. 1998a) although anti-p59S antibodies do recognize centrosomes and midbodies of mitotic cells. αB-crystallin phosphorylated at Ser-59 was thus observed at microtubule organizing centers by means of double staining with anti-β-tubulin antibody on aster formation analysis. In addition, αB-crystallin phosphorylated at Ser-59 co-localizes with γ-tubulin in centrosomes (Inaguma et al., unpublished data). These results suggest that αB-crystallin might be involved in the machinery of microtubule polymerization.

The physiological significance of the phosphorylation of αB-crystallin remains to be elucidated in detail. In rat lens, phosphorylated αB-crystallin, preferentially at Ser-45, increases after birth (Ito et al. 1999). Phosphorylation of αB-crystallin at Ser-59 has recently been reported to be apparently related to a protective effect against ischemic stress in perfused rat hearts (Eaton et al. 2000) and sorbitol-induced apoptosis in cultured cardiac myocytes (Hoover et al. 2000). The fact that αB-crystallin phosphorylated at Ser-59 accumulates in brains of patients with Alexander's disease and also in aged normal brains (Kato et al. 2001a) is interesting in this context.

References

Alessi DR, Cuenda A, Cohen P, Dudley DT, Saltiel AR (1995) PD 098059 is a specific inhibitor of the activation of mitogen- activated protein kinase kinase in vitro and in vivo. J Biol Chem 270:27489–27494

Amici C, Sistonen L, Santoro MG, Morimoto RI (1992) Antiproliferative prostaglandins activate heat shock transcription factor. Proc Natl Acad Sci USA 89:6227–6231

Arrigo AP (1990) Tumor necrosis factor induces the rapid phosphorylation of the mammalian heat shock protein hsp28 [published erratum appears in Mol Cell Biol 1990 July;10(7):3857]. Mol Cell Biol 10:1276–1280

Arrigo AP, Welch WJ (1987) Characterization and purification of the small 28,000-dalton mammalian heat shock protein. J Biol Chem 262:15359–15369

Beall A, Bagwell D, Woodrum D, Stoming TA, Kato K, Suzuki A, Rasmussen H, Brophy CM (1999) The small heat shock-related protein, HSP20, is phosphorylated on serine 16 during cyclic nucleotide-dependent relaxation [published erratum appears in J Biol Chem 1999 Sept 24;274(39):28058]. J Biol Chem 274:11344–11351

Beall AC, Kato K, Goldenring JR, Rasmussen H, Brophy CM (1997) Cyclic nucleotide-dependent vasorelaxation is associated with the phosphorylation of a small heat shock-related protein. J Biol Chem 272:11283–11287

Beckmann RP, Lovett M, Welch WJ (1992) Examining the function and regulation of hsp 70 in cells subjected to metabolic stress. J Cell Biol 117:1137–1150

Blake MJ, Udelsman R, Feulner GJ, Norton DD, Holbrook NJ (1991) Stress-induced heat shock protein 70 expression in adrenal cortex: an adrenocorticotropic hormone-sensitive, age-dependent response. Proc Natl Acad Sci USA 88:9873–9977

Brophy CM, Dickinson M, Woodrum D (1999a) Phosphorylation of the small heat shock-related protein, HSP20, in vascular smooth muscles is associated with changes in the macromolecular associations of HSP20. J Biol Chem 274:6324–6329

Brophy CM, Lamb S, Graham A (1999b) The small heat shock-related protein-20 is an actin-associated protein. J Vasc Surg 29:326–333

Bruce JL, Price BD, Coleman CN, Calderwood SK (1993) Oxidative injury rapidly activates the heat shock transcription factor but fails to increase levels of heat shock proteins. Cancer Res 53:12–15

Cairns J, Qin S, Tan YH, Guy GR (1994) Dephosphorylation of the small heat shock protein Hsp27 in vivo by protein phosphatase 2A. J Biol Chem 269:9176–9183

Caltabiano MM, Koestler TP, Poste G, Greig RG (1986) Induction of 32- and 34-kDa stress proteins by sodium arsenite, heavy metals, and thiol-reactive agents. J Biol Chem 261:13381–13386

Caspers GJ, Leunissen JA, de Jong WW (1995) The expanding small heat-shock protein family, and structure predictions of the conserved "α-crystallin domain". J Mol Evol 40:238–248

Chen Q, Yu K, Stevens JL (1992) Regulation of the cellular stress response by reactive electrophiles. The role of covalent binding and cellular thiols in transcriptional activation of the 70-kilodalton heat shock protein gene by nephrotoxic cysteine conjugates. J Biol Chem 267: 24322–24327

Chiesa R, Gawinowicz-Kolks MA, Kleiman NJ, Spector A (1987) The phosphorylation sites of the B2 chain of bovine alpha-crystallin. Biochem Biophys Res Commun 144:1340–1347

Choi HS, Li B, Lin Z, Huang E, Liu AY (1991) cAMP and cAMP-dependent protein kinase regulate the human heat shock protein 70 gene promoter activity. J Biol Chem 266:11858–11865

Cotto JJ, Kline M, Morimoto RI (1996) Activation of heat shock factor 1 DNA binding precedes stress-induced serine phosphorylation. Evidence for a multistep pathway of regulation. J Biol Chem 271:3355–3358

De Jong WW, Caspers GJ, Leunissen JA (1998) Genealogy of the α-crystallin–small heat-shock protein superfamily. Int J Biol Macromol 22:151–162

Domin J, Higgins T, Rozengurt E (1994) Preferential inhibition of platelet-derived growth factor-stimulated DNA synthesis and protein tyrosine phosphorylation by nordihydroguaiaretic acid. J Biol Chem 269:8260–8267

Eaton P, Awad WI, Miller JI, Hearse DJ, Shattock MJ (2000) Ischemic preconditioning: a potential role for constitutive low molecular weight stress protein translocation and phosphorylation?. J Mol Cell Cardiol 32:961–971

Edington BV, Hightower LE (1990) Induction of a chicken small heat shock (stress) protein: evidence of multilevel posttranscriptional regulation. Mol Cell Biol 10:4886–4898

Edington BV, Whelan SA, Hightower LE (1989) Inhibition of heat shock (stress) protein induction by deuterium oxide and glycerol: additional support for the abnormal protein hypothesis of induction. J Cell Physiol 139:219–228

Edwards DP, Adams DJ, Savage N, McGuire WL (1980) Estrogen induced synthesis of specific proteins in human breast cancer cells. Biochem Biophys Res Commun 93:804–812

Favata MF, Horiuchi KY, Manos EJ et al. (1998) Identification of a novel inhibitor of mitogen-activated protein kinase kinase. J Biol Chem 273:18623–18632

Freeman ML, Borrelli MJ, Syed K, Senisterra G, Stafford DM, Lepock JR (1995) Characterization of a signal generated by oxidation of protein thiols that activates the heat shock transcription factor. J Cell Physiol 164:356–366

Fukushima M (1992) Biological activities and mechanisms of action of PGJ2 and related compounds: an update. Prostaglandins Leukot Essent Fatty Acids 47:1–12

Gaestel M, Schroder W, Benndorf R, Lippmann C, Buchner K, Hucho F, Erdmann VA, Bielka H (1991) Identification of the phosphorylation sites of the murine small heat shock protein hsp25. J Biol Chem 266:14721–14724

Gaestel M, Benndorf R, Hayess K, Priemer E, Engel K (1992) Dephosphorylation of the small heat shock protein hsp25 by calcium/calmodulin-dependent (type 2B) protein phosphatase. J Biol Chem 267:21607–21611

Gaestel M, Gotthardt R, Muller T (1993) Structure and organisation of a murine gene encoding small heat-shock protein Hsp25. Gene 128:279–283

Griffith OW (1982) Mechanism of action, metabolism, and toxicity of buthionine sulfoximine and its higher homologs, potent inhibitors of glutathione synthesis. J Biol Chem 257:13704–13712

Gschwendt M, Dieterich S, Rennecke J, Kittstein W, Mueller HJ, Johannes FJ (1996) Inhibition of protein kinase C mu by various inhibitors. Differentiation from protein kinase c isoenzymes. FEBS Lett 392:77–80

Guy GR, Cao X, Chua SP, Tan YH (1992) Okadaic acid mimics multiple changes in early protein phosphorylation and gene expression induced by tumor necrosis factor or interleukin-1. J Biol Chem 267:1846–18452

Hamazaki S, Okada S, Toyokuni S, Midorikawa O (1989) Thiobarbituric acid-reactive substance formation of rat kidney brush border membrane vesicles induced by ferric nitrilotriacetate. Arch Biochem Biophys 274:348–354

Head MW, Hurwitz L, Goldman JE (1996) Transcription regulation of αB-crystallin in astrocytes: analysis of HSF and AP1 activation by different types of physiological stress. J Cell Sci 109: 1029–1039

Hickey E, Brandon SE, Potter R, Stein G, Stein J, Weber LA (1986) Sequence and organization of genes encoding the human 27 kDa heat shock protein [published erratum appears in Nucleic Acids Res 1986 Oct 24;14(20):8230]. Nucleic Acids Res 14:4127–4145

Holbrook NJ, Udelsman R (1994) Heat shock protein gene expression in response to physiologic stress and aging, In: Morimoto RI, Tisseieres A, Georgopoulos C (eds) The biology of heat shock proteins and molecular chaperones. Cold Spring Harbor Laboratory Press, New York, pp 577–593

Holbrook NJ, Carlson SG, Choi AM, Fargnoli J (1992) Induction of HSP70 gene expression by the antiproliferative prostaglandin PGA2: a growth-dependent response mediated by activation of heat shock transcription factor. Mol Cell Biol 12:1528–1534

Hoover HE, Thuerauf DJ, Martindale JJ, Glembotski CC (2000) Alpha B-crystallin gene induction and phosphorylation by MKK6-activated p38. A potential role for alpha B-crystallin as a target of the p38 branch of the cardiac stress response. J Biol Chem 275:23825–23833

Huang LE, Zhang H, Bae SW, Liu AY (1994) Thiol reducing reagents inhibit the heat shock response. Involvement of a redox mechanism in the heat shock signal transduction pathway. J Biol Chem 269:30718–30725

Huang MT, Lysz T, Ferraro T, Abidi TF, Laskin JD, Conney AH (1991) Inhibitory effects of curcumin on in vitro lipoxygenase and cyclooxygenase activities in mouse epidermis. Cancer Res 51:813–819

Inaguma Y, Hasegawa K, Goto S, Ito H, Kato K (1995) Induction of the synthesis of hsp27 and αB crystallin in tissues of heat-stressed rats and its suppression by ethanol or an α1- adrenergic antagonist. J Biochem (Tokyo) 117:1238–1243

Ireland RC, Berger EM (1982) Synthesis of low molecular weight heat shock peptides stimulated by ecdysterone in a cultured Drosophila cell line. Proc Natl Acad Sci USA 79:855–859

Ireland RC, Berger E, Sirotkin K, Yund MA, Osterbur D, Fristrom J (1982) Ecdysterone induces the transcription of four heat-shock genes in Drosophila S3 cells and imaginal discs. Dev Biol 93:498–507

Ito H, Hasegawa K, Inaguma Y, Kozawa O, Asano T, Kato K (1995) Modulation of the stress-induced synthesis of stress proteins by a phorbol ester and okadaic acid. J Biochem (Tokyo) 118: 629–634

Ito H, Hasegawa K, Inaguma Y, Kozawa O, Kato K (1996) Enhancement of stress-induced synthesis of hsp27 and αB crystallin by modulators of the arachidonic acid cascade. J Cell Physiol 166:332–339

Ito H, Okamoto K, Kato K (1997a) Prostaglandins stimulate the stress-induced synthesis of hsp27 and αB crystallin. J Cell Physiol 170:255–262

Ito H, Okamoto K, Nakayama H, Isobe T, Kato K (1997b) Phosphorylation of αB-crystallin in response to various types of stress. J Biol Chem 272:29934–29941

Ito H, Okamoto K, Kato K (1998) Enhancement of expression of stress proteins by agents that lower the levels of glutathione in cells. Biochim Biophys Acta 1397:223–230

Ito H, Iida K, Kamei K, Iwamoto I, Inaguma Y, Kato K (1999) αB-crystallin in the rat lens is phosphorylated at an early post- natal age. FEBS Lett 446:269–272

Jacquier-Sarlin MR, Jornot L, Polla BS (1995) Differential expression and regulation of hsp70 and hsp90 by phorbol esters and heat shock. J Biol Chem 270:14094–14099

Jurivich DA, Sistonen L, Kroes RA, Morimoto RI (1992) Effect of sodium salicylate on the human heat shock response. Science 255:1243–1245

Jurivich DA, Sistonen L, Sarge KD, Morimoto RI (1994) Arachidonate is a potent modulator of human heat shock gene transcription. Proc Natl Acad Sci USA 91:2280–2284

Kaida T, Kozawa O, Ito T et al. (1999) Vasopressin stimulates the induction of heat shock protein 27 and αB-crystallin via protein kinase C activation in vascular smooth muscle cells. Exp Cell Res 246:327–337

Kantorow M, Horwitz J, van Boekel MA, de Jong WW, Piatigorsky J (1995) Conversion from oligomers to tetramers enhances autophosphorylation by lens αA-crystallin. Specificity between αA- and αB-crystallin subunits. J Biol Chem 270:17215–17220

Kantorow M, Piatigorsky J (1994) α-crystallin/small heat shock protein has autokinase activity. Proc Natl Acad Sci USA 91:3112–3116

Kato K, Goto S, Hasegawa K, Inaguma Y (1993) Coinduction of two low-molecular-weight stress proteins, αB crystallin and HSP28, by heat or arsenite stress in human glioma cells. J Biochem (Tokyo) 114:640–647

Kato K, Goto S, Inaguma Y, Hasegawa K, Morishita R, Asano T (1994a) Purification and characterization of a 20-kDa protein that is highly homologous to αB crystallin. J Biol Chem 269:15302–15309

Kato K, Hasegawa K, Goto S, Inaguma Y (1994b) Dissociation as a result of phosphorylation of an aggregated form of the small stress protein, hsp27. J Biol Chem 269:11274–11278

Kato K, Ito H, Hasegawa K, Inaguma Y, Suzuki A, Kozawa O, Asano T (1995) Enhancement of stress-induced synthesis of stress proteins by mastoparan in C6 rat glioma cells. J Biochem (Tokyo) 118:149–153

Kato K, Ito H, Hasegawa K, Inaguma Y, Kozawa O, Asano T (1996a) Modulation of the stress-induced synthesis of hsp27 and αB-crystallin by cyclic AMP in C6 rat glioma cells. J Neurochem 66:946–950

Kato K, Ito H, Inaguma Y, Okamoto K, Saga S (1996b) Synthesis and accumulation of αB crystallin in C6 glioma cells is induced by agents that promote the disassembly of microtubules. J Biol Chem 271:26989–26994

Kato K, Ito H, Okamoto K (1997) Modulation of the arsenite-induced expression of stress proteins by reducing agents. Cell Stress Chaperones 2:199–209

Kato K, Ito H, Kamei K, Inaguma Y, Iwamoto I, Saga S (1998a) Phosphorylation of αB-crystallin in mitotic cells and identification of enzymatic activities responsible for phosphorylation. J Biol Chem 273:28346–28354

Kato K, Ito H, Kamei K, Iwamoto I (1998b) Stimulation of the stress-induced expression of stress proteins by curcumin in cultured cells and in rat tissues in vivo. Cell Stress Chaperones 3:152–160

Kato K, Ito H, Kamei K, Iwamoto I (1999) Selective stimulation of Hsp27 and αB-crystallin but not Hsp70 expression by p38 MAP kinase activation. Cell Stress Chaperones 4:94–101

Kato K, Inaguma Y, Ito H, Iida K, Iwamoto I, Kamei K, Ochi N, Ohta H, Kishikawa M (2001a) Ser-59 is the major phosphorylation site in αB-crystallin accumulated in the brain of patients with Alexander's disease. J Neurochem 76:730–736

Kato K, Ito H, Iwamoto I, Iida K, Inaguma Y (2001b) Protein kinase inhibitors can suppress stress-induced dissociation of Hsp27 Cell Stress Chaperones 6:16–20

Kim K, Rhee SG, Stadtman ER (1985) Nonenzymatic cleavage of proteins by reactive oxygen species generated by dithiothreitol and iron. J Biol Chem 260:15394–15397

Kim SH, Kim JH, Erdos G, Lee YJ (1993) Effect of staurosporine on suppression of heat shock gene expression and thermotolerance development in HT-29 cells. Biochem Biophys Res Commun 193:759–763

Klemenz R, Frohli E, Aoyama A, Hoffmann S, Simpson RJ, Moritz RL, Schafer R (1991) αB crystallin accumulation is a specific response to Ha-ras and v- mos oncogene expression in mouse NIH 3T3 fibroblasts. Mol Cell Biol 11:803–812

Klemperer NS, Pickart CM (1989) Arsenite inhibits two steps in the ubiquitin-dependent proteolytic pathway. J Biol Chem 264:19245–19252

Kozawa O, Niwa M, Matsuno H, Tokuda H, Miwa M, Ito H, Kato K, Uematsu T (1999a) Sphingosine 1-phosphate induces heat shock protein 27 via p38 mitogen-activated protein kinase activation in osteoblasts. J Bone Miner Res 14:1761–1767

Kozawa O, Tanabe K, Ito H, Matsuno H, Niwa M, Kato K, Uematsu T (1999b) Sphingosine 1-phosphate regulates heat shock protein 27 induction by a p38 MAP kinase-dependent mechanism in aortic smooth muscle cells. Exp Cell Res 250:376–380

Kozawa O, Tokuda H, Miwa M, Ito H, Matsuno H, Niwa M, Kato K, Uematsu T (1999c) Involvement of p42/p44 mitogen-activated protein kinase in prostaglandin F2α-stimulated induction of heat shock protein 27 in osteoblasts. J Cell Biochem 75:610–619

Krief S, Faivre JF, Robert P et al. (1999) Identification and characterization of cvHsp. A novel human small stress protein selectively expressed in cardiovascular and insulin-sensitive tissues. J Biol Chem 274:36592–36600

Ku RH, Billings RE (1986) The role of mitochondrial glutathione and cellular protein sulfhydryls in formaldehyde toxicity in glutathione-depleted rat hepatocytes. Arch Biochem Biophys 247:183–189

Kummer JL, Rao PK, Heidenreich KA (1997) Apoptosis induced by withdrawal of trophic factors is mediated by p38 mitogen-activated protein kinase. J Biol Chem 272:20490–20494

Lam WY, Wing Tsui SK, Law PT, Luk SC, Fung KP, Lee CY, Waye MM (1996) Isolation and characterization of a human heart cDNA encoding a new member of the small heat shock protein family–HSPL27. Biochim Biophys Acta 1314:120–124

Landry J, Chretien P, Laszlo A, Lambert H (1991) Phosphorylation of HSP27 during development and decay of thermotolerance in Chinese hamster cells. J Cell Physiol 147:93–101

Landry J, Lambert H, Zhou M, Lavoie JN, Hickey E, Weber LA, Anderson CW (1992) Human HSP27 is phosphorylated at serines 78 and 82 by heat shock and mitogen-activated kinases that recognize the same amino acid motif as S6 kinase II. J Biol Chem 267:794–803

Lee JC, Laydon JT, McDonnell PC et al. (1994) A protein kinase involved in the regulation of inflammatory cytokine biosynthesis. Nature 372:739–746

Liu H, Lightfoot R, Stevens JL (1996) Activation of heat shock factor by alkylating agents is triggered by glutathione depletion and oxidation of protein thiols. J Biol Chem 271:4805–4812

Ludwig S, Engel K, Hoffmeyer A, Sithanandam G, Neufeld B, Palm D, Gaestel M, Rapp UR (1996) 3pK, a novel mitogen-activated protein (MAP) kinase-activated protein kinase, is targeted by three MAP kinase pathways. Mol Cell Biol 16:6687–6697

Maizels ET, Peters CA, Kline M, Cutler JRE, Shanmugam M, Hunzicker-Dunn M (1998) Heat-shock protein-25/27 phosphorylation by the delta isoform of protein kinase C. Biochem J 332:703–712

McLaughlin MM, Kumar S, McDonnell PC, Van Horn S, Lee JC, Livi GP, Young PR (1996) Identification of mitogen-activated protein (MAP) kinase-activated protein kinase-3, a novel substrate of CSBP p38 MAP kinase. J Biol Chem 271:8488–8492

Mehlen P, Kretz-Remy C, Preville X, Arrigo AP (1996) Human hsp27, Drosophila hsp27 and human αB-crystallin expression- mediated increase in glutathione is essential for the protective activity of these proteins against TNFα-induced cell death. EMBO J 15:2695–2706

Mehlen P, Hickey E, Weber LA, Arrigo AP (1997) Large unphosphorylated aggregates as the active form of hsp27 which controls intracellular reactive oxygen species and glutathione levels and generates a protection against TNFalpha in NIH-3T3-ras cells. Biochem Biophys Res Commun 241:187–192

Meier R, Rouse J, Cuenda A, Nebreda AR, Cohen P (1996) Cellular stresses and cytokines activate multiple mitogen-activated- protein kinase kinase homologues in PC12 and KB cells. Eur J Biochem 236:796–805

Milarski KL, Morimoto RI (1986) Expression of human HSP70 during the synthetic phase of the cell cycle. Proc Natl Acad Sci USA 83:9517–9521

Morimoto RI, Tissieres A, Georgopoulos C (1990) The stress response, function of the proteins, and perspectives. In: Morimoto RI, Tissieres A, Georgopoulos C (eds) Stress proteins in biology and medicine. Cold Spring Harbor Laboratory Press, New York, pp 1–36

Narumiya S, Fukushima M (1986) Site and mechanism of growth inhibition by prostaglandins. I. Active transport and intracellular accumulation of cyclopentenone prostaglandins, a reaction leading to growth inhibition. J Pharmacol Exp Ther 239:500–505

Narumiya S, Ohno K, Fujiwara M, Fukushima M (1986) Site and mechanism of growth inhibition by prostaglandins. II. Temperature-dependent transfer of a cyclopentenone prostaglandin to nuclei. J Pharmacol Exp Ther 239:506–511

Ohno K, Fukushima M, Fujiwara M, Narumiya S (1988) Induction of 68,000-dalton heat shock proteins by cyclopentenone prostaglandins. Its association with prostaglandin-induced G1 block in cell cycle progression. J Biol Chem 263:19764–19770

Petronini PG, Alfieri R, Campanini C, Borghetti AF (1995) Effect of an alkaline shift on induction of the heat shock response in human fibroblasts. J Cell Physiol 162:322–329

Pizurki L, Polla BS (1994) cAMP modulates stress protein synthesis in human monocytes-macrophages. J Cell Physiol 161:169–177

Preville X, Schultz H, Knauf U, Gaestel M, Arrigo AP (1998) Analysis of the role of Hsp25 phosphorylation reveals the importance of the oligomerization state of this small heat shock protein in its protective function against TNFα- and hydrogen peroxide-induced cell death. J Cell Biochem 69:436–452

Rogalla T, Ehrnsperger M, Preville X et al. (1999) Regulation of Hsp27 oligomerization, chaperone function, and protective activity against oxidative stress/tumor necrosis factor alpha by phosphorylation. J Biol Chem 274:18947–18956

Saklatvala J, Kaur P, Guesdon F (1991) Phosphorylation of the small heat-shock protein is regulated by interleukin 1, tumour necrosis factor, growth factors, bradykinin and ATP. Biochem J 277:635–642

Santoro MG, Garaci E, Amici C (1989) Prostaglandins with antiproliferative activity induce the synthesis of a heat shock protein in human cells. Proc Natl Acad Sci USA 86:8407–8411

Sheikh-Hamad D, Di Mari J, Suki WN, Safirstein R, Watts BA, Rouse D (1998) p38 kinase activity is essential for osmotic induction of mRNAs for HSP70 and transporter for organic solute betaine in Madin-Darby canine kidney cells. J Biol Chem 273:1832–1837

Shimizu T, Wolfe LS (1990) Arachidonic acid cascade and signal transduction. J Neurochem 55:1–15

Smith WL (1989) The eicosanoids and their biochemical mechanisms of action. Biochem J 259:315–324

Spector A, Chiesa R, Sredy J, Garner W (1985) cAMP-dependent phosphorylation of bovine lens α-crystallin. Proc Natl Acad Sci USA 82:4712–4716

Srivastava KC, Bordia A, Verma SK (1995) Curcumin, a major component of food spice turmeric (Curcuma longa) inhibits aggregation and alters eicosanoid metabolism in human blood platelets. Prostaglandins Leukot Essent Fatty Acids 52:223–227

Stokoe D, Engel K, Campbell DG, Cohen P, Gaestel M (1992) Identification of MAPKAP kinase 2 as a major enzyme responsible for the phosphorylation of the small mammalian heat shock proteins. FEBS Lett 313:307–313

Suzuki A, Sugiyama Y, Hayashi Y, Nyu-i N, Yoshida M, Nonaka I, Ishiura S, Arahata K, Ohno S (1998) MKBP, a novel member of the small heat shock protein family, binds and activates the myotonic dystrophy protein kinase. J Cell Biol 140:1113–1124

Tokuda H, Oiso Y, Kozawa O (1992) Protein kinase C activation amplifies prostaglandin F2 alpha-induced prostaglandin E2 synthesis in osteoblast-like cells. J Cell Biochem 48:262–268

Toullec D, Pianetti P, Coste H et al. (1991) The bisindolylmaleimide GF 109203X is a potent and selective inhibitor of protein kinase C. J Biol Chem 266:15771–15781

Udelsman R, Blake MJ, Stagg CA, Li DG, Putney DJ, Holbrook NJ (1993) Vascular heat shock protein expression in response to stress. Endocrine and autonomic regulation of this age-dependent response. J Clin Invest 91:465–473

Voorter CE, Mulders JW, Bloemendal H, de Jong WW (1986) Some aspects of the phosphorylation of α-crystallin A. Eur J Biochem 160:203–210

Welch WJ (1985) Phorbol ester, calcium ionophore, or serum added to quiescent rat embryo fibroblast cells all result in the elevated phosphorylation of two 28,000-dalton mammalian stress proteins. J Biol Chem 260:3058–3062

Yim MB, Chae HZ, Rhee SG, Chock PB, Stadtman ER (1994) On the protective mechanism of the thiol-specific antioxidant enzyme against the oxidative damage of biomacromolecules. J Biol Chem 269:1621–1626

shHsp-Phosphorylation: Enzymes, Signaling Pathways and Functional Implications

M. Gaestel[1]

1
Introduction

Several posttranslational modifications of shHsps have been detected, including phosphorylation (Kim et al. 1983; Voorter et al. 1986), deamidation, acylation as well as mixed intermolecular disulfide formation, oxidation and glycation (for a review of the latter modifications, see Groenen et al. 1994). In this chapter, one of the most prominent modifications, the shHsp phosphorylation, the enzymes responsible for this modification as well as the possible functional implications will be reviewed in detail. However, it might well be that phosphorylation regulates shHsp structure and function in a complex interplay with some of the other modifications mentioned above.

A recent methodological description of the analysis of shHsp phosphorylation is given elsewhere (Benndorf et al. 2000).

2
Identification of Phosphorylated shHsps

Initially, phosphorylation of a 25-kDa rat stress protein was described in muscle cells (Kim et al. 1983) and it has been demonstrated that rat Hsp25/28 is rapidly phosphorylated in embryonic fibroblasts following heat shock or addition of either the phorbol ester (PMA), the calcium ionophore A23187 or serum in embryonic fibroblasts (Welch 1985). Meanwhile, covalent modification of shHsps by phosphorylation has been detected for bovine αA-crystallin (Voorter et al. 1986) and αB-crystallin (Chiesa et al. 1987), human Hsp27 (Arrigo and Welch 1987), mouse Hsp25 (Benndorf et al. 1988), hamster Hsp27 (Chretien and Landry 1988) as well as for human and bovine Hsp20 (Beall et al. 1999). In all these proteins, phosphoserines are the exclusively phosphorylated amino acid residues identified so far. However, in specific cells other residues of shHsps might also be phosphorylated (Butt and Gaestel, unpubl.).

[1] Institut für Pharmazeutische Biologie, Martin-Luther-Universität Halle/Wittenberg, Hoher Weg 8, 06120 Halle, Germany
Present address: M. Gaestel, Center Biochemistry/Institute of Biochemistry, Medical School Hannover, Carl-Neuberg-Str. 1, 30625 Hannover, Germany

Progress in Molecular and Subcellular Biology, Vol. 28
A.-P. Arrigo and W.E.G. Müller (Eds.)
© Springer-Verlag Berlin Heidelberg 2002

Phosphorylation of α-crystallin was observed in lens fiber cells (Chiesa et al. 1989), in Alexander-diseased brain (Mann et al. 1991), in human U373 MG glioma cells (Ito et al. 1997; Kato et al. 1998) and in different rat tissues including heart and diaphragm even under normal, non-stressed conditions (Ito et al. 1997). While relatively stable in the mature lens fiber, α-crystallin phosphorylation could be reversed by dephosphorylation in the lens epithelium (Chiesa and Spector 1989).

In vitro, αA-crystallin is phosphorylated at S122 (Voorter et al. 1986; phosphorylation sites of sHsp are summarized in Table 1). Phosphorylations of αB-crystallin could be detected at S45 and S59 when labeled in organ culture (Chiesa et al 1987) and also at S19 and S45 after in vivo labeling in the eye lens (Voorter et al. 1989). The serine residues S122 in αA- and S19,45,59 in αB-crystallin can be phosphorylated in vitro by protein kinase A (PKA) (Spector et al. 1985), an enzyme which is present in the lens. For αA-crystallin, a weak autophosphorylation at a site between amino acids 131 and 145, which can be enhanced by dissociation of oligomers, has also been reported (Kantorov and Piatigorsky 1994; Kantorow et al. 1995). In human U373 MG glioma cells, phosphorylation of αB-crystallin at all three sites S19, S45 and S59 is stimulated to a certain extent by heat shock, arsenite, PMA, ocadaic acid, H_2O_2, anisomycin and high osmolarity. The phosphorylation induced by both mitogenic and stress stimuli in these cells has been elegantly shown using phosphorylation site specific antibodies (Ito et al. 1997). As a result of possible phosphorylation at three sites, eight differentially phosphorylated isoforms of αB-crystallin can exist which migrate in four different spots in isoelectric focusing (IEF).

Mouse Hsp25 is phosphorylated at two sites, S15 and S86 (Gaestel et al. 1991). Hence, Hsp25 can exist in four different isoforms (S15,86, S15-P, S86-P, S15,86-P) which are represented by three different spots in isoelectric focussing (non-, mono- and bis-phosphorylated). Human Hsp27 possesses

Table 1. Phosphorylation sites of sHsp and identification of protein kinases. Amino acid residues which are part of the conserved kinase recognition consensus motif are underlined

sHsp	Phosphorylation site in sequence	Responsible enzymes	References
Mouse Hsp25	LLRSPS$_{15}$WEP LNRQLS$_{86}$SGV	MK 2/3, MK5/PRAK?, PKC-δ?	Gaestel et al. (1991); Stokoe et al. (1992)
Human Hsp27	LLRGPS$_{15}$WDP YSRALS$_{78}$RQL LSRQLS$_{82}$SGV	MK 2/3, MK5/PRAK?	Landry et al. (1992); Stokoe et al. (1992)
Bovine Hsp20	WLRRAS$_{16}$APL	PKG?, PKA?	Beall et al. (1999)
Bovine αA-Cry	RYRLPS$_{122}$NVD	PKA?	Voorter et al. (1986)
Bovine αB-Cry	FFPFHS$_{19}$PSR TSTSLS$_{45}$PFY FLRAPS$_{59}$WFD	? p42/44 MAPK/ERK1/2 MK2/3, MK5/PRAK?	Kato et al. (1998)

three phosphorylation sites, S15, S78 and S82 (Landry et al. 1992). Recently, phosphorylation site-specific antibodies against human Hsp27 have been successfully used (Eyers et al. 1999).

Although sHsps are also phosphorylated in response to mitogenic signals such as platelet-derived growth factor (PDGF), acidic fibroblast growth factor (FGF) (Saklatvala et al. 1991) and basic FGF (Zhou et al. 1993), the most striking feature is that human Hsp27 or mouse Hsp25 were rapidly phosphorylated after treatment of cells with the pro-inflammatory cytokines IL-1 and TNFα (Kaur et al. 1989; Arrigo 1990) and stimuli-inducing chemical or oxidative stress such as arsenite, H_2O_2 (Oesterreich et al. 1990; Huot et al. 1995) or UVB light (Nozaki et al. 1997). Since pretreatment of mammalian cells with antioxidants could completely inhibit TNFα-, H_2O_2- or arsenite-induced sHsp-phosphorylation, it has been suggested that reactive oxygen species are involved in this process (Huot et al. 1995). Differentiation-inducing agents such as phorbol ester and leukemia inhibitory factor/D-factor lead to an increased sHsp-phosphorylation which correlates in some cases with the differentiation process (Regazzi et al. 1988; Michishita et al. 1991; Spector et al. 1993; Minowada and Welch 1995). sHsp-phosphorylation has also been described as the result of a wide variety of other stimuli including bradykinin (Saklatvala et al. 1991), thrombin (Mendelsohn et al. 1991), histamine (Santell et al. 1992), ifosfamide (Issels et al. 1993), ocadaic acid (Kasahara et al. 1993), vascular endothelial growth factor (Rousseau et al. 1997), cholecystokinin (Groblewski et al. 1997), Sindbis virus infection (Nakatsue et al. 1998) and cyclosporin A (Paslaru et al. 2000).

A special situation seems to apply to bovine and human Hsp20: Phosphorylation of this protein in smooth muscle cells is stimulated by the adenylate cyclase activator forskolin, the guanylate cyclase activator nitroprusside and the phosphodiesterase inhibitor isobutylmethylxanthine (Brophy et al. 1999). Although several phosphorylated isoforms of Hsp20 exist in smooth muscles during relaxation, only one phosphorylation site, S16, has been localized so far (Beall et al. 1999).

The other human muscle-specific sHsps, HSPB3 (Boehlens et al. 1998) and MKBP (Suzuki et al. 1998a), are not known to be phosphorylated and do not carry the phosphorylation consensus motifs known for Hsp27 or αB-crystallin.

A heat shock- and ecdysterone-dependent appearance of two phosphorylated isoforms of Drosophila Hsp27 has been detected (Rollet and Best-Belpomme 1986). The phosphorylated amino acids and the phosphorylation site in Dm-Hsp27 have not been identified so far. Recently, however, it has been shown that Dm-Hsp27 could be phosphorylated by the purified Hsp27-kinase from hamster (Marin et al. 1996), indicating similar phosphorylation sites.

Cytosolic and chloroplast sHsps of plants are not phosphorylated in vivo (Suzuki et al. 1998b).

3

The Major Pathway for sHsp Phosphorylation Via p38 MAPK and MAPKAP Kinase 2

Phosphorylation in response to a wide variety of extracellular stimuli summarized as "stress" is an intriguing feature of mammalian sHsps. In the last few years, major progress has been achieved by identifying a stress-stimulated kinase-pathway which leads to sHsp-phosphorylation.

Initially, several attempts to characterize and purify the sHsp-kinase in mammals were undertaken, which showed that the sHsp-kinase was distinct from PKA, PKC, p90 ribosomal S6-kinase (RSK), calcium/calmodulin-dependent PKs (Benndorf et al. 1992; Zhou et al. 1993) and MAPK and that the human enzyme has an estimated molecular mass of about 45 kDa (Guesdon et al. 1993). A first direct indication of the kinase involved in sHsp-phosphorylation came from the observation that rabbit MAP kinase-activated protein (MAPKAP) kinase 2 (MK2) very efficiently phosphorylates Hsp25 and Hsp27 in vitro and that this kinase co-purifies with the Hsp25 kinase activity during the whole purification process from rabbit muscle (Stokoe et al. 1992a). MK2 shares all properties described so far for the sHsp-kinases (Stokoe et al. 1992b; Cano et al. 1996). It is activated by heat shock (Engel et al. 1995a) and possesses a phosphorylation site recognition motif HyXRXXS(P) (Hy stands for a bulky hydrophobic residue as F, L, Y, V; Stokoe et al. 1993), which is in agreement with the sites in Hsp25 and Hsp27. The major breakthrough in understanding this signaling pathway for sHsp phosphorylation was the identification of the main activator of MK2 as a stress-activated MAP kinase, the p38 MAPK (Freshney et al. 1994; Rouse et al 1994). Meanwhile we know that two of the four isoforms of p38 MAPK, p38α and p38β also designated as SAPK2a and b, are the main activators of MK2 which phosphorylate MK2 at two or three regulatory sites (Stokoe et al. 1992b; Ben-Levy et al. 1995; Cuenda et al. 1995; Engel et al. 1995b; Cohen 1997). Upstream of p38 MAPK, two stress-activated MAP kinase kinases, MKK6 and MKK3, have been described which in turn could be activated by MAP kinase kinase kinases such as TPL2, MEKK, MLK/DLK, TAK or ASK (for review, see Kyriakis and Avruch 1996; Schaeffer and Weber 1999).

The finding that p38 MAPK is a central component of a stress- and inflammatory cytokine-activated signaling cascade which is activated by IL-1, TNFα, heat shock, osmotic stress, UV-light, H_2O_2, anisomycin, bacterial lipopolysaccharide (LPS) and arsenite (reviewed in Kyriakis and Avruch 1996) and that it activates the sHsp kinase MK2 explains sHsp-phosphorylation as being a result of most of the stimuli and stress conditions mentioned above. Only PMA-induced Hsp-phosphorylation could not be explained by this cascade per se. However, a cross talk between PMA-induced isoforms of PKC and upstream components of the p38 cascade seems likely (Schultz et al. 1997) and a direct phosphorylation of Hsp25 in the *corpus luteum* of rats by PKC-δ is highly probable (Maizels et al. 1998). The finding that both Hsp27 and αB-crystallin are phosphorylated in HeLa cells by similar stress-stimuli (IJssel et al. 1998) is probably the result of the existence of MK2 phosphorylation sites in both proteins (S59 in αB-crystallin and S15,78,82 in Hsp27). This notion is supported by the fact that MKK6-activated p38 MAPK leads to activation of MK2 and an increased phosphorylation of αB-crystallin at serine 59 (Hoover et al. 2000).

Two distinct isoforms of MK2, both of which can phosphorylate sHsps, were detected in mouse embryonic fibroblasts (Kotlyarov et al. 1999) and in rat cardiac myocytes (Chevalier and Allen 2000). Since deletion of the MK2 gene in mice leads to the disappearance of both isoforms (Kotlyarov et al. 1999), it is supposed that both isoforms result from the same MK2 gene by differential splicing, probably altering the C-terminus of the enzyme (Zu et al. 1994; K. Engel and C. Schubert, pers. comm.; L.-L. Lin, pers. comm.). Three other kinases very similar to MK2 have been identified – MAPKAP kinases 3, 4 and 5 (MK3, 4, 5). MK3 (McLaughlin et al. 1996), also known as 3pK (Sithanandam et al. 1996), shows 75% sequence identity to MK2 at the amino acid level and is widely expressed in mammalian tissues, especially in heart and skeletal muscle. It is also activated by p38 MAPKα,β, and it phosphorylates Hsp25 and Hsp27 very efficiently in vitro and probably also in vivo at the same sites as MAPKAP kinase 2 (Clifton et al. 1996; Ludwig et al. 1996; McLaughlin et al. 1996). MAPKAP kinase 4 has been identified in sea urchin and is 65% identical to MK2 (Komatsu et al. 1997). So far, no mammalian homologue for MK4 has been identified. MK5 (Ni et al. 1998), also described as p38-regulated/activated protein kinase (PRAK; New et al. 1998), is about 40% identical to MK2, strictly regulated by p38 MAPKα,β and ubiquitously expressed in human and mouse. MK5/PRAK is also able to phosphorylate sHsps in vitro at the same sites identified for MK2 and 3 (New et al. 1998). So far, it is not possible to estimate the relative contribution of MK2, 3 and 5 to stress-dependent sHsp phosphorylation in vivo. The first information about the contribution of MK2 comes from analysis of Hsp25 phosphorylation in MK2 knock out mice (Kotlyarov et al. 1999). In heart tissue of these animals, no bis-phosphorylated isoform of Hsp25 could be detected after LPS-stress treatment indicating a major role of MK2 for sHsp phosphorylation in these cells under these specific conditions. However, the existence of mono-phosphorylated Hsp25 in MK2 knockout animals indicates that there are further Hsp25 kinases relevant in vivo. This idea is supported by the finding that in other tissues of the MK2-minus animals, no reduction of stress-induced Hsp25 phosphorylation could be detected at all (Kotlyarov and Gaestel, unpubl. result).

A specific inhibitor for p38 MAPKα,β, the pyridyl imidazole SB203580 (Lee et al. 1994), has been successfully used in many publications to inhibit the p38 MAPK cascade (reviewed in Cohen, 1997). Since p38 MAPKα,β plays the central role in the signaling processes leading to sHsp-phosphorylation and since MK2 activation and sHsp phosphorylation are obviously obligatorily coupled to activation of this pathway, this inhibitor also suppresses sHsp-phosphorylation in vivo (e.g. Cuenda et al. 1995; Schultz et al. 1996). However, it has emerged that other SB203580-sensitive enzymes such as protein kinase

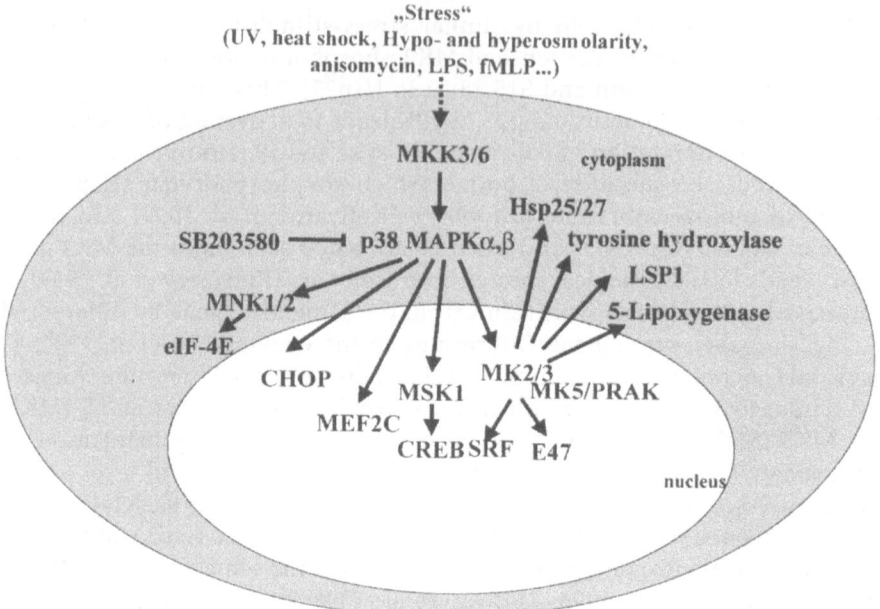

Fig. 1. The "p38MAPK/Hsp27 pathway". As seen here, the stress-activated p38 MAPK/Hsp27 pathway is far from linear but is strongly branched. There are several targets for p38 MAPKα,β, which are phosphorylated and activated by this enzymes: Transcription factors CHOP (Wang and Ron 1996) and MEF2C (Han et al. 1997) and protein kinases MK2, 3, 5 (Rouse et al. 1994; McLaughlin et al. 1996; New et al. 1998) as well as MNK1, 2 (Fukunaga and Hunter 1997) and MSK1, 2 (Deak et al. 1998). MK2, 3, 5 and MSK1 in turn can also phosphorylate transcription factors CREB (Deak et al. 1998), SRF (Heidenreich et al. 1999) and E47 (Neufeld et al. 2000). MNK1 in turn phosphorylates initiation factor 4B (Waskiewicz et al. 1999). Interestingly, as a consequence of phosphorylation and activation, MK2, 3, 5 are translocated to the cytoplasm (Engel et al. 1998; Ugo Moens, pers. comm.). In the cytoplasm of the cell besides sHsps, further substrates for MK2, 3 exist: tyrosine hydroxylase (Thomas et al. 1997), leukocyte-specific protein 1 (LSP1, Huang et al. 1997) and 5-lipoxygenase. (Werz et al. 2000)

B (Lali et al. 2000) and cyclo-oxygenases 1 and 2 (Borsch-Haubold et al. 1998) exist. On the other hand, inhibition of p38 MAPK blocks a wide variety of downstream events ranging from phosphorylation of transcription factors such as MEF2 C (Han et al. 1997) and CHOP (Wang and Ron 1996) to phosphorylation and activation of further protein kinases such as MNKs (Fukunaga and Hunter 1997; Waskiewicz et al. 1997) and MSKs (Deak et al. 1998) (see Fig. 1). Hence, the use of SB203580 and its related compounds can best be useful for demonstrating that sHsp phosphorylation, besides all other events downstream of p38 MAPK and the other targets of SB, is not involved in a certain biological phenomenon or process (Schultz et al. 1997; Preville et al. 1998; Martin et al. 1999; Davidson and Morange 2000). To specifically demonstrate the involvement of sHsp phosphorylation, in addition to the SB compounds, the intracellular overexpression or microinjection of shHsp phosphorylation site mutants was used in several experimental approaches (Lavoie et al. 1995; Schneider et al. 1998; Hedges et al. 1999; Lasa et al. 2000). However, even in these approaches, it is not clear whether a direct mimicry effect or the blocking shHsps phosphorylation is measured or whether the effects observed result from negatively interfering upstream properties of the non-phosphorylatable mutants which could inhibit the enzymatic activity of MK2, 3 or 5 and block phosphorylation of other substrates of these enzymes. Hence, so far there is no experimental system available which allows definitive conclusions to be drawn, on the function of shHsp phosphorylation in vivo. In future, we expect to learn more about the general function of these proteins from the targeted disruption of the Hsp25 gene in mice and, hopefully, we will develop new ideas to test the involvement of Hsp25 phosphorylation in this function, too.

4
Other Pathways

Besides the stress-activated p38 MAPK cascade, several other signal transduction pathways are involved in shHsp phosphorylation. It has been demonstrated that the classical mitogen-activated p42/44 MAPK cascade is responsible for increased phosphorylation of serine residue 45 in αB-crystallin in mitotic human cells (Kato et al. 1998) and that phosphorylation of αB-crystallin in rat lens is due to developmental activation of p42/44 MAPK in the early post-natal period (Ito et al. 1999). The classical p42/44 MAPK pathway is also able to activate MK2 and MK3 in cells where specific components of this signaling pathway are overexpressed (Ludwig et al. 1996). However, in vivo p38 MAPK has been identified as the major activator of MK2 and MK3 and SB203580 (in contrast to the MKK1-inhibitor PD098059) was able to prevent activation of MK2/3 and phosphorylation of shHsps at the appropriate sites in all cells examined so far (reviewed in Cohen 1997). So, the only shHsp targeted by the mitogene-stimulated p42/44 MAPK cascade in vivo known so far, is αB-crystallin with its serine 45.

Hsp25 is highly phosphorylated in the rat *corpus luteum* in late pregnancy. Under these conditions no activation of p38 MAPK, but a specific and prominent induction of the Ca^{2+}-independent PKC-isoform PKC-δ is observed (Maizels et al. 1998). Since PKC-δ and -α are also able to phosphorylate Hsp25 at the same sites as MK2 in vitro (Gaestel et al. 1991; Maizels et al. 1998), it is well possible that increased luteal Hsp25 phosphorylation in late pregnancy is the result of direct phosphorylation by PKC-δ in vivo.

S16 of Hsp20 is located in a sequence motif RRXS (cf. Table 1) which is a characteristic consensus motif for phosphorylation by cyclic nucleotide-(cAMP, cGMP)-activated protein kinases (PKA, PKG). Since activation of cyclic nucleotide-dependent signaling pathways leads to phosphorylation of Hsp20 at S16 in various smooth muscles cells (A.C. Beall et al. 1997; A. Beall et al.

1999), it is supposed that these pathways are responsible for this specific phosphorylation of Hsp20.

It is probable that the degree of sHsp phosphorylation is not only regulated by phosphorylating enzymes but also by specific protein phosphatases. For mouse Hsp25 an in vitro dephosphorylation by the calcium-dependent protein phosphatase 2B (calcineurin) has been described (Gaestel et al. 1992). Since it is known that heat shock increases the intracellular calcium level, this enzyme could be of physiological relevance for the regulation of sHsp-phosphorylation. Using specific inhibitors for several classes of protein phosphatases the Hsp27 dephosphorylating enzyme in MRC-5 cells has been characterized to belong to the protein phosphatase 2 A group which itself is regulated by signal-dependent phosphorylation (Cairns et al. 1994). Hence, these findings indicate that other mechanisms may be involved in regulation of sHsp-phosphorylation by extracellular signals.

5
Functional Aspects of·sHsps-Phosphorylation

In order to identify functional consequences of sHsp phosphorylation, it would be very helpful to have a well-defined sHsp function, the modulation of which could be analyzed by the post-translational modification. Unfortunately, sHsps show a wide variety of different, seemingly unrelated functions in different experimental systems (reviewed in Ehrnsperger et al. 1998). These range from action as ATP-independent molecular chaperones (Horwitz 1992; Jakob et al. 1993; Merck et al. 1993; Ehrnsperger et al. 1997, see Chap. 3, this Vol.), specific interactions with certain components of the cytoskeleton such as actin (Miron et al. 1991; Lavoie et al. 1993; Benndorf et al. 1994; Zhu et al. 1994) or intermediate filaments (Nicholl and Quinlan 1994; Wisniewski and Goldman 1998; Perng et al. 1999; see Chap. 12, this Vol.), regulation of intracellular glutathione level (Mehlen et al. 1996a), stabilization of RNA (Nover et al. 1989; Lasa et al. 1999), inhibition of apoptosis (Mehlen et al. 1996b; Garrido et al. 1999; Bruey et al. 2000; Pandey et al. 2000; see Chap. 10, this Vol.) up to the inhibition of elastase (Voorter et al. 1994). Very recently, another very interesting function for sHsps under stress-conditions has been characterized: Hsp27 is able to inhibit protein translation during heat shock by binding initiation factor eIF4G and facilitating dissociation of cap-initiation complexes (Cuesta et al. 2000).

The diversity in function (discussed elsewhere in this book) makes a definition for the role of sHsp phosphorylation rather complicated, especially regarding a unifying concept. In addition, the experimental approaches using the inhibitor SB203580 or phosphorylation site mutants of sHsps harbor the limitations discussed in section 3. However, taking into account these reservations, some potential effects of sHsp phosphorylation emerge which are discussed below.

5.1
Phosphorylation and Oligomerisation

An influence of sHsp-phosphorylation on the size of the oligomeric sHsp particles has been described: Depending on the experimental conditions and cellular systems, phosphorylation of Hsp27 correlates with either the dissociation or a further increase in size of the oligomers. In U251 MG human glioma cells PMA, IL-1, TNFα, okadaic acid and chemical stresses such as arsenite induce phosphorylation and, in parallel, dissociation of Hsp27 from 300–400 kDa oligomers to structures smaller than 70 kDa (Kato et al. 1994). In hamster CCL39 cells, a reduction of the multimeric size of transfected human Hsp27 in response to mitogenic stimuli and arsenite is only observed for the wild-type protein but not for a phosphorylation site mutant (Lavoie et al. 1995), indicating that phosphorylation is necessary for dissociation. In contrast, in starved HeLa cells, hsp27 is dephosphorylated and quantitatively recovered in the form of small structures (<200 kDa) whereas the protein is phosphorylated after serum stimulation leading to an increased molecular mass (up to 700 kDa) (Mehlen and Arrigo 1994). Following TNFα-treatment of HeLa cells, Hsp27 is rapidly phosphorylated and detected in larger oligomers (>700 kDa), but after 4 h of treatment the large complexes dissociate into smaller species (<200 kDa). During the whole process, the phosphorylated isoforms of Hsp27 remained in the smaller oligomers of the protein (<300 kDa) (Mehlen et al. 1995). Interestingly, in stably transfected NIH 3T3 or L929 cells, non-phosphorylatable mutants of Hsp25 or Hsp27 are also shifted to large oligomeric complexes after TNFα-treatment, but, in contrast to the wild-type protein, they stay constitutively hyperaggregated over the whole time of treatment (Mehlen et al. 1997; Preville et al. 1998). Similarly, mutants of hamster Hsp27 lacking the phosphorylatable serine 90 (homologue to S82 in human and S86 in mouse sHsp), but still carrying the second phosphorylation site serine 15, are no longer able to dissociate under stress conditions, indicating that phosphorylation of S90 regulates oligomerization of this protein probably by modulating intermolecular interaction in the N-terminus of the molecule (Lambert et al. 1999). Mimicry of human Hsp27 phosphorylation using replacement of the phosporylatable serine residues S15, S78 and S82 by negatively charged aspartate residues results in dissociation of Hsp27 in vitro (Rogalla et al. 1999). In smooth muscle cells Hsp20 dissociates as a result of stimulation of cyclic nucleotide-dependent kinases, probably as the result of phosphorylation of S16 (Brophy et al. 1999). From the data above one may conclude that phosphorylation is necessary but, at least at the cellular level, not always sufficient for dissociation of the sHsp-oligomers and that other, unknown mechanisms could also contribute to the regulation of the sHsp oligomeric size.

5.2
Speculations About the Function of sHsp Phosphorylation

Since distinct biochemical activities have been described for small and large oligomers of sHsps, phosphorylation may regulate these functions in a complementary manner. Monomers of sHsps were reported as F-actin cap-binding proteins which inhibit actin polymerization whereas oligomeric complexes do not show this effect (Miron et al. 1991; Benndorf et al. 1994). In contrast, oligomeric complexes of sHsps exhibit chaperone properties in vitro (Ehrnsperger et al. 1997; Leroux et al. 1997) and dissociation of human Hsp27 achieved by mimicry of serine phosphorylation by aspartate correlates with a decrease in chaperone activity of the mutants (Rogalla et al. 1999). Since constitutively hyper-aggregated phosphorylation site mutants of sHsps are still able to protect cells against TNF-α (Mehlen et al. 1997; Preville et al. 1998) or against ischemic damage (Martin et al. 1999), the chaperone function of sHsps could be more relevant for this protection than the actin-binding function. It may well be that sHsp phosphorylation can act as a switch between these both functions, stimulating actin-binding and suppressing chaperone function of sHsps. However, the in vitro chaperone function of sHsps is not always restricted to large oligomeric structures, since in the yeast *S. cerevisiae*, where no phosphorylation of sHsps has been detected so far, a temperature-induced dissociation of Hsp26 oligomers to tetramers is reported to increase the chaperone activity (Haslbeck et al. 1999).

In addition, the phosphorylation-dependent actin-modulating function of Hsp27 suggests a role of at least some mammalian sHsps in signaling to the cytoskeleton (reviewed in Landry and Huot 1999). In this regard it is interesting to note that the enzymes responsible for sHsp-phosphorylation, MK2,3 and 5 are downstream components of the p38 MAPK cascade, which could be activated through the P21-activated kinase 1 (Pak1) by the Rho family GTPases Rac and Cdc42 (Zhang et al. 1995). Interestingly, these small GTPases are also found to be responsible for actin filament accumulation at the plasma membrane forming membrane ruffles (Ridley et al. 1992) and the formation of actin-based filopodia (Nobes and Hall 1995) suggesting a possible involvement of sHsp-phosphorylation in this processes. This idea is supported by the observation that dominant negative Rho inhibits ceramide- and endothelin-induced intracellular redistribution of Hsp27 in rabbit smooth muscle (Wang and Bitar 1998).

The potential importance of sHsp phosphorylation for the function of the cytoskeleton is further supported by several independent observations: In human platelets only phosphorylated Hsp27 associates with the activation-dependent cytoskeleton (Zhu et al. 1994). In ROS 17/2.8 osteosarcoma cells, restoration of stress fibers and focal adhesions in the heat shock recovery was inhibited by microinjection of non-phosphorylated Hsp27 but not of PKA-phosphorylated Hsp27 (Schneider et al. 1998). In addition to that, wild-type Hsp27, but not its phosphorylation site mutants stabilize the cytoskeleton and protect cells from the effects of cytochalasin D (Lavoie et al. 1995) as well as from high doses of Cholecystokinin (CCK), a stimulus which activates the p38 MAPK cascade in rat acini (Schäfer et al. 1999).

Because of its effect on the cytoskeleton, sHsp phosphorylation could also be involved in regulation of the motility of cells. Migration of bovine endothelial cells is stimulated about two-fold by Hsp27 and inhibited by a non-phosphorylatable mutant of Hsp27 probably by affecting the generation of lamellipodia microfilaments (Piotrowicz et al. 1998). Similarly, the non-phosphorylatable mutant of Hsp27 and dominant negative p38 MAPKα also blocks PDGF-stimulated migration of smooth muscle cells, while a constitutively active MKK6 increased cell migration (Hedges et al. 1999).

Phosphorylation of two different sHsps, Hsp27 and Hsp20, seem to regulate smooth muscle cell contraction in different directions. Phosphorylation of Hsp27 is correlated with muscle cell and mesangial cell contraction (Müller et al. 1999; Meloche et al. 2000) and inhibition of the p38 MAPK cascade, and Hsp27 antibodies attenuate endothelin-1 induced contraction (Yamboliev et al. 2000). In contrast, Hsp20 phosphorylation is increased during vascular smooth muscle relaxation induced by nitroglycerin (Rembold and O'Connor 2000) and cAMP/cGMP-dependent inhibition of contraction of smooth muscle correlates with increased phosphorylation of Hsp20 (Woodrum et al. 1999). The antagonistic relation of phosphorylation of Hsp20 and Hsp27 in smooth muscle contraction is further supported by the observation, that phosphorylated Hsp27 could inhibit Hsp20 phosphorylation in vitro (Fuchs et al. 2000).

It is interesting to note that sHsp phosphorylation is also influenced by ischemia and under conditions of ischemic preconditioning. Ischemia in rat kidney leads to decreased glomerular Hsp25 phosphorylation (Smoyer et al. 2000) while a 10-fold increase of S59 phosphorylation of αB-crystallin is observed after ischemic preconditioning in rat hearts (Eaton et al. 2000).

Finally, it should also be mentioned that there is increasing evidence for a role of the p38 MAPK cascade in regulating mRNA stability and translation (Lee et al. 1994; Kotlyarov et al. 1999; Lasa et al. 1999; Winzen et al. 1999; Pages et al. 2000). Since sHsps have also been detected in specific mRNA-protein particles, so-called stress granula (Nover et al. 1989), and since sHsp specifically interact with a component of the translation initiation complex (Cuesta et al. 2000), it cannot be excluded that sHsps are involved in this regulation as well. The observations that a mutant of Hsp27 which mimics phosphorylation is able to partially stabilize a reporter mRNA carrying the AUUUA-rich 3'UTR of cyclooxygenase 2 (Lasa et al. 1999) and that MK2 targets the AUUUA-rich region of TNF-α (Kotlyarov et al. 1999; Kollias and Gaestel, unpublished) could give a first indication of a function of sHsps in this context. However, as mentioned before in section 3, the experimental approaches do not exclude that other targets of p38 MAPK and MK2,3 or 5 could be involved in this regulation. The existence of such hypothetical targets would permit the observed effects to be explained without the necessity for the involvement of sHsps and their phosphorylation.

Acknowledgements. I would like to thank Rainer R. Benndorf (Ann Arbor) and Armin Neininger (Halle) for critical reading of the manuscript and for their valuable comments.

References

Arrigo AP (1990) Tumor necrosis factor induces the rapid phosphorylation of the mammalian heat shock protein hsp28. Mol Cell Biol 10:1276–1280

Arrigo AP, Welch WJ (1987) Characterization and purification of the small 28,000-dalton mammalian heat shock protein. J Biol Chem 262:15359–15369

Beall A, Bagwell D, Woodrum D, Stoming TA, Kato K, Suzuki A, Rasmussen H, Brophy CM (1999) The small heat shock-related protein, HSP20, is phosphorylated on serine 16 during cyclic nucleotide-dependent relaxation. J Biol Chem 274:11344–11351

Beall AC, Kato K, Goldenring JR, Rasmussen H, Brophy CM (1997) Cyclic nucleotide-dependent vasorelaxation is associated with the phosphorylation of a small heat shock-related protein. J Biol Chem 272:11283–11287

Ben-Levy R, Leighton IA, Doza YN, Attwood P, Morrice N, Marshall CJ, Cohen P (1995) Identification of novel phosphorylation sites required for activation of MAPKAP kinase-2. EMBO J 14:5920–5930

Benndorf R, Kraft R, Otto A, Stahl J, Bohm H, Bielka H (1988) Purification of the growth-related protein p25 of the Ehrlich ascites tumor and analysis of its isoforms. Biochem Int 17:225–234

Benndorf R, Hayess K, Stahl J, Bielka H (1992) Cell-free phosphorylation of the murine small heat-shock protein hsp25 by an endogenous kinase from Ehrlich ascites tumor cells. Biochim Biophys Acta 1136:203–207

Benndorf R, Hayess K, Ryazantsev S, Wieske M, Behlke J, Lutsch G (1994) Phosphorylation and supramolecular organization of murine small heat shock protein HSP25 abolish its actin polymerization-inhibiting activity. J Biol Chem 269:20780–20784

Benndorf R, Engel K, Gaestel M (2000) Analysis of small Hsp phosphorylation. Methods Mol Biol 99:431–445

Boelens WC, Van Boekel MA, De Jong WW (1998) HspB3, the most deviating of the six known human small heat shock proteins. Biochim Biophys Acta 1388:513–516

Borsch-Haubold AG, Pasquet S, Watson SP (1998) Direct inhibition of cyclooxygenase-1 and -2 by the kinase inhibitors SB 203580 and PD 98059. SB203580 also inhibits thromboxane synthase. J Biol Chem 273:28766–28772

Brophy CM, Dickinson M, Woodrum D (1999) Phosphorylation of the small heat shock-related protein, HSP20, in vascular smooth muscles is associated with changes in the macromolecular associations of HSP20. J Biol Chem 274:6324–6329

Bruey JM, Ducasse C, Bonniaud P, Ravagnan L, Susin SA, Diaz-Latoud C, Gurbuxani S, Arrigo AP, Kroemer G, Solary E, Garrido C (2000) Hsp27 negatively regulates cell death by interacting with cytochrome c. Nat Cell Biol 2:645–652

Cairns J, Qin S, Philp R, Tan YH, Guy GR (1994) Dephosphorylation of the small heat shock protein Hsp27 in vivo by protein phosphatase 2A. J Biol Chem 269:9176–9183

Cano E, Doza YN, Ben-Levy R, Cohen P, Mahadevan LC (1996) Identification of anisomycin-activated kinases p45 and p55 in murine cells as MAPKAP kinase-2. Oncogene 12:805–812

Chevalier D, Allen BG (2000) Two distinct forms of MAPKAP kinase-2 in adult cardiac ventricular myocytes. Biochemistry 39:6145–6156

Chiesa R, Spector A (1989) The dephosphorylation of lens alpha-crystallin A chain. Biochem Biophys Res Commun 162:1494–1501

Chiesa R, Gawinowicz-Kolks MA, Kleiman NJ, Spector A (1987) The phosphorylation sites of the B2 chain of bovine alpha-crystallin. Biochem Biophys Res Commun 144:1340–1347

Chiesa R, McDermott MJ, Spector A (1989) Differential synthesis and phosphorylation of the alpha-crystallin A and B chains during bovine lens fiber cell differentiation. Curr Eye Res 8:151–158

Chretien P, Landry J (1988) Enhanced constitutive expression of the 27-kDa heat shock proteins in heat-resistant variants from Chinese hamster cells. J Cell Physiol 137:157–166

Clifton AD, Young PR, Cohen P (1996) A comparison of the substrate specificity of MAPKAP kinase-2 and MAPKAP kinase-3 and their activation by cytokines and cellular stress. FEBS Lett 392:209–214

Cohen P (1997) The search for physiological substrates of MAP and SAP kinases in mammalian cells. Trends Cell Biol 7:353–361

Cuenda A, Rouse J, Doza YN, Meier R, Cohen P, Gallagher TF, Young PR, Lee JC (1995) SB 203580 is a specific inhibitor of a MAP kinase homologue which is stimulated by cellular stresses and interleukin-1. FEBS Lett 364:229–233

Cuesta R, Laroia G, Schneider RJ (2000) Chaperone hsp27 inhibits translation during heat shock by binding eIF4G and facilitating dissociation of cap-initiation complexes. Genes Dev 14: 1460–1470

Davidson SM, Morange M (2000) Hsp25 and the p38 MAPK pathway are involved in differentiation of cardiomyocytes. Dev Biol 218:146–160

Deak M, Clifton AD, Lucocq LM, Alessi DR (1998) Mitogen- and stress-activated protein kinase-1 (MSK1) is directly activated by MAPK and SAPK2/p38, and may mediate activation of CREB. EMBO J 17:4426–4441

Eaton P, Awad WI, Miller JI, Hearse DJ, Shattock MJ (2000) Ischemic preconditioning: a potential role for constitutive low molecular weight stress protein translocation and phosphorylation? J Mol Cell Cardiol 32:961–971

Ehrnsperger M, Graber S, Gaestel M, Buchner J (1997) Binding of non-native protein to Hsp25 during heat shock creates a reservoir of folding intermediates for reactivation. EMBO J 16:221–229

Ehrnsperger M, Buchner J, Gaestel M (1998) Structure and function of small heat shock proteins. In: Fink AL, Goto Y (eds) Molecular chaperones in the life cycle of proteins. Marcel Dekker, New York, pp 533–566

Engel K, Ahlers A, Brach MA, Herrmann F, Gaestel M (1995a) MAPKAP kinase 2 is activated by heat shock and TNF-alpha: in vivo phosphorylation of small heat shock protein results from stimulation of the MAP kinase cascade. J Cell Biochem 57:321–330

Engel K, Schultz H, Martin F, Kotlyarov A, Plath-K, Hahn M, Heinemann U, Gaestel M (1995b) Constitutive activation of mitogen-activated protein kinase-activated protein kinase 2 by mutation of phosphorylation sites and an A-helix motif. J Biol Chem 270:27213–27221

Engel K, Kotlyarov A, Gaestel M (1998) Leptomycin B-sensitive nuclear export of MAPKAP kinase 2 is regulated by phosphorylation. EMBO J 17:3363–3371

Eyers PA, van den IJssel P, Quinlan RA, Goedert M, Cohen P (1999) Use of a drug-resistant mutant of stress-activated protein kinase 2a/p38 to validate the in vivo specificity of SB 203580. FEBS Lett 451:191–196

Freshney NW, Rawlinson L, Guesdon F, Jones E, Cowley S, Hsuan J, Saklatvala J (1994) Interleukin-1 activates a novel protein kinase cascade that results in the phosphorylation of Hsp27. Cell 78:1039–1049

Fuchs LC, Giulumian AD, Knoepp L, Pipkin W, Dickinson M, Hayles C, Brophy C (2000) Stress causes decrease in vascular relaxation linked with altered phosphorylation of heat shock proteins. Am J Physiol Regul Integr Comp Physiol 279:R492–R498

Fukunaga R, Hunter T (1997) MNK1, a new MAP kinase-activated protein kinase, isolated by a novel expression screening method for identifying protein kinase substrates. EMBO J 16:1921–1933

Gaestel M, Schroder W, Benndorf R, Lippmann C, Buchner K, Hucho F, Erdmann VA, Bielka H (1991) Identification of the phosphorylation sites of the murine small heat shock protein hsp25. J Biol Chem 266:14721–14724

Gaestel M, Benndorf R, Hayess K, Priemer E, Engel K (1992) Dephosphorylation of the small heat shock protein hsp25 by calcium/calmodulin-dependent (type 2B) protein phosphatase. J Biol Chem 267:21607–21611

Garrido C, Bruey JM, Fromentin A, Hammann A, Arrigo AP, Solary E (1999) HSP27 inhibits cytochrome c-dependent activation of procaspase-9. FASEB J 13:2061–2070

Groblewski GE, Grady T, Mehta N, Lambert H, Logsdon CD, Landry J, Williams JA (1997) Cholecystokinin stimulates heat shock protein 27 phosphorylation in rat pancreas both in vivo and in vitro. Gastroenterology 112:1354–1361

Groenen PJ, Merck KB, de Jong WW, Bloemendal H (1994) Structure and modifications of the junior chaperone alpha-crystallin. From lens transparency to molecular pathology. Eur J Biochem 225:1–19

Guesdon F, Freshney N, Waller RJ, Rawlinson L, Saklatvala J (1993) Interleukin 1 and tumor necrosis factor stimulate two novel protein kinases that phosphorylate the heat shock protein hsp27 and beta-casein. J Biol Chem 268:4236–4243

Han J, Jiang Y, Li Z, Kravchenko VV, Ulevitch RJ (1997) Activation of the transcription factor MEF2 C by the MAP kinase p38 in inflammation. Nature 386:296–299

Haslbeck M, Walke S, Stromer T, Ehrnsperger M, White HE, Chen S, Saibil HR, Buchner J (1999) Hsp26: a temperature-regulated chaperone. EMBO J 18:6744–6751

Hedges JC, Dechert MA, Yamboliev IA, Martin JL, Hickey E, Weber LA, Gerthoffer WT (1999) A role for p38(MAPK)/HSP27 pathway in smooth muscle cell migration. J Biol Chem 274: 24211–24219

Heidenreich O, Neininger A, Schratt G, Zinck R, Cahill MA, Engel K, Kotlyarov A, Kraft R, Kostka S, Gaestel M, Nordheim A (1999) MAPKAP kinase 2 phosphorylates serum response factor in vitro and in vivo. J Biol Chem 274:14434–14443

Hoover HE, Thuerauf DJ, Martindale JJ, Glembotski CC (2000) Alpha B-crystallin gene induction and phosphorylation by MKK6-activated p38. A potential role for alpha B-crystallin as a target of the p38 branch of the cardiac stress response. J Biol Chem 275:23825–23833

Horwitz J (1992) Alpha-crystallin can function as a molecular chaperone. Proc Natl Acad Sci USA 89:10449–10453

Huang CK, Zhan L, Ai Y, Jongstra J (1997) LSP1 is the major substrate for mitogen-activated protein kinase-activated protein kinase 2 in human neutrophils. J Biol Chem 272:17–19

Huot J, Lambert H, Lavoie JN, Guimond A, Houle F, Landry J (1995) Characterization of 45-kDa/54-kDa HSP27 kinase, a stress-sensitive kinase which may activate the phosphorylation-dependent protective function of mammalian 27-kDa heat-shock protein HSP27. Eur-J-Biochem 227:416–427

Van den IJssel PR, Overkamp P, Bloemendal H, de Jong WW (1998) Phosphorylation of alphaB-crystallin and HSP27 is induced by similar stressors in HeLa cells. Biochem Biophys Res Commun 247:518–523

Issels RD, Meier TH, Muller E, Multhoff G, Wilmanns W (1993) Ifosfamide induced stress response in human lymphocytes. Mol Aspects Med 14:281–286

Ito H, Okamoto K, Nakayama H, Isobe T, Kato K (1997) Phosphorylation of alphaB-crystallin in response to various types of stress. J Biol Chem 272:29934–29941

Ito H, Iida K, Kamei K, Iwamoto I, Inaguma Y, Kato K (1999) AlphaB-crystallin in the rat lens is phosphorylated at an early post-natal age. FEBS Lett 446:269–272

Jakob U, Gaestel M, Engel K, Buchner J (1993) Small heat shock proteins are molecular chaperones. J Biol Chem 268:1517–1520

Kantorow M, Piatigorsky J (1994) Alpha-crystallin/small heat shock protein has autokinase activity. Proc Natl Acad Sci USA 91:3112–3116

Kantorow M, Horwitz J, van Boekel MA, de Jong WW, Piatigorsky J (1995) Conversion from oligomers to tetramers enhances autophosphorylation by lens alpha A-crystallin. Specificity between alpha A- and alpha B-crystallin subunits. J Biol Chem 270:17215–17220

Kasahara K, Ikuta T, Chida K, Asakura R, Kuroki T (1993) Rapid phosphorylation of 28-kDa heat-shock protein by treatment with okadaic acid and phorbol ester of BALB/MK-2 mouse keratinocytes. Eur J Biochem 213:1101–1107

Kato K, Hasegawa K, Goto S, Inaguma Y (1994) Dissociation as a result of phosphorylation of an aggregated form of the small stress protein, hsp27. J Biol Chem 269:11274–11278

Kato K, Ito H, Kamei K, Inaguma Y, Iwamoto I, Saga S (1998) Phosphorylation of alphaB-crystallin in mitotic cells and identification of enzymatic activities responsible for phosphorylation. J Biol Chem 273:28346–28354

Kaur P, Welch WJ, Saklatvala J (1989) Interleukin 1 and tumour necrosis factor increase phosphorylation of the small heat shock protein. Effects in fibroblasts, Hep G2 and U937 cells. FEBS Lett 258:269–273

Kim YJ, Shuman J, Sette M, Przybyla A (1983) Phosphorylation pattern of a 25 Kdalton stress protein from rat myoblasts. Biochem Biophys Res Commun 117:682–687

Komatsu S, Murai N, Totsukawa G, Abe M, Akasaka K, Shimada H, Hosoya H (1997) Identification of MAPKAPK homolog (MAPKAPK-4) as a myosin II regulatory light-chain kinase in sea urchin egg extracts. Arch Biochem Biophys 343:55–62

Kotlyarov A, Neininger A, Schubert C, Eckert R, Birchmeier C, Volk HD, Gaestel M (1999) MAPKAP kinase 2 is essential for LPS-induced TNF-alpha biosynthesis. Nat Cell Biol 1:94–97

Kyriakis JM, Avruch J (1996) Protein kinase cascades activated by stress and inflammatory cytokines. BioEssays 18:567–577

Lali FV, Hunt AE, Turner SJ, Foxwell BM (2000) The pyridinyl imidazole inhibitor SB203580 blocks phosphoinositide-dependent protein kinase activity, protein kinase B phosphorylation, and retinoblastoma hyperphosphorylation in interleukin-2-stimulated T cells independently of p38 mitogen-activated protein kinase. J Biol Chem 275:7395–7402

Lambert H, Charette SJ, Bernier AF, Guimond A, Landry J (1999) HSP27 multimerization mediated by phosphorylation-sensitive intermolecular interactions at the amino terminus. J Biol Chem 274:9378–9385

Landry J, Huot J (1999) Regulation of actin dynamics by stress-activated protein kinase 2 (SAPK2)-dependent phosphorylation of heat-shock protein of 27 kDa (Hsp27). Biochem Soc Symp 64:79–89

Landry J, Lambert H, Zhou M, Lavoie JN, Hickey E, Weber LA, Anderson CW (1992) Human HSP27 is phosphorylated at serines 78 and 82 by heat shock and mitogen-activated kinases that recognize the same amino acid motif as S6 kinase II. J Biol Chem 267:794–803

Lasa M, Mahtani KR, Finch A, Brewer G, Saklatvala J, Clark AR (2000) Regulation of cyclooxygenase 2 mRNA stability by the mitogen-activated protein kinase p38 signaling cascade. Mol Cell Biol 20:4265–4274

Lavoie JN, Gingras-Breton G, Tanguay RM, Landry J (1993) Induction of Chinese hamster HSP27 gene expression in mouse cells confers resistance to heat shock. HSP27 stabilization of the microfilament organization. J Biol Chem 268:3420–3429

Lavoie JN, Lambert H, Hickey E, Weber LA, Landry J (1995) Modulation of cellular thermoresistance and actin filament stability accompanies phosphorylation-induced changes in the oligomeric structure of heat shock protein 27. Mol Cell Biol 15:505–516

Lee JC, Laydon JT, McDonnell PC, Gallagher TF, Kumar S, Green D, McNulty D, Blumenthal MJ, Heys JR, Landvatter SW et al. (1994) A protein kinase involved in the regulation of inflammatory cytokine biosynthesis. Nature 372:739–746

Leroux MR, Melki R, Gordon B, Batelier G, Candido EP (1997) Structure-function studies on small heat shock protein oligomeric assembly and interaction with unfolded polypeptides. J Biol Chem 272:24646–24656

Ludwig S, Engel K, Hoffmeyer A, Sithanandam G, Neufeld B, Palm D, Gaestel M, Rapp UR (1996) 3pK, a novel mitogen-activated protein (MAP) kinase-activated protein kinase, is targeted by three MAP kinase pathways. Mol Cell Biol 16:6687–6697

Maizels ET, Peters CA, Kline M, Cutler RE Jr, Shanmugam M, Hunzicker-Dunn M (1998) Heat-shock protein-25/27 phosphorylation by the delta isoform of protein kinase C. Biochem J 332:703–712

Mann E, McDermott MJ, Goldman J, Chiesa R, Spector A (1991) Phosphorylation of alpha-crystallin B in Alexander's disease brain. FEBS Lett 294:133–136

Marin R, Landry J, Tanguay RM (1996) Tissue-specific posttranslational modification of the small heat shock protein HSP27 in Drosophila. Exp Cell Res 223:1–8

Martin JL, Hickey E, Weber LA, Dillmann WH, Mestril R (1999) Influence of phosphorylation and oligomerization on the protective role of the small heat shock protein 27 in rat adult cardiomyocytes. Gene Expr 7:349-355

McLaughlin MM, Kumar S, McDonnell PC, Van-Horn S, Lee JC, Livi GP, Young-PR (1996) Identification of mitogen-activated protein (MAP) kinase-activated protein kinase-3, a novel substrate of CSBP p38 MAP kinase. J Biol Chem 271:8488-8492

Mehlen P, Arrigo AP (1994) The serum-induced phosphorylation of mammalian hsp27 correlates with changes in its intracellular localization and levels of oligomerization. Eur J Biochem 221:327-334

Mehlen P, Mehlen A, Guillet D, Preville X, Arrigo AP (1995) Tumor necrosis factor-alpha induces changes in the phosphorylation, cellular localization, and oligomerization of human hsp27, a stress protein that confers cellular resistance to this cytokine. J Cell Biochem 58:248-259

Mehlen P, Kretz-Remy C, Preville X, Arrigo AP (1996a) Human hsp27, Drosophila hsp27 and human alphaB-crystallin expression-mediated increase in glutathione is essential for the protective activity of these proteins against TNFalpha-induced cell death. EMBO-J 15:2695-2706

Mehlen P, Schulze-Osthoff K, Arrigo AP (1996b) Small stress proteins as novel regulators of apoptosis. Heat shock protein 27 blocks Fas/APO-1- and staurosporine-induced cell death. J Biol Chem 271:16510-16514

Mehlen P, Hickey E, Weber LA, Arrigo AP (1997) Large unphosphorylated aggregates as the active form of hsp27 which controls intracellular reactive oxygen species and glutathione levels and generates a protection against TNFalpha in NIH-3T3-ras cells. Biochem Biophys Res Commun 241:187-192

Meloche S, Landry J, Huot J, Houle F, Marceau F, Giasson E (2000) p38 MAP kinase pathway regulates angiotensin II-induced contraction of rat vascular smooth muscle. Am J Physiol Heart Circ Physiol 279:H741-H751

Mendelsohn ME, Zhu Y, O'Neill S (1991) The 29-kDa proteins phosphorylated in thrombin-activated human platelets are forms of the estrogen receptor-related 27-kDa heat shock protein. Proc Natl Acad Sci USA 88:11212-11216

Merck KB, Groenen PJ, Voorter CE, de Haard-Hoekman WA, Horwitz J, Bloemendal H, de Jong WW (1993) Structural and functional similarities of bovine alpha-crystallin and mouse small heat-shock protein. A family of chaperones. J Biol Chem 268:1046-1052

Michishita M, Satoh M, Yamaguchi M, Hirayoshi K, Okuma M, Nagata K (1991) Phosphorylation of the stress protein hsp27 is an early event in murine myelomonocytic leukemic cell differentiation induced by leukemia inhibitory factor/D-factor. Biochem Biophys Res Commun 176:979-984

Minowada G, Welch W (1995) Variation in the expression and/or phosphorylation of the human low molecular weight stress protein during in vitro cell differentiation. J Biol Chem 270:7047-7054

Miron T, Vancompernolle K, Vandekerckhove J, Wilchek M, Geiger B (1991) A 25-kD inhibitor of actin polymerization is a low molecular mass heat shock protein. J Cell Biol 114:255-261.

Müller E, Burger-Kentischer A, Neuhofer W, Fraek ML, Marz J, Thurau K, Beck FX (1999) Possible involvement of heat shock protein 25 in the angiotensin II-induced glomerular mesangial cell contraction via p38 MAP kinase. J Cell Physiol 181:462-469

Nakatsue T, Katoh I, Nakamura S, Takahashi Y, Ikawa Y, Yoshinaka Y (1998) Acute infection of Sindbis virus induces phosphorylation and intracellular translocation of small heat shock protein HSP27 and activation of p38 MAP kinase signaling pathway. Biochem Biophys Res Commun 253:59-64

Neufeld B, Grosse-Wilde A, Hoffmeyer A, Jordan BW, Chen P, Dinev D, Ludwig S, Rapp UR (2000) Serine/Threonine kinases 3pK and MAPK-activated protein kinase 2 interact with the basic helix-loop-helix transcription factor E47 and repress its transcriptional activity. J Biol Chem 275:20239-20242

New L, Jiang Y, Zhao M, Liu K, Zhu W, Flood LJ, Kato Y, Parry GC, Han J (1998) PRAK, a novel protein kinase regulated by the p38 MAP kinase. EMBO J 17:3372-3384

Ni H, Wang XS, Diener K, Yao Z (1998) MAPKAPK5, a novel mitogen-activated protein kinase (MAPK)-activated protein kinase, is a substrate of the extracellular-regulated kinase (ERK) and p38 kinase. Biochem Biophys Res Commun 243:492–496

Nicholl ID, Quinlan RA (1994) Chaperone activity of alpha-crystallins modulates intermediate filament assembly. EMBO J 13:945–953

Nobes CD, Hall A (1995) Rho, rac, and cdc42 GTPases regulate the assembly of multimolecular focal complexes associated with actin stress fibers, lamellipodia, and filopodia. Cell 81:53–62

Nover L, Scharf KD, Neumann D (1989) Cytoplasmic heat shock granules are formed from precursor particles and are associated with a specific set of mRNAs. Mol Cell Biol 9:1298–1308

Nozaki J, Takehana M, Kobayashi S (1997) UVB irradiation induces changes in cellular localization and phosphorylation of mouse HSP27. Photochem Photobiol 65:843–848

Oesterreich S, Benndorf R, Bielka H (1990) The expression of the growth-related 25 kDa protein (p25) of Ehrlich ascites tumor cells is increased by hyperthermic treatment (heat shock). Biomed Biochim Acta 49:219–226

Pages G, Berra E, Milanini J, Levy AP, Pouyssegur J (2000) Stress-activated protein kinases (JNK and p38/HOG) are essential for vascular endothelial growth factor mRNA stability. J Biol Chem 275:26484–26491

Pandey P, Farber R, Nakazawa A, Kumar S, Bharti A, Nalin C, Weichselbaum R, Kufe D, Kharbanda S (2000) Hsp27 functions as a negative regulator of cytochrome c-dependent activation of procaspase-3. Oncogene 19:1975–1981

Paslaru L, Rallu M, Manuel M, Davidson S, Morange M (2000) Cyclosporin A induces an atypical heat shock response. Biochem Biophys Res Commun 269:464–469

Perng MD, Cairns L, van den IJssel P, Prescott A, Hutcheson AM, Quinlan RA (1999) Intermediate filament interactions can be altered by HSP27 and alphaB-crystallin. J Cell Sci 112:2099–2112

Piotrowicz RS, Hickey E, Levin EG (1998) Heat shock protein 27 kDa expression and phosphorylation regulates endothelial cell migration. FASEB J 12:1481–1490

Preville X, Schultz H, Knauf U, Gaestel M, Arrigo AP (1998) Analysis of the role of Hsp25 phosphorylation reveals the importance of the oligomerization state of this small heat shock protein in its protective function against TNFalpha- and hydrogen peroxide-induced cell death. J Cell Biochem 69:436–452

Regazzi R, Eppenberger U, Fabbro D (1988) The 27,000 daltons stress proteins are phosphorylated by protein kinase C during the tumor promoter-mediated growth inhibition of human mammary carcinoma cells. Biochem Biophys Res Commun 152:62–68

Rembold CM, O'Connor M (2000) Caldesmon and heat shock protein 20 phosphorylation in nitroglycerin- and magnesium-induced relaxation of swine carotid artery. Biochim Biophys Acta 1500:257–264

Ridley AJ, Paterson HF, Johnston CL, Diekmann D, Hall A (1992) The small GTP-binding protein rac regulates growth factor-induced membrane ruffling. Cell 70:401–410

Rogalla T, Ehrnsperger M, Preville X, Kotlyarov A, Lutsch G, Ducasse C, Paul C, Wieske M, Arrigo AP, Buchner J, Gaestel M (1999) Regulation of Hsp27 oligomerization, chaperone function, and protective activity against oxidative stress/tumor necrosis factor alpha by phosphorylation. J Biol Chem 274:18947–18956

Rollet E, Best-Belpomme M (1986) HSP 26 and 27 are phosphorylated in response to heat shock and ecdysterone in Drosophila melanogaster cells. Biochem Biophys Res Commun 141:426–433

Rouse J, Cohen P, Trigon S, Morange M, Alonso-Llamazares A, Zamanillo D, Hunt T, Nebreda AR (1994) A novel kinase cascade triggered by stress and heat shock that stimulates MAPKAP kinase-2 and phosphorylation of the small heat shock proteins. Cell 78:1027–1037

Rousseau S, Houle F, Landry J, Huot J (1997) p38 MAP kinase activation by vascular endothelial growth factor mediates actin reorganization and cell migration in human endothelial cells. Oncogene 15:2169–2177

Saklatvala J, Kaur P, Guesdon F (1991) Phosphorylation of the small heat-shock protein is regulated by interleukin 1, tumour necrosis factor, growth factors, bradykinin and ATP. Biochem J 27:635–642

Santell L, Bartfeld NS, Levin EG (1992) Identification of a protein transiently phosphorylated by activators of endothelial cell function as the heat-shock protein HSP27. A possible role for protein kinase C. Biochem J 284:705–710

Schaeffer HJ, Weber MJ (1999) Mitogen-activated protein kinases: specific messages from ubiquitous messengers. Mol Cell Biol 19:2435–2444

Schafer C, Clapp P, Welsh MJ, Benndorf R, Williams JA (1999) HSP27 expression regulates CCK-induced changes of the actin cytoskeleton in CHO-CCK-A cells. Am J Physiol 277:C1032–C1043

Schneider GB, Hamano H, Cooper LF (1998) In vivo evaluation of hsp27 as an inhibitor of actin polymerization: hsp27 limits actin stress fiber and focal adhesion formation after heat shock. J Cell Physiol 177:575–584

Schultz H, Engel K, Gaestel M (1997) PMA-induced activation of the p42/44ERK- and p38RK-MAP kinase cascades in HL-60 cells is PKC dependent but not essential for differentiation to the macrophage-like phenotype. J Cell Physiol 173:310–318

Schultz H, Rogalla T, Engel K, Lee JC, Gaestel M (1997) The protein kinase inhibitor SB203580 uncouples PMA-induced differentiationof HL-60 cells from phosphorylation of Hsp27. Cell Stress Chaperones 2:41–49

Sithanandam G, Latif F, Duh FM, Bernal R, Smola U, Li H, Kuzmin I, Wixler V, Geil L, Shrestha S (1996) 3pK, a new mitogen-activated protein kinase-activated protein kinase located in the small cell lung cancer tumor suppressor gene region. Mol Cell Biol 16:868–876

Smoyer WE, Ransom R, Harris RC, Welsh MJ, Lutsch G, Benndorf R (2000) Ischemic acute renal failure induces differential expression of small heat shock proteins. J Am Soc Nephrol 11:211–221

Spector A, Chiesa R, Sredy J, Garner W (1985) cAMP-dependent phosphorylation of bovine lens alpha-crystallin. Proc Natl Acad Sci USA 82:4712–4716

Spector NL, Ryan C, Samson W, Levine H, Nadler LM, Arrigo AP (1993) Heat shock protein is a unique marker of growth arrest during macrophage differentiation of HL-60 cells. J Cell Physiol 156:619–625

Stokoe D, Engel K, Campbell DG, Cohen P, Gaestel M (1992a) Identification of MAPKAP kinase 2 as a major enzyme responsible for the phosphorylation of the small mammalian heat shock proteins. FEBS Lett 313:307–313

Stokoe D, Campbell DG, Nakielny S, Hidaka H, Leevers SJ, Marshall C, Cohen P (1992b) MAPKAP kinase-2; a novel protein kinase activated by mitogen-activated protein kinase. EMBO J 11:3985–3994

Stokoe D, Caudwell B, Cohen PT, Cohen P (1993) The substrate specificity and structure of mitogen-activated protein (MAP) kinase-activated protein kinase-2. Biochem J 296:843–849

Suzuki A, Sugiyama Y, Hayashi Y, Nyu-i N, Yoshida M, Nonaka I, Ishiura S, Arahata K, Ohno S (1998a) MKBP, a novel member of the small heat shock protein family, binds and activates the myotonic dystrophy protein kinase. J Cell Biol 140:1113–1124

Suzuki TC, Krawitz DC, Vierling E (1998b) The chloroplast small heat-shock protein oligomer is not phosphorylated and does not dissociate during heat stress in vivo. Plant Physiol 116:1151–1161

Thomas G, Haavik J, Cohen P (1997) Participation of a stress-activated protein kinase cascade in the activation of tyrosine hydroxylase in chromaffin cells. Eur J Biochem 247:1180–1189

Voorter CE, Mulders JW, Bloemendal H, de Jong WW (1986) Some aspects of the phosphorylation of alpha-crystallin A. Eur J Biochem 160:203–210

Voorter CE, de Haard-Hoekman WA, Roersma ES, Meyer HE, Bloemendal H, de Jong WW (1989) The in vivo phosphorylation sites of bovine alpha B-crystallin. FEBS Lett 259:50–52

Voorter CE, de Haard-Hoekman W, Merck KB, Bloemendal H, de Jong WW (1994) Elastase inhibition by the C-terminal domains of alpha-crystallin and small heat-shock protein. Biochim Biophys Acta 1204:43–47

Wang P, Bitar KN (1998) Rho A regulates sustained smooth muscle contraction through cytoskeletal reorganization of HSP27. Am J Physiol 275:G1454–G1462

Wang XZ, Ron D (1996) Stress-induced phosphorylation and activation of the transcription factor CHOP (GADD153) by p38 MAP kinase. Science 272:1347–1349

Waskiewicz AJ, Flynn A, Proud CG, Cooper JA (1997) Mitogen-activated protein kinases activate the serine/threonine kinases Mnk1 and Mnk2. EMBO J 16:1909–1920

Waskiewicz AJ, Johnson JC, Penn B, Mahalingam M, Kimball SR, Cooper JA (1999) Phosphorylation of the cap-binding protein eukaryotic translation initiation factor 4E by protein kinase Mnk1 in vivo. Mol Cell Biol 19:1871–1880

Welch WJ (1985) Phorbol ester, calcium ionophore, or serum added to quiescent rat embryo fibroblast cells all result in the elevated phosphorylation of two 28,000-dalton mammalian stress proteins. J Biol Chem 260:3058–3062

Werz O, Klemm J, Samuelsson B, Radmark O (2000) 5-lipoxygenase is phosphorylated by p38 kinase-dependent MAPKAP kinases. Proc Natl Acad Sci USA 97:5261–5266

Winzen R, Kracht M, Ritter B, Wilhelm A, Chen CY, Shyu AB, Muller M, Gaestel M, Resch K, Holtmann H (1999) The p38 MAP kinase pathway signals for cytokine-induced mRNA stabilization via MAP kinase-activated protein kinase 2 and an AU-rich region-targeted mechanism. EMBO J 18:4969–4980

Wisniewski T, Goldman JE (1998) Alpha B-crystallin is associated with intermediate filaments in astrocytoma cells. Neurochem Res 23:385–392

Woodrum DA, Brophy CM, Wingard CJ, Beall A, Rasmussen H (1999) Phosphorylation events associated with cyclic nucleotide-dependent inhibition of smooth muscle contraction. Am J Physiol 277:H931–H939

Yamboliev IA, Hedges JC, Mutnick JL, Adam LP, Gerthoffer WT (2000) Evidence for modulation of smooth muscle force by the p38 MAP kinase/HSP27 pathway. Am J Physiol Heart Circ Physiol 278:H1899–H1907

Zhang S, Han J, Sells MA, Chernoff J, Knaus UG, Ulevitch RJ, Bokoch GM (1995) Rho family GTPases regulate p38 mitogen-activated protein kinase through the downstream mediator Pak1. J Biol Chem 270:23934–23936

Zhou M, Lambert H, Landry J (1993) Transient activation of a distinct serine protein kinase is responsible for 27-kDa heat shock protein phosphorylation in mitogen-stimulated and heat-shocked cells. J Biol Chem 268:35–43

Zhu Y, O'Neill S, Saklatvala J, Tassi L, Mendelsohn ME (1994) Phosphorylated HSP27 associates with the activation-dependent cytoskeleton in human platelets. Blood 84:3715–3723

Zu YL, Wu F, Gilchrist A, Ai Y, Labadia ME, Huang CK (1994) The primary structure of a human MAP kinase activated protein kinase 2. Biochem Biophys Res Commun 200:1118–1124

Small Stress Proteins: Modulation of Intracellular Redox State and Protection Against Oxidative Stress

André-Patrick Arrigo, Catherine Paul, Cécile Ducasse, Olivier Sauvageot, and Carole Kretz-Remy[1]

1
Introduction

Small stress proteins (also denoted small heat shock proteins: sHsp) are oligomeric phospho-polypeptides (Arrigo and Welch 1987; Arrigo et al. 1988; de Jong et al. 1993; Buchner et al. 1998) which increase the cell resistance to different types of stress, including heat shock and oxidative stress (reviewed in Arrigo and Landry 1994; Arrigo 1998, 2000; Arrigo and Préville 1999). In vitro, these proteins have been described as ATP-independent chaperones which counteract protein denaturation and help in the refolding of misfolded polypeptides (Jakob et al. 1993; Jakob and Buchner 1994; Ehrnsperger et al. 2000). Except for a role in maintaining cytoskeletal architecture, little information was available concerning the mode of action of these proteins in vivo. Recently, it has been proposed that large sHsp oligomers bind to misfolded polypeptides (Ehrnsperger et al. 1997; Lee et al. 1997) and present them to ATP-dependent protein chaperones (Hsp70, Hsp40, Hsp90 and co-chaperones) (see Haslbeck and Buchner, Chap. 3, this Vol.).

In recent years, there has been renewed interest in mammalian sHsp because they confer protection against stress-induced apoptosis (Mehlen et al. 1996b; Samali and Cotter 1996; Samali and Orrenius 1998; Arrigo and Préville 1999; Arrigo 2000) and the cytotoxic effects induced by a variety of toxic chemicals, particularly those used in cancer chemotherapy, i.e. cisplatin and doxorubicin (Huot et al. 1991; Oesterreich et al. 1993; Garrido et al. 1996, 1997; Richards et al. 1996). Moreover, sHsp expression also appears to be linked to the oncogenic status of the cell (Têtu et al. 1992, 1995; see also Ciocca and Vargas-Roig, Chap. 11, this Vol.). This phenomenon, which may be related to the protection mediated by sHsp against inflammatory mediators such as tumor necrosis factor (TNFα) (Mehlen et al. 1995c; Wang et al. 1996; Park et al. 1998), suggests that in vivo these stress proteins modulate the immuno-surveillance mediated by this cytokine (Arrigo 1998, 2000). Mammalian sHsp are also known to modulate the intracellular redox state in several different cells and protect against the deleterious effects induced by oxidative stress.

[1] Laboratoire Stress Oxydant, Chaperons et Apoptose, Centre de Génétique Moléculaire et Cellulaire, CNRS UMR-5534, Université Claude Bernard Lyon-I, 69622 Villeurbanne, France

Progress in Molecular and Subcellular Biology, Vol. 28
A.-P. Arrigo and W.E.G. Müller (Eds.)
© Springer-Verlag Berlin Heidelberg 2002

Oxidative stress is a common phenomenon in eukaryotic cells. Aerobic existence is coupled with the use of molecular oxygen leading to respiratory energy production. However, this thirst for electrons also generates side products such as reactive oxygen species (ROS) which can modulate cell physiology. The expression of several genes is under the control of intracellular redox state (Arrigo 1999). For example, ROS stimulate the expression of genes which are under the control of the redox-sensitive transcription factor NF-κB (Kretz-Remy et al. 1996). ROS can also be cytotoxic when they are produced at a high concentration, as it is the case during oxidative stress (Wong et al. 1989). Hence, oxidative stress is frequently observed when cells are exposed to peroxides, glutathione-depriving drugs, toxins, radiation, inflammatory cytokines and also heat shock. Oxidative stress is also observed in diseases such as cancer, atherosclerosis, rheumatoid arthritis, AIDS, and neurodegenerative disease.

Cells have developed sophisticated mechanisms to protect themselves in the face of an extracellular oxidizing environment (Powis et al. 1995). They include detoxifying enzymes (i.e. superoxide dismutase, catalase and glutathione peroxidase), vitamins C and E and thiol-containing molecules, such as glutathione. The anti-apoptotic protein Bcl2 can also provides protection against oxidative stress (Hockenbery et al. 1993; Kane et al. 1993). The present chapter summarizes recent observations concerning the protective activity of mammalian sHsp against the cytotoxicity induced by oxidative stress.

2
Cell Death Induced in Response to Oxidative Stress

Two types of cell death can be induced by agents or conditions that are deleterious to the cell: (1) necrosis or death by explosion of cells following a toxic stimuli (murder of the cell) and (2) apoptosis, which corresponds to a more physiological elimination of cells following a definite program (suicide of the cell). Necrotic and apoptotic processes differ by several points such as: (1) apoptosis requires ATP production but necrosis does not (Lelli et al. 1998; Nicotera et al. 1998); (2) apoptosis is induced by the activation of cysteine proteases, referred to as caspases; (3) necrosis releases intracellular material outside the cell, a phenomenon which, in vivo, induces inflammation. In contrast, apoptotic cells, which do not release cellular material are phagocytosed by macrophages without inducing any inflammation process.

In tissue culture cells exposed to oxidative stress, the type of death occurring depends on the intensity of the stress. Apoptosis is usually induced in the case of moderate oxidative stress. In contrast, drastic exposure to oxidant conditions induces a necrotic type of death. Hence, a stressful agent can induce either apoptosis or necrosis depending on its concentration (Vayssier et al. 1998). In the case of apoptosis, endogenous ROS induce a glutathione depletion which triggers the mitochondrial apoptotic pathway (Zucker et al. 1997; see Arrigo et al., Chap. 10, this Vol.). In contrast, high levels of ROS induce an intense oxidation of macromolecules such as proteins and lipids which results in a necrotic type of death (Jacobson 1996). In this type of cell death, the high level of ROS inhibits caspase activity and decreases ATP levels (Samali et al. 1999). In the next section we will discuss how sHsp negatively modulate ROS levels and as a consequence protect the cell against necrotic or apoptotic death.

3
sHsp Expression Protects Against Oxidative Stress-Induced Cell Death

One interesting feature of sHsp concerns their cytoprotector effect in cells exposed to oxidative stress (Huot et al. 1991, 1996; Mehlen et al. 1993, 1996a; Preville et al. 1999; see Fig. 1A). Oxidative stress can be induced by agents or conditions such as hydrogen peroxide, menadione, doxorubicin, tumor necrosis factor (TNFα) and even heat shock (Gorman et al. 1999; Katschinski et al. 2000). We have been particularly interested by TNFα produced in vivo primarily by activated macrophages. This cytokine is a fundamental mediator of inflammation which can also trigger the necrotic elimination of many tumor cells (Fiers 1991). The binding of TNFα to its cell surface receptors rapidly generates ROS, whose appearance induces the phosphorylation of 27-kDa polypeptides (Kaur and Saklatvala 1988; Robaye et al. 1989) later identified as Hsp27 phospho-isoforms (Arrigo 1990). Using murine L929 or NIH 3T3-ras fibroblasts, which are highly sensitive to oxidative stress and devoid of endogenous sHsp expression, we and others have demonstrated that the expression of different sHsp (human Hsp27, murine Hsp25, *Drosophila* Hsp27 and mammalian αB-crystallin) interfered with TNFα-induced cell death (Mehlen et al. 1995c; Park et al. 1998; Wang et al. 1996).

3.1
Expression of sHsp Can Modulate the Intracellular Redox State

We have investigated whether the protective activity generated by sHsp expression was linked to a modulation of intracellular ROS level. In murine L929 (Mehlen et al. 1995c), NIH 3T3-ras (Mehlen et al. 1997) and human colorectal cancer HT-29 (Garrido et al. 1997) cells, we observed that the expression of several sHsp, such as human Hsp27, *Drosophila* Hsp27 or mammalian αB-crystallin, significantly decreased the basal level of intracellular ROS (superoxide anion and peroxides). Interestingly, the expression of these proteins also strongly decreased the intracellular burst of ROS generated by the binding of TNFα to its receptor (Mehlen et al. 1995b,c, 1996a, 1997; Rogalla et al. 1999). Consequently, ROS-dependent phenomena, such as lipid peroxidation, protein oxidation, and NF-κB activation were impaired by the expression of Hsp27 (Mehlen et al. 1996a). Moreover, disruption of F-actin architecture, a

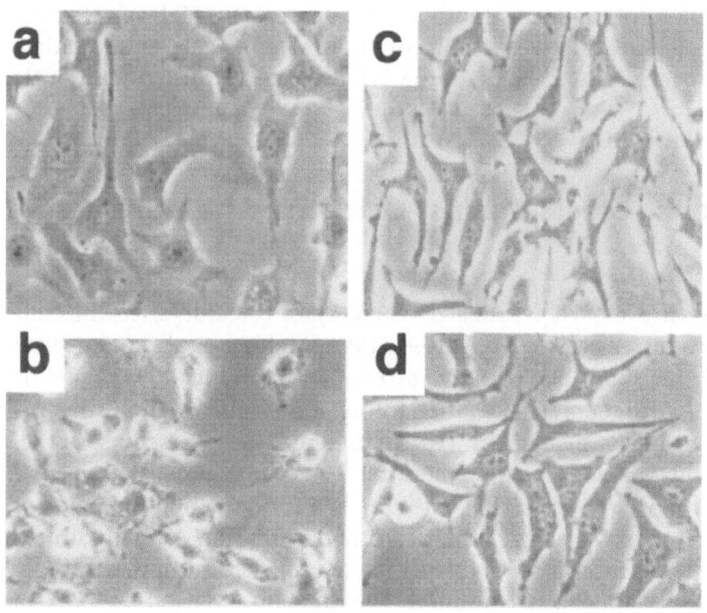

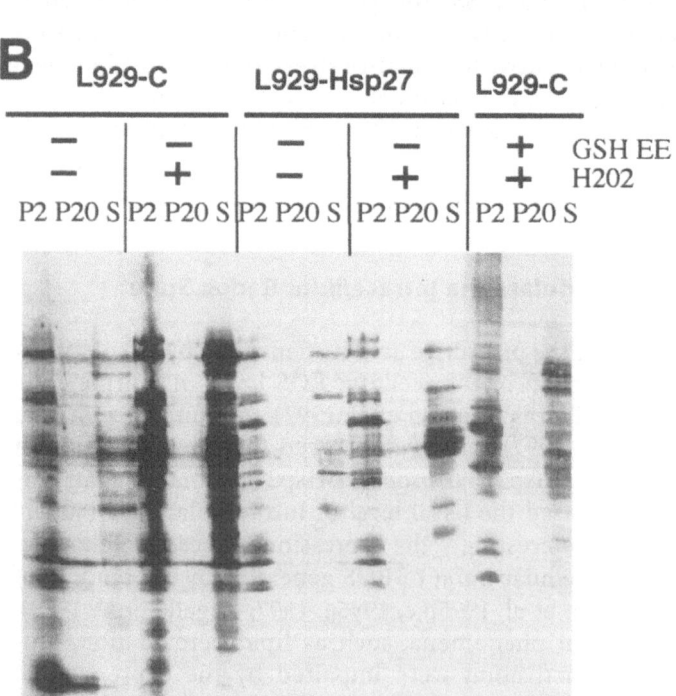

very early event that occurs in cells exposed to oxidative stress, was less intense in cells expressing human Hsp27 (Huot et al. 1996, 1997; Guay et al. 1997; Préville et al. 1998a). This protective activity may reflect the ability of Hsp27 to act as an F-actin specific protein chaperone (Lavoie et al. 1993) and/or be a consequence of the lower level of ROS produced in Hsp27-expressing cells (Préville et al. 1998a).

We have compared the anti-ROS activity of Hsp27 to that mediated by the well-known anti-apoptotic protein Bcl-2 (Reed 1997). Both Bcl-2 and Hsp27 can act as anti-oxidant proteins that decrease intracellular ROS levels (Hockenbery et al. 1993; Kane et al. 1993; Mehlen et al. 1996a). When similar numbers of Bcl-2 and Hsp27 molecules were considered, it was concluded that Bcl-2 decreases the intracellular level of ROS and induces protection against oxidative stress more efficiently than Hsp27 (Guenal et al. 1997; Paul and Arrigo 2000). It has also been reported that the expression of Hsp27 increases the production of extracellular superoxide in CCL39 fibroblasts (Souren et al. 1996). This phenomenon, which may be related to the protection of NADPH-oxidase activity by Hsp27, is compatible with the generation of a pro-reduced state inside the cell.

Two questions then arise: what is the mechanism induced by sHsp expression that decreases intracellular ROS levels in murine cells normally devoid of sHsp expression (see Sect. 3.2) and what is the mechanism of action of sHsp in cells exposed to oxidative injuries (see Sects. 3.3–3.4)?

◄ ───

Fig. 1. Protective activity of Hsp27 against oxidative stress. A Morphological analysis of control and Hsp27-expressing L929 cells exposed to H_2O_2. Control L929 cells devoid of endogenous sHsp expression (a, b) and cells overexpressing human Hsp27 (c, d) were treated with or without 400 µM H_2O_2. Cells were analyzed by phase-contrast microscopy prior to treatment (a, c) and after 4 h (b, d) of oxidative challenge. Note that control cells are rapidly damaged by oxidative stress, whereas no morphological changes can be detected in Hsp27-expressing cells. B Hsp27 expression interferes with protein oxidation as detected by immunoblot analysis of oxidized proteins (oxiblot). Control and human Hsp27-expressing cells were either left untreated or submitted to 1-h oxidative stress mediated by 400 µM H_2O_2. Where indicated, control cells were pretreated overnight with 5 mM GSH ethyl ester (*GSH EE*) prior to oxidative challenge. This treatment increased the level of reduced glutathione inside the cell to the same extent as that induced by Hsp27 expression. Cells were then lyzed and subcellular fractions (*P2, P20* and *S*) were treated with 2,4-dinitrophenylhydrazine (DNPH) as referenced in Preville et al (1999). Note that (1) human Hsp27 expression strongly attenuates the formation of carbonyl residues in response to a strong oxidative stress, (2) an increased intracellular GSH content also buffers protein oxidation but the pattern of the protected proteins is different from that induced by Hsp27 expression. This suggests that Hsp27 acts as a specific chaperone which interferes with the oxidation and/or degradation of specific proteins. (Adapted from Préville et al. 1999))

3.2
In Murine Fibroblasts Hsp27 Modulates the Glutathione Level and Intracellular Redox State Through Upregulation of Glucose 6-Phosphate Dehydrogenase Activity

In cells that are normally devoid of endogenous Hsp27 expression (i.e. murine L929 or NIH 3T3-ras cells), the protective activity of mammalian Hsp27 against oxidative stress was found to depend on the tripeptide glutathione (Mehlen et al. 1996a). In these cells, Hsp27 expression upregulates the glutathione level and upholds this redox modulator in its reduced form (Baek et al. 2000; Preville et al. 1999; Paul and Arrigo 2000). However, in cells that constitutively express endogenous Hsp27 (i.e. HeLa cells), an increased expression of this protein (mediated by expression-vector transfection) was not found to modify the level of glutathione. This phenomenon is probably related to the fact that, in L929 cells, the level of glutathione reaches a plateau when the level of Hsp27 is about 1 ng/mg of total cellular proteins (Mehlen et al. 1996a).

Glutathione is the major source of cellular thiol (Meister and Anderson 1983) which participates in the maintenance of redox state homeostasis. Glutathione is also a potent detoxicant that protects cells from oxidative stress (Yamauchi et al. 1990). As a consequence of its glutathione dependence, Hsp27 did not protect against the oxidative stress generated by either buthionine sulfoximine (BSO) (a specific and essentially irreversible inhibitor of γ-glutamyl-cysteine synthetase), or diethyl maleate (DEM) which binds the free sulfhydryl groups of glutathione (Mehlen et al. 1996a). Interestingly, Bcl-2 over-expression also upregulates glutathione level (Hockenbery et al. 1993; Kane et al. 1993; Paul and Arrigo 2000). Similarly, as described in the case of ROS (see Sect. 3.1), Bcl-2 modulated glutathione levels more efficiently than Hsp27 (Paul and Arrigo 2000). However, since Hsp27 is active in the form of a large oligomer (see Sect. 3.4), this structure appears more active than the well described Bcl-2 homodimeric structure. An intriguing property of Bcl-2 concerns its ability to induce the redistribution of glutathione into the nucleus (Voehringer et al. 1998). It is not yet known whether a similar property exists for Hsp27.

Similarly to Bcl-2 (Hockenbery et al. 1993; Kane et al. 1993), Hsp27 expression also decreases the oxidation of glutathione (GSH oxidized in GSSG) during oxidative stress (Preville et al. 1999). This phenomenon may be related to the fact that Hsp27 expression can modulate the expression and/or the activity of enzymes involved in the ROS-glutathione pathway. Interestingly, among the enzymes analyzed, we noticed that the expression of human Hsp27 in L929 cells significantly increased the activity of glucose-6-phosphate dehydrogenase, glutathione reductase and glutathione transferase (Preville et al. 1999). These changes are probably responsible of the pro-reducing state observed in Hsp27-expressing L929 cells. This hypothesis is supported by the fact that genetically manipulated L929 and HeLa cells which express a high level of glucose-6-phosphate dehydrogenase displayed an increased level of reduced glutathione (Preville et al. 1999). The possibility therefore exists that sHsp may act as specialized chaperones towards glucose-6-phosphate dehydrogenase. It is also noteworthy that, in vitro, αB-crystallin interferes with the glycation inactivation of glucose-6-phosphate dehydrogenase and the carbamylation inactivation of 6-phosphogluconate deshydrogenase (Ganea and Harding 1995; Ganea and Harding 1996).

3.3
Hsp27 Expression Protects Against Protein Oxidation and the Inactivation of Key Anti-Oxidant Enzymes

In murine L929 cells exposed to oxidative stress, the rise in glutathione as well as the upholding of this redox modulator in its reduced form are directly responsible of the protection observed at the level of the cell morphology, cytoskeletal architecture (Préville et al. 1998a, 1999) and mitochondrial membrane potential (Preville et al. 1999; Paul and Arrigo 2000).

Mammalian Hsp27 expression also buffers the increase in protein oxidation following oxidative stress and protects several key enzymes (such as glutathione transferase) against inactivation (Preville et al. 1999). In this case, however, the protection necessitated both an increase in GSH and the presence of Hsp27 per se. Indeed, the pattern of proteins which are oxidized (carbonyl residues induced by hydroxyl radical accumulation) in response to H_2O_2 treatment in cells that are pretreated with the anti-oxidant drug glutathione ethyl ester is different from that observed in cells expressing Hsp27 (see the oxiblot presented in Fig. 1B). This suggests an in vivo chaperone activity of Hsp27 that protects specific proteins against oxidative modifications.

3.4
Oligomerization as a Key Property Regulating sHsp Protective Activity Against Oxidative Stress

Drastic changes in the cellular locale, structural organization and phosphorylation of Hsp27 are induced by ROS. For example, dynamic changes in the Hsp27 oligomerization profile are observed in cells exposed to oxidative stress. During the first hour of TNFα treatment, Hsp27 native molecular weight drastically increases (up to 800 kDa). This phenomenon is followed by a rapid reduction in the native size of Hsp27 (about 200 kDa) (Mehlen et al. 1995b). This transient shift in the size of Hsp27 oligomers is ROS dependent (Mehlen et al. 1995a). As mentioned in Section 2, a rapid increase in Hsp27 phosphorylation is induced in response to TNFα. The phosphorylated Hsp27 isoforms are essentially recovered as small or medium-sized oligomers (<300 kDa) (Mehlen et al. 1995b). Analysis of non-phosphorylatable Hsp27 mutants led to the conclusion that the large unphosphorylated oligomers of Hsp27 represent the active form of the protein which modulates ROS and glutathione levels (Mehlen et al. 1997; Rogalla et al. 1999) and displays in vitro chaperone activity (Rogalla et al. 1999). Moreover, the use of the P38 kinase inhibitor SB203580 led to the conclusion that phosphorylation concentrates Hsp27 in the form of small oligomers (Préville et al. 1998b). This effect may either inactivate Hsp27 chaperone activity or induce its recycling through dynamic deaggregation-oligomerization steps of the protein.

It is rather improbable that Hsp27 large oligomers which transiently accumulate during oxidative stress act as reservoirs that facilitate the detoxification and renaturation of oxidized proteins by other chaperones. Indeed, mammalian cells exhibit only limited direct repair mechanisms and most oxidized proteins undergo selective proteolysis (Grune et al. 1997). In this particular context, one can imagine that the Hsp27 containing reservoirs activate the presentation of oxidized proteins to the ubiquitin-independent 20S proteasome, a protein degradation machine with a high affinity for oxidized proteins (Grune and Davies 1997; Grune et al. 1997; Sitte et al. 1998). The Hsp27-dependent formation of reservoirs containing oxidized proteins may then counteract the deleterious accumulation of proteolysis-resistant large aggregates of oxidized proteins (formation of lipofuscin (Sitte et al. 2000)) (see Fig. 2).

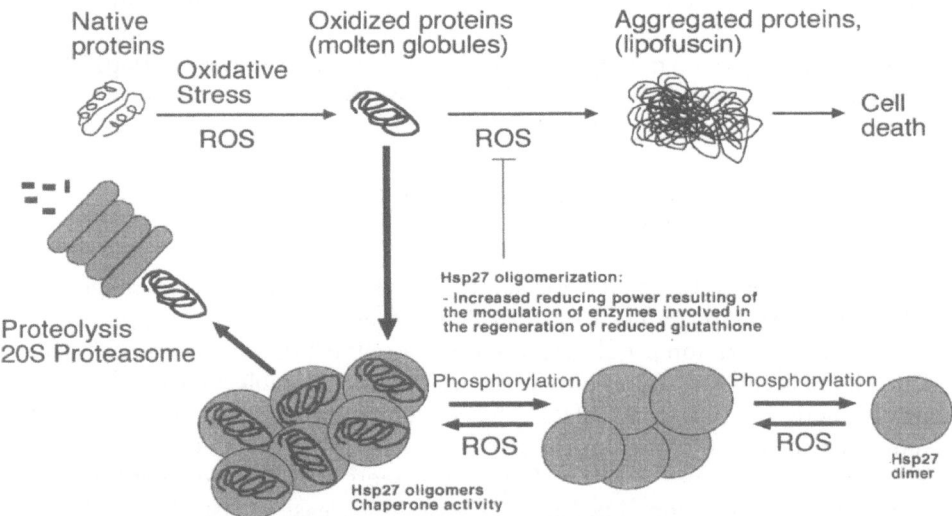

Fig. 2. Schematic illustration of Hsp27 putative mode of action during oxidative stress. Large Hsp27 oligomers modulate the activity of enzymes involved in the glutathione-ROS pathway and modify their activity. This leads to a pro-reduced state that interferes with ROS accumulation and protein oxidation. This phenomenon also counteracts the formation of lipofuscin. Hsp27 large oligomers are supposed to bind oxidized proteins, hence creating a reservoir of polypeptides that will be presented to the 20S proteasome pathway. (Adapted in part from Ehrnsperger et al. 1997; Lee et al. 1997; Arrigo and Préville 1999)

We have also observed that a decrease in the intracellular content of glutathione induced Hsp27 to concentrate in the form of small oligomers. In contrast, high levels of this redox modulator triggered the formation of large Hsp27 oligomers (Mehlen et al. 1997). A model describing the putative role of Hsp27 against oxidative stress postulates that glutathione depletion inhibits Hsp27 protective activity because this stress dissociates the high molecular weight complexes formed by Hsp27, probably in association with oxidized proteins. In contrast, a rise in glutathione should increase the activity of these chaperone complexes by requiring more Hsp27 in the form of large oligomers. The rise in the level of glutathione and/or the upholding of glutathione in its reduced form induced by Hsp27 expression, may be required because of the increased need of reductant to maintain the complexes formed by Hsp27 with oxidized proteins. This reducing activity may also stimulate the presentation of oxidized proteins to the 20S proteasome. These glutathione-dependent effects may be related to the fact that Hsp27 dimers are the building bricks that assemble to form oligomeric complexes (Zavialov et al. 1998).

4
sHsp Expression Can Also Enhance Oxidative Stress Induced Cell Death

As described above, the ectopic expression of Hsp27 confers resistance to heat shock, oxidative stress and other types of stress. However, reports have been published describing a sensitization of the cells to oxidative stress by Hsp27 overexpression. One can cite, for example, the KMST-6 human immortalized fibroblasts (Arata et al. 1995), a squamous carcinoma cell line (Trautinger et al. 1997) and subspecies of L929 fibrosarcoma cells (Mairesse et al. 1998). Interestingly, in these cells lines, Hsp27 still confers heat shock resistance, hence suggesting that different Hsp27 mechanisms trigger resistance to heat shock and oxidative stress. These observations also suggest that the intra cellular context in which Hsp27 is expressed is important. In this regard, it is interesting to mention that in the absence of Hsp27 expression, the above mentioned cell lines were originally quite resistant to oxidative stress (i.e. the L929 subcell line described above is far more resistant to oxidative stress than what has been generally observed for the classical L929 cell line described above in the previous sections). Production of Hsp27 in these resistant cells led to a marked decrease in growth rate associated with a series of phenotypic changes, including cell spreading and accumulation of actin stress fibers. By contrast, in cells that become resistant to oxidative stress because of Hsp27 expression (see Fig. 1A), a more fusiform appearance of the cell was induced by Hsp27 expression. Another important parameter to consider is the level of expression of Hsp27. Indeed, in spite of the fact that a low level of expression of Hsp27 generates protection against oxidative stress, a high level of this protein may be toxic and could sensitize cells to oxidative stress.

5
Conclusions and Future Perspectives

Several studies have described the ability of mammalian Hsp27 to protect cells against the cytotoxicity induced by agents and/or conditions that generate an oxidative stress, and particularly the pro-inflammatory cytokine TNFα. This phenomenon may have important consequences in vivo by modulating the inflammation process and diseases such as cancer, atherosclerosis, rheumatoid arthritis, AIDS, and neurodegenerative diseases.

From a mechanistic point of view, Hsp27 expression drives murine fibroblasts towards a pro-reduced state which is characterized by a reduced level of ROS and an upregulation of glutathione cellular content. This effect, however, is not observed when Hsp27 bearing cDNA vectors are transfected in cells that already constitutively express a high level of endogenous sHsp. Moreover, the activity of several enzymes such as glucose-6-phosphate dehydrogenase, glutathione reductase and glutathione transferase is modified by Hsp27 expression (Preville et al. 1999). More work will be required to determine whether such effects are related to a specific chaperone activity of Hsp27 toward these enzymes. Concerning the protective activity mediated by Hsp27 against oxidative stress, it appears to depend on the ability of this protein to uphold glutathione in its reduced form. This led to better detoxification of ROS; the protection observed resembled that mediated by anti-oxidants (particularly at the level of F-actin network and mitochondrial membrane potential) (Préville et al. 1998a). A reduced level of ROS also decreased the level of oxidized proteins, lipids and other macromolecules. In addition to these effects, we have also noticed that GSH is used by Hsp27 as an essential parameter of its chaperone activity (Preville et al. 1999). Indeed, a comparison of protein carbonyls patterns in Hsp27-expressing cells and anti-oxidant treated normal cells revealed that the protection mediated by Hsp27 is not only GSH dependent but also relies on the presence of Hsp27 per se. Hence, through its ability to decrease the level of protein oxidation, Hsp27 prevents the inactivation of key enzymes and maintains the detoxifying abilities of the cell at an optimal level.

During oxidative stress and similarly to what has been observed during heat shock, the large Hsp27 oligomers represent the active, GSH dependent, chaperone form of Hsp27. Phosphorylation interferes with the formation of these large oligomers; a phenomenon which induces the inactivation and/or the recycling of Hsp27 protective activity (Préville et al. 1998b). Although this has not yet been demonstrated, oxidized proteins may bind these oligomeric structures. Indeed, the presence of carbonyl groups produces modified polypeptides with properties similar to those of the 'molten-globule' intermediates of the protein folding-unfolding pathway. This favors the hypothesis that as a result of its chaperone activity, Hsp27 prevents further oxidative modifications of proteins and avoids the formation of lipofuscin (Sitte et al. 2000). However, it is rather improbable that the large Hsp27 oligomers act by presenting the oxidized proteins to the refolding chaperone machinery (Hsp70 and co-chaperones) since most oxidized proteins cannot be refolded and undergo selective proteolysis (Grune et al. 1997). One attractive hypothesis to be tested is that the Hsp27 oligomers stimulate the presentation of oxidized proteins to the 20S proteasome (Grune and Davies 1997; Grune et al. 1997; Sitte et al. 1998). The model describing such a putative role of Hsp27 is presented in Fig. 2.

Acknowledgements. This work was supported by the Association pour la Recherche sur le Cancer (grant # 5204) and the Région Rhône-Alpes.

References

Arata S, Hamaguchi S, Nose K (1995) Effects of the overexpression of the small heat shock protein, Hsp27, on the sensitivity of human fibroblast cells exposed to oxidative stress. J Cell Physiol 163:458–465

Arrigo A-P (1990) Tumor necrosis factor induces the rapid phosphorylation of the mammalian heat shock protein hsp28. Mol Cell Biol 10:1276–1280

Arrigo AP (1998) Small stress proteins: chaperones that act as regulators of intracellular redox state and programmed cell death. Biol Chem 379:19–26

Arrigo AP (1999) Gene expression and the thiol redox state. Free Radic Biol Med 27:936–944

Arrigo AP (2000) sHsp as novel regulators of programmed cell death and tumorigenicity. Pathol Biol (Paris) 48:280–288

Arrigo A-P, Landry J (1994) Expression and Function of the Low-molecular-weight Heat Shock Proteins. In: Morimoto RI, Tissieres A, Georgopoulos C (eds) The biology of heat shock proteins and molecular chaperones. Cold Spring Harbor Laboratory Press, Cold Spring Harbor, pp 335–373

Arrigo A-P, Préville X (1999) Role of Hsp27 and related proteins. In: Latchman DS (ed) Stress Proteins. Springer, Berlin Heidelberg New York, pp 101–132

Arrigo A-P, Welch W (1987) Characterization and purification of the small 28,000-dalton mammalian heat shock protein. J Biol Chem 262:15359–15369

Arrigo A-P, Suhan JP, Welch WJ (1988) Dynamic changes in the structure and intracellular locale of the mammalian low-molecular-weight heat shock protein. Mol Cell Biol 8:5059–5071

Baek SH, Min JN, Park EM, Han MY, Lee YS, Lee YJ, Park YM (2000) Role of small heat shock protein HSP25 in radioresistance and glutathione-redox cycle. J Cell Physiol 183:100–107

Buchner J, Ehrnsperger M, Gaestel M, Walke S (1998) Purification and characterization of small heat shock proteins. Methods Enzymol 290:339–349

De Jong W, Leunissen J, Voorter C (1993) Evolution of the alpha-crystallin/small heat-shock protein family. Mol Biol Evol 10:103–126

Ehrnsperger M, Graber S, Gaestel M, Buchner J (1997) Binding of non-native protein to Hsp25 during heat shock creates a reservoir of folding intermediates for reactivation. EMBO J 16: 221–229

Ehrnsperger M, Gaestel M, Buchner J (2000) Analysis of chaperone properties of small Hsp's. Methods Mol Biol 99:421–429

Fiers W (1991) Tumor necrosis factor. Characterization at the molecular, cellular and in vivo level. FEBS Lett 285:199–212

Ganea E, Harding J (1996) Inhibition of 6-phosphogluconate dehydrogenase by carbamylation and protection by alpha-crystallin, a chaperone-like protein. Biochem Biophys Res Commun 222:626–631

Ganea E, Harding JJ (1995) Molecular chaperones protect against glycation-induced inactivation of glucose-6-phosphate dehydrogenase. Eur J Biochem 231:181–187

Garrido C, Mehlen P, Fromentin A, Hammann A, Assem M, Arrigo A-P, Chauffert B (1996) Inconstant association between 27-kDa heat-shock protein (Hsp27) content and doxorubicin resistance in human colon cancer cells. The doxorubicin-protecting effect of Hsp27. Eur J Biochem 237:653–659

Garrido C, Ottavi P, Fromentin A, Hammann A, Arrigo AP, Chauffert B, Mehlen P (1997) HSP27 as a mediator of confluence-dependent resistance to cell death induced by anticancer drugs. Cancer Res 57:2661–2667

Gorman AM, Heavey B, Creagh E, Cotter TG, Samali A (1999) Antioxidant-mediated inhibition of the heat shock response leads to apoptosis. FEBS Lett 445:98–102

Grune T, Davies KJ (1997) Breakdown of oxidized proteins as a part of secondary antioxidant defenses in mammalian cells. Biofactors 6:165–172

Grune T, Reinheckel T, Davies KJ (1997) Degradation of oxidized proteins in mammalian cells. FASEB J 11:526–534

Guay J, Lambert H, Gingras Breton G, Lavoie JN, Huot J, Landry J (1997) Regulation of actin filament dynamics by p38 map kinase-mediated phosphorylation of heat shock protein 27. J Cell Sci 110:357–368

Guenal I, Sidoti-de Fraisse C, Gaumer S, Mignotte B (1997) Bcl-2 and Hsp27 act at different levels to suppress programmed cell death. Oncogene 15:347–360

Hockenbery DM, Oltvai ZN, Yin XM, Milliman CL, Korsmeyer SJ (1993) Bcl-2 functions in an antioxidant pathway to prevent apoptosis. Cell 75:241–251

Huot J, Roy G, Lambert H, Chretien P, Landry J (1991) Increased survival after treatments with anticancer agents of Chinese hamster cells expressing the human 27,000 heat shock protein. Cancer Res 51:5245–5252

Huot J, Houle F, Spitz DR, Landry J (1996) HSP27 phosphorylation-mediated resistance against actin fragmentation and cell death induced by oxidative stress. Cancer Res 56:273–279

Huot J, Houle F, Marceau F, Landry J (1997) Oxidative stress-induced actin reorganization mediated by the p38 mitogen-activated protein kinase/heat shock protein 27 pathway in vascular endothelial cells. Circ Res 80:383–392

Jacobson MD (1996) Reactive oxygen species and programmed cell death. Trends Biochem Sci 21:83–86

Jakob U, Buchner J (1994) Assisting spontaneity: the role of Hsp90 and small Hsps as molecular chaperones. Trends Biochem Sci 19:205–211

Jakob U, Gaestel M, Engels K, Buchner J (1993) Small heat shock proteins are molecular chaperones. J Biol Chem 268:1517–1520

Kane DJ, Sarafian TA, Anton R, Hahn H, Gralla EB, Valentine JS, Ord T, Bredesen DE (1993) Bcl-2 inhibition of neural death: decreased generation of reactive oxygen species. Science 262:1274–1277

Katschinski DM, Boos K, Schindler SG, Fandrey J (2000) Pivotal role of reactive oxygen species as intracellular mediators of hyperthermia-induced apoptosis. J Biol Chem 275:21094–21098

Kaur P, Saklatvala J (1988) Interleukin 1 and tumour necrosis factor increase phosphorylation of fibroblast proteins. FEBS Lett 241:6–10

Kretz-Remy C, Mehlen P, Mirault ME, Arrigo A-P (1996) Inhibition of I kappa B-alpha phosphorylation and degradation and subsequent NF-kappa B activation by glutathione peroxidase overexpression. J Cell Biol 133:1083–1093

Lavoie JN, Hickey E, Weber LA, Landry J (1993) Modulation of actin microfilament dynamics and fluid phase pinocytosis by phosphorylation of Heat Shock Protein 27. J Biol Chem 268:24210–24214

Lee GJ, Roseman AM, Saibil HR, Vierling E (1997) A small heat shock protein stably binds heat-denatured model substrates and can maintain a substrate in a folding-competent state. EMBO J 16:659–671

Lelli JL Jr, Becks LL, Dabrowska MI, Hinshaw DB (1998) ATP converts necrosis to apoptosis in oxidant-injured endothelial cells. Free Radic Biol Med 25:694–702

Mairesse N, Bernaert D, Del Bino G, Horman S, Mosselmans R, Robaye B, Galand P (1998) Expression of HSP27 results in increased sensitivity to tumor necrosis factor, etoposide, and H2O2 in an oxidative stress-resistant cell line. J Cell Physiol 177:606–617

Mehlen P, Briolay J, Smith L, Diaz-Latoud C, Pauli D, Arrigo A-P (1993) Analysis of the resistance to heat and hydrogen peroxide stresses in COS cells transiently expressing wild type or deletion mutants of the *Drosophila* 27-kDa heat-shock protein. Eur J Biochem 215:277–284

Mehlen P, Kretzremy C, Briolay J, Fostan P, Mirault ME, Arrigo AP (1995a) Intracellular reactive oxygen species as apparent modulators of heat-shock protein 27 (hsp27) structural organization and phosphorylation in basal and tumour necrosis factor alpha-treated T47D human carcinoma cells. Biochem J 312:367–375

Mehlen P, Mehlen A, Guillet D, Préville X, Arrigo A-P (1995b) Tumor necrosis factor-a induces changes in the phosphorylation, cellular localization, and oligomerization of human hsp27, a stress protein that confers cellular resistance to this cytokine. J Cell Biochem 58:248–259

Mehlen P, Préville X, Chareyron P, Briolay J, Klemenz R, Arrigo A-P (1995c) Constitutive expression of human hsp27, *Drosophila* hsp27, or human alpha B-crystallin confers resistance to TNF- and oxidative stress-induced cytotoxicity in stably transfected murine L929 fibroblasts. J Immunol 154:363–374

Mehlen P, Préville X, Kretz-Remy C, Arrigo A-P (1996a) Human hsp27, *Drosophila* hsp27 and human αB-crystallin expression-mediated increase in glutathione is essential for the protective activity of these protein against TNFα-induced cell death. EMBO J 15:2695–2706

Mehlen P, Schulze-Osthoff K, Arrigo A-P (1996b) Small stress proteins as novel regulators of apoptosis – heat shock protein 27 blocks Fas/APO-1- and staurosporine-induced cell death. J Biol Chem 271:16510–16514

Mehlen P, Hickey E, Weber L, Arrigo A-P (1997) Large unphosphorylated aggregates as the active form of hsp27 which controls intracellular reactive oxygen species and glutathione levels and generates a protection against TNFa in NIH-3T3-ras cells. Biochem Biophys Res Commun 241:187–192

Meister A, Anderson ME (1983) Glutathione. Annu Rev Biochem 52:711–760

Nicotera P, Leist M, Ferrando-May E (1998) Intracellular ATP, a switch in the decision between apoptosis and necrosis. Toxicol Lett 102–103:139–142

Oesterreich S, Weng C-N, Qiu M, Hilsenbeck SG, Osborne CK, Fuqua SW (1993) The small heat shock protein hsp27 is correlated with growth and drug resistance in human breast cancer cell lines. Cancer Res 53:4443–4448

Park YM, Han MY, Blackburn RV, Lee YJ (1998) Overexpression of HSP25 reduces the level of TNF alpha-induced oxidative DNA damage biomarker, 8-hydroxy-2'-deoxyguanosine, in L929 cells. J Cell Physiol 174:27–34

Paul C, Arrigo AP (2000) Comparison of the protective activities generated by two survival proteins: Bcl-2 and Hsp27 in L929 murine fibroblasts exposed to menadione or staurosporine. Exp Gerontol 35:757–766

Powis G, Briehl M, Oblong J (1995) Redox signalling and the control of cell growth and death. Pharmacol Ther 68:149–173

Préville X, Gaestel M, Arrigo AP (1998a) Phosphorylation is not essential for protection of L929 cells by Hsp25 against H2O2-mediated disruption actin cytoskeleton, a protection which appears related to the redox change mediated by Hsp25. Cell Stress Chaperones 3:177–187

Préville X, Schultz H, Knauf U, Gaestel M, Arrigo AP (1998b) Analysis of the role of Hsp25 phosphorylation reveals the importance of the oligomerization state of this small heat shock protein in its protective function against TNFalpha- and hydrogen peroxide-induced cell death. J Cell Biochem 69:436–452

Preville X, Salvemini F, Giraud S, Chaufour S, Paul C, Stepien G, Ursini MV, Arrigo AP (1999) Mammalian small stress proteins protect against oxidative stress through their ability to increase glucose-6-phosphate dehydrogenase activity and by maintaining optimal cellular detoxifying machinery. Exp Cell Res 247:61–78

Reed JC (1997) Double identity for proteins of the Bcl-2 family. Nature 387:773–776

Richards EH, Hickey E, Weber LA, Master JR (1996) Effect of overexpression of the small heat shock protein HSP27 on the heat and drug sensitivities of human testis tumor cells. Cancer Res 56:2446–2451

Robaye B, Hepburn A, Lecocq R, Fiers W, Boeynaems JM, Dumont JE (1989) Tumor necrosis factor-a induces the phosphorylation of 28 kDa stress proteins in endothelial cells: Possible role in protection against cytotoxicity? Biochem Biophys Res Commun 163:301–308

Rogalla T, Ehrnsperger M, Preville X, Kotlyarov A, Lutsch G, Ducasse C, Paul C, Wieske M, Arrigo AP, Buchner J, Gaestel M (1999) Regulation of Hsp27 oligomerization, chaperone function, and protective activity against oxidative stress/tumor necrosis factor alpha by phosphorylation. J Biol Chem 274:18947–18956

Samali A, Cotter TG (1996) Heat shock proteins increase resistance to apoptosis. Exp Cell Res 223:163–170

Samali A, Orrenius S (1998) Heat shock proteins: regulators of stress response and apoptosis. Cell Stress Chaperones 3:228–236

Samali A, Nordgren H, Zhivotovsky B, Peterson E, Orrenius S (1999) A comparative study of apoptosis and necrosis in HepG2 cells: oxidant- induced caspase inactivation leads to necrosis. Biochem Biophys Res Commun 255:6–11

Sitte N, Merker K, Grune T (1998) Proteasome-dependent degradation of oxidized proteins in MRC-5 fibroblasts. FEBS Lett 440:399–402

Sitte N, Huber M, Grune T, Ladhoff A, Doecke WD, Von Zglinicki T, Davies KJ (2000) Proteasome inhibition by lipofuscin/ceroid during postmitotic aging of fibroblasts. FASEB J 14:1490–1498

Souren JE, Van Aken H, Van Wijk R (1996) Enhancement of superoxide production and protection against heat shock by HSP27 in fibroblasts. Biochem Biophys Res Commun 227:816–821

Têtu B, Lacasse B, Bouchard H-L, Lagacé R, Huot J, Landry J (1992) Prognostic influence of HSP-27 expression in malignant fibrous histiocytoma: a clinicopathological and immunohisto-chemical study. Cancer Res 52:2325–2328

Têtu B, Brisson J, Landry J, Huot J (1995) Prognostic significance of heat-shock protein-27 in node-positive breast carcinoma: an immunohistochemical study. Breast Cancer Res Treat 36:93–97

Trautinger F, Kokesch C, Herbacek I, Knobler RM, Kindas-Mugge I (1997) Overexpression of the small heat shock protein, hsp27, confers resistance to hyperthermia, but not to oxidative stress and UV-induced cell death, in a stably transfected squamous cell carcinoma cell line. J Photochem Photobiol 39:90–95

Vayssier M, Banzet N, Francois D, Bellmann K, Polla BS (1998) Tobacco smoke induces both apoptosis and necrosis in mammalian cells: differential effects of HSP70. Am J Physiol 275:771–779

Voehringer DW, McConkey DJ, McDonnell TJ, Brisbay S, Meyn RE (1998) Bcl-2 expression causes redistribution of glutathione to the nucleus. Proc Natl Acad Sci USA 95:2956–2960

Wang G, Klostergaard J, Khodadadian M, Wu J, Wu TW, Fung KP, Carper SW, Tomasovic SP (1996) Murine cells transfected with human Hsp27 cDNA resist TNF-induced cytotoxicity. J Immunother Emphasis Tumor Immunol 19:9–20

Wong GHW, Elwell JE, Oberby LW, Goeddel D (1989) Manganous superoxide dismutase is essential for cellular resistance to tumor necrosis factor. Cell 58:923–931

Yamauchi N, Kuriyama YH, Watanabe N, Neda H, Maeda M, Himeno T, Tsuji Y (1990) Suppressive effects of intracellular glutathione on hydroxyl radical production induced by tumor necrosis factor. Int J Cancer 46:884–888

Zavialov AV, Gaestel M, Korpela T, Zav'yalov VP (1998) Thiol/disulfide exchange between small heat shock protein 25 and glutathione. Biochim Biophys Acta 1388:123–132

Zucker B, Hanusch J, Bauer G (1997) Glutathione depletion in fibroblasts is the basis for apoptosis-induction by endogenous reactive oxygen species. Cell Death Differ 4:388–395

Small Stress Proteins: Novel Negative Modulators of Apoptosis Induced Independently of Reactive Oxygen Species

André-Patrick Arrigo, Catherine Paul, Cécile Ducasse, Florence Manero, Carole Kretz-Remy, Sophie Virot, Etienne Javouhey, Nicole Mounier, and Chantal Diaz-Latoud[1]

1
Introduction

The execution phase of the apoptotic cell death process occurs throughout the proteolytic activation of proteolytic enzymes called caspases (Nicholson and Thornberry 1997; Thornberry and Lazebnik 1998). Several different pathways can lead to the activation of caspases, among them, one can cite the death receptors (i.e. Fas) (Scaffidi et al. 1998) and mitochondria pathways (Reed 1997; Green and Reed 1998). When activated by ligand binding, death receptors (i.e. Fas) recruit adapter polypeptides (i.e. FADD) that interact with and subsequently activate pro-caspases (i.e. pro-caspase 8) or trigger a signal transduction pathway that activates specific genes (i.e. the DAXX/ASK1/JNK pathway). In contrast, in the mitochondria pathway, different inducers have the ability to induce the release in the cytoplasm of different proteins present in mitochondria, such as cytochrome c, apoptosis-inducing factor (AIP), Hsp60, Hsp10, adenylate kinase, Smac/Diablo as well as the fraction of pro-caspase 2, 3, 8, and 9 present in mitochondria (Kluck et al. 1997; Reed 1997; Yang et al. 1997; Kohler et al. 1999; Samali et al. 1999a; Susin et al. 1999a,b; Xanthoudakis et al. 1999; Du et al. 2000; Verhagen et al. 2000). Once it has been released from the mitochondria, cytochrome c interacts with Apaf-1 in the presence of ATP/dATP. This results in the formation of the apoptosome complex which recruits and activates pro-caspase 9 which subsequently activates pro-caspase 3 (Li et al. 1997; Saleh et al. 1999). The release of cytochrome c from the mitochondria is a caspase-independent early phenomenon that precedes mitochondrial membrane potential loss (Bossy-Wetzel et al. 1998). This phenomenon may be induced by conformational changes of Bax (Desagher et al. 1999) or by BAK oligomerization induced by tBID, a membrane-targeted death ligand (Wei et al. 2000). Stress-induced apoptosis usually occurs through the activation of the mitochondria pathway. This is particularly the case when mild oxidative or heat stresses are considered (Samali et al. 2000).

---

[1] Laboratoire Stress Oxydant, Chaperons et Apoptose, Centre de Génétique Moléculaire et Cellulaire, CNRS UMR-5534, Université Claude Bernard Lyon-I, 69622 Villeurbanne, France

Several proteins have been described as modulators of apoptosis. Among them are members of the Bcl-2 (Kluck et al. 1997) and IAP (inhibitor of apoptosis proteins) (Deveraux et al. 1998) families. Bcl-2 is a constitutively expressed anti-apoptotic protein localized in the mitochondria and endoplasmic reticulum membranes, which inhibits the release of apoptogenic cytochrome c from the mitochondria [probably through an inhibition of the voltage-dependent anion channels (VDAC) formation] (Kluck et al. 1997; Shimizu et al. 1999). Bcl-2 may also inhibit caspase activation downstream of cytochrome c release (Rosse et al. 1998). In contrast, other members of the Bcl-2 family such as Bax promote apoptosis by inhibiting Bcl-2 action and by triggering mitochondria channel formation.

Recent studies have demonstrated that chaperone proteins control specific steps of the apoptotic pathway. For example, it has been shown that the major stress protein Hsp70 negatively regulates apoptosis by interfering with the release of cytochrome c from the mitochondria (Mosser et al. 2000). The chaperone function of Hsp70 is required for this type of protection. Hsp70 is also active downstream of cytochrome c release and upstream of caspase 3 activation (Chun-Ying Li 2000), probably by altering the formation of the apoptosome complex through its interaction with the apoptosome subunit Apaf-1 (Beere et al. 2000; Saleh et al. 2000). Moreover, Hsp70 can interact with BAG-1 (Takayama et al. 1997) and therefore stimulates Bcl-2 action, or exerts its anti-apoptotic activities downstream of caspase-3 activation (Jäättelä et al. 1998). As a consequence of its anti-apoptotic properties, Hsp70 stimulates cell tumorigenicity (Jäättelä 1995). Interestingly, Hsp90 has also been reported to act as a negative modulator of cytochrome c-dependent apoptosis through its binding to Apaf-1 (Pandey et al. 2000a).

In contrast to Hsp70 and Hsp90, the mitochondrial Hsp60 and Hsp10 act as pro-apoptotic chaperones. Indeed, these proteins stimulate the conversion of pro-caspase 3 in active caspase 3, once they are released from the mitochondria together with cytochrome c (Samali et al. 1999a; Xanthoudakis et al. 1999).

The present chapter summarizes recent observations concerning the protective activity of mammalian small stress proteins (Hsp) (particularly human and murine Hsp27) against apoptotic cell death induced by agents and/or conditions that act independently of reactive oxygen species (ROS).

2
sHsp Expression Protects Against Apoptosis Induced Independently of ROS Formation

Recently, the role played by sHsp has been analyzed in the context of apoptosis generated independently of intracellular ROS burst. In this respect, we and others have reported that mammalian Hsp27, *Drosophila* Hsp27 as well as alpha-B crystallin overexpression exerts an anti-apoptotic effect in cells exposed to the kinase C inhibitor staurosporine. Human Hsp27 was also shown to protect against apoptosis induced by actinomycin D, camptothecin, Fas/ APO-1 receptor (Mehlen et al. 1996), the topoisomerase II inhibitor etoposide (Garrido et al. 1999; Samali and Orrenius 1998), cisplatin (Garrido et al. 1997) and doxorubicin (Hansen et al. 1999). In contrast, Hsp27 protects less efficiently against T antigen/p53-mediated cell death (Guenal et al. 1997) and appears to promote the apoptotic cell death mediated by cytotoxic T-lymphocyte (CTL) cells (Beresford et al. 1998).

Apoptotic cells release reduced glutathione (GSH) and/or induce its oxidation independently of ROS production (Beaver and Waring 1995; Slater et al. 1995; Van den Dobbelsteen et al. 1996; Ghibelli et al. 1998). Hence, in spite of the fact that oxidative alterations may not be necessary for the development of physiological apoptosis, redox imbalance may be required for the stress-induced apoptotic pathway. This is suggested by the finding that GSH depletion, a common event in stress-induced apoptosis, is necessary and sufficient to induce cytochrome c release, the key event of this pathway (Coppola and Ghibelli 2000). Consequently, there are some reports that an increase in GSH levels and/or the upholding of this redox modulator in its reduced form could delay the apoptotic process (Van den Dobbelsteen et al. 1996). However, the redox modulation by Hsp27 (see Arrigo et al., Chap. 9, this Vol.) has not yet been found to play a crucial role in the protective activity of this protein against ROS-independent apoptosis.

2.1
Hsp27 Interferes with the Apoptotic Machinery Downstream of Cytochrome c Release from the Mitochondria

In a recent report (Garrido et al. 1999), we have shown that, in U937 cells exposed to etoposide, human Hsp27 overexpression reduced apoptosis by counteracting pro-caspase 9 activation. As a consequence, this effect inhibited the activation of pro-caspase 3 downstream of caspase 9. As seen in Fig. 1A,B in murine L929 cells exposed to the kinase inhibitor staurosporine, the expression of murine Hsp27 also downregulated the activation of pro-caspase 9 and pro-caspase 3. Moreover, in an in vitro system allowing caspases to be activated by exogenous cytochrome c and dATP, the presence of recombinant Hsp27 interfered with pro-caspase 3 activation. In contrast, the presence of Hsp60 and Hsp10 induced the opposite effect confirming their pro-apoptotic properties (Fig. 1C). Further analysis revealed that the inhibition of pro-caspase 9 activation by Hsp27 correlated with the binding of Hsp27 to cytochrome c, once this apoptogenic polypeptide is released from the mitochondria (Bruey et al. 2000a; see also Fig. 2A). The binding of Hsp27 to cytochrome c is thought to inhibit the ability of this apoptogenic agent to trigger apoptosome formation. The putative binding site of cytochrome c to Hsp27 lies in the N-terminal part of the protein between amino acids 51 and 88 (Bruey et al. 2000a), a domain which does not appear to be related to the chaperone activity of Hsp27 (C. Ducasse, O. Sauvageot and A.-P. Arrigo, in prep.). However, this mechanism of action of

In Vivo

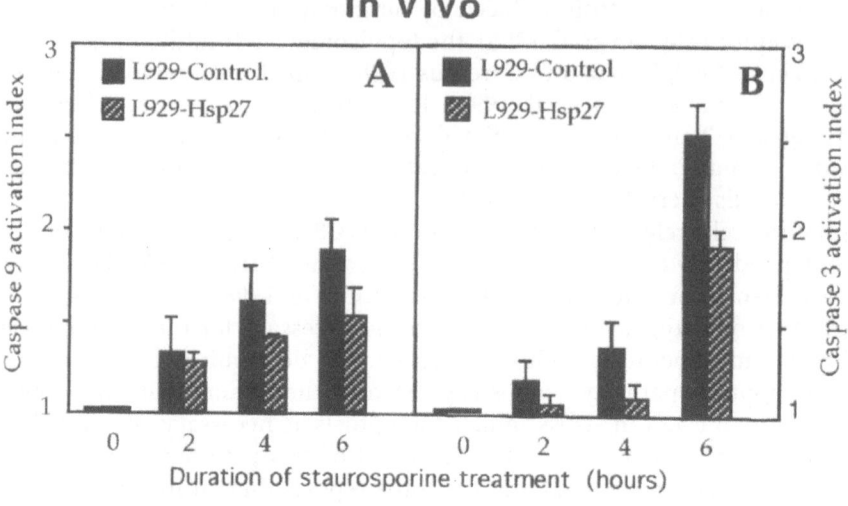

Duration of staurosporine treatment (hours)

In Vitro

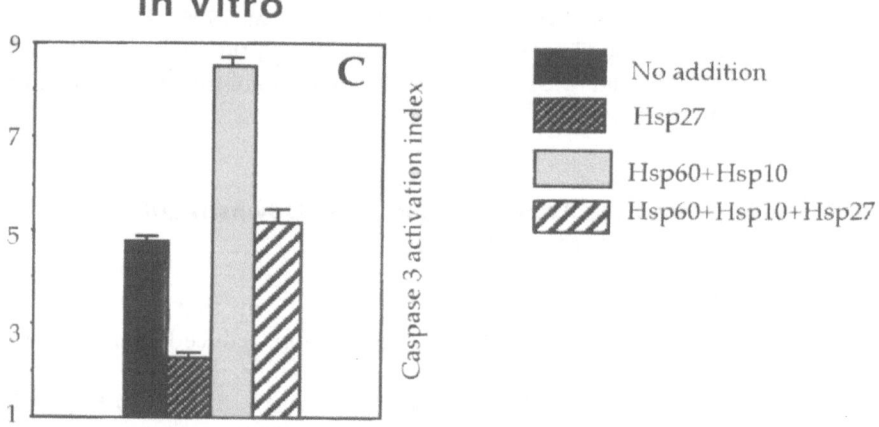

Fig. 1. Modulation of caspase activation by Hsp27. Analysis of in vivo pro-caspases 9 (A)and 3 (B) activation. Caspase activity was measured in control or Hsp27-expressing cells treated for the indicated time period with 1 μM of staurosporine. 10^6 cells were harvested and DEVD-dependent caspase activity (essentially caspase 3) was determined using the Apo Alert CPP32 fluorometric assay kit (Clontech, Montigny les Bretonneux, France). The determination of pro-caspase 9 activation was performed using the caspase 9 fluorometric assay kit (R&D systems, Abingdon, UK) using 2×10^6 cells as starting material. The activation index was determined as the ratio between the activity in extracts of treated cells to that measured in extracts of untreated cells. Standard deviations are presented ($n = 3$). C In vitro activation of DEVD-dependent pro-caspases. A post-mitochondrial extract derived from L929 cells was obtained and activated by the addition of cytochrome c and dATP. This cell-free system was kept on ice for 30 min at 4 °C in the absence or presence of 1 ng (per μg of total proteins) of either *Hsp27, Hsp60 + Hsp10* or *Hsp27 + Hsp60 + Hsp10*. Cytochrome c and dATP were then added to the extracts. After an incubation of 1 h at 37 °C, the activation of DEVD-dependent pro-caspases was determined using the fluorometric technique described above. The presented histograms are representative of three identical experiments, standard deviations are presented ($n = 3$)

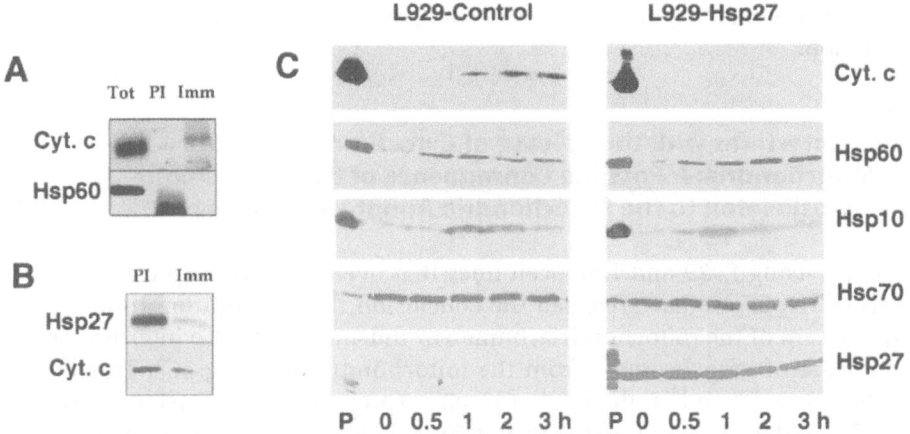

Fig. 2. Hsp27 interacts with cytochrome c and interferes with the release of this apoptogenic protein from mitochondria. **A** Cytochrome c co-immunoprecipitates with Hsp27. HeLa cells, treated for 2 h with 0.125 µM of staurosporine, were lyzed in the presence of Triton X-100 in order to destroy the mitochondria, leading most cytochrome c to be recovered in the soluble cell lysate. Immunoprecipitations were carried out with non-immune (*PI*) or Hsp27 antibody (*Imm*). Total cellular proteins (*Tot*) as well as immunoprecipitated proteins were then probed with antibodies specific for cytochrome c and Hsp60. Note that cytochrome c, but not Hsp60, co-immunoprecipitates with Hsp27. **B** Analysis of the relative fraction of endogenous cytochrome c that co-immunoprecipitates with Hsp27. In this case, the proteins present in the supernatants obtained from the immunoprecipitation step described in **A** were analyzed in immunoblots probed with both Hsp27 and cytochrome c antisera. Note only a weak decrease in the level of endogenous cytochrome c in conditions where Hsp27 is quantitatively immunodepleted. **C** Immunoblot analysis of the release of cytochrome c from the mitochondria. Control and Hsp27 expressing L929 cells were either kept untreated or treated for various time periods (0.5–3 h) with 1 µM staurosporine before being prepared according to Bossy-Wetzel et al. (1998). In brief, cells were lyzed and the pellet (*P* pellet from untreated cells) and supernatant fractions were analyzed in immunoblots probed with specific antibodies. Cytochrome c becomes detectable in the soluble fraction when it is released from the mitochondria. Soluble fractions isolated from either untreated cells (*0*) or cells treated for the indicated hours with staurosporine (0.5–3 h). Note the inhibition of cytochrome c release by Hsp27 expression and the lack of effect of this protein on the release of Hsp60 and Hsp10. Analysis of Hsp27 and Hsc70 in the different fractions was used as control

Hsp27 is far from being understood, since we have recently observed that only a small fraction of total endogenous cytochrome c interacts with Hsp27 (see the immunoprecipitation and immunodepletion analysis presented in Fig. 2A,B, respectively). This suggests that most cytochrome c molecules do not interact with this chaperone protein and/or that only a very minor and particular form of Hsp27 can interact with this polypeptide. An interaction of Hsp27 with pro-caspase 3 has also been documented (Pandey et al. 2000b), therefore suggesting that Hsp27 can act at different levels of the apoptotic execution machinery. We have also shown that Hsp27 does not interfere with the caspase-independent apoptosis inducing fraction (AIF) pathway (Bruey et al. 2000a) which triggers chromatin condensation and large-scale DNA fragmentation.

2.2
Hsp27 Interferes with the Release of Cytochrome c from the Mitochondria: A Possible Consequence of the Modulation of a Cytoskeleton to the Mitochondria Apoptotic Pathway

Recently, using L929 and HeLa cell lines that over- or underexpress different levels of Hsp27, we have reached the conclusion that, in addition to its effect downstream of the mitochondria, human or murine Hsp27 also interferes with the release of cytochrome c from the mitochondria (see Fig. 2C). This effect was found to depend on the level of Hsp27 expression. In contrast, no effect of Hsp27 was detected at the level of Hsp60 and Hsp10, which are released concomitantly with cytochrome c. This upstream activity necessitated a higher level of Hsp27 expression compared to the activity (described above in Sect. 2.1) that interferes with pro-caspase activation downstream of cytochrome c release. This study also shows that the high level of endogenous Hsp27 expressed in HeLa cells protects them against the eventuality of apoptotic death. Indeed, a decrease in the Hsp27 level, which sensitized HeLa cells to apoptosis, reduced the delay required for cytochrome c release and pro-caspase-3 activation.

How could high levels of Hsp27 expression modulate the kinetics of cytochrome c release from the mitochondria? Hsp27 is not specifically mitochondria-associated, hence suggesting that this protein acts upstream of this organelle. In this respect, we have investigated the role of Hsp27 as a modulator of F-actin architecture. It was observed that the F-actin depolymerization agent cytochalasin D rapidly induces the release of cytochrome c from the mitochondria and that this phenomenon is inhibited by Hsp27 expression. Hence, the negative effect of Hsp27 towards mitochondrial cytochrome c loss may result as an effect of the protective activity mediated by this stress protein against F-actin damage induced by cytochalasin D or other apoptosis inducers (staurosporine, ROS, oxidative stress, heat shock, etc.). This study also led to the discovery of an apoptotic-signaling pathway linking cytoskeleton damage to the mitochondria (C. Paul et al., submitted).

2.3
Hsp27 Interferes with Apoptosis Induced by Death Ligands

As mentioned above, Hsp27 expression negatively regulates cell death induced by death ligands such as TNFα or Fas ligand (Arrigo 1998). A recent report also indicates that αAB-crystallin expression in lens epithelial cells generates protection against Fas-induced cell death (Andley et al. 2000). The protection against the necrotic cell death induced by TNFα is probably a consequence of the protective effect of Hsp27 against oxidative stress (see Arrigo et al., Chap.

9, this Vol.). However, it cannot be excluded that Hsp27 may also interfere with TNFα-induced caspase activation, particularly when the apoptotic cell death induced by this cytokine is considered. Interestingly, a recent report describes that phosphorylated dimers of Hsp27 interact with DAXX, a mediator of Fas-induced apoptosis, preventing the interaction of DAXX with both Fas and apoptotic signaling kinase 1 (Ask1) and therefore blocking DAXX-mediated apoptosis (Charette et al. 2000).

2.4
Hsp27 Expression Generates Protection Against Heat Shock-Induced Apoptosis

Heat stresses that are not too drastic generate an apoptotic process which is blocked by the broad caspase inhibitor z-VAD-fmk (Samali and Cotter 1996; Samali and Orrenius 1998; Samali et al. 1999b). It was also shown that this type of cell death is dependent on cytochrome c release from the mitochondria and caspase 3 activation. Furthermore, these studies suggest that the lack of Hsp induction during heat shock is the cause rather than the consequence of cell death and that the enhanced cell survival and resistance to apoptosis during thermotolerance is due to an increased expression of heat shock proteins. An anti-sense approach allowed us to conclude that the inhibition of apoptosis during thermotolerance is due, at least in part, to Hsp27 which preferentially blocks mitochondrial cytochrome c release and to Hsp70 which interferes with apoptosomal caspase activation (Samali et al. 2000).

2.5
Role of Hsp27 Phosphorylation and Oligomerization in the Protection Against Apoptosis

The protection against the cell death induced by heat shock (Arrigo et al. 1988), oxidative stress and TNFα (Mehlen et al. 1997a; Préville et al. 1998; Rogalla et al. 1999) requires the transient formation of unphosphorylated large Hsp27 oligomers. In contrast, in the case of Fas ligand-mediated cell death, preliminary evidence indicates that this phenomenon does not correlate with any change in Hsp27 phosphorylation and/or oligomerization status. In the case of staurosporine or cytochalasin D treatment, a drastic change in the oligomerization and phosphorylation of Hsp27 has also been observed. These apoptotic stresses induced the transient formation of two subpopulations of Hsp27 molecules that are characterized by large and small molecular masses (C. Paul, F. Manero, S. Virot and A.-P. Arrigo, in prep.). As for etoposide treatment, this drug only induces the transient formation of large oligomers. Analysis performed with unphosphorylatable Hsp27 polypeptides revealed that phosphorylation inhibits the protective activity of Hsp27 against etoposide and that only the large oligomers formed by this protein can protect against this apoptotic agent (Bruey et al. 2000b). In contrast, as mentioned above in the case of

DAXX-mediated cell death, only the unphosphorylated dimers of Hsp27 are active (Charette et al. 2000). Additional experiments will then be required to better understand the structural organization of Hsp27 which is active against ROS-independent apoptosis. In this respect it will be of particular interest to determine which oligomeric form of Hsp27 interacts with cytochrome c, once it is released from the mitochondria.

3
sHsp as Essential Anti-Apoptotic Proteins During Early Cell Differentiation

Numerous studies have reported that sHsp are transiently expressed during the early differentiation of different cell types. This phenomenon, which was first reported in *Drosophila* (Pauli et al. 1990), has been observed in all the organisms analyzed so far (Arrigo and Landry 1994, Arrigo 1995). For example, Hsp27 is transiently expressed during the early differentiation of numerous mammalian cells such as Ehrlich ascites cells (Gaestel et al. 1989), embryonal carcinoma and stem cells (Stahl et al. 1992; Mehlen et al. 1997b), normal B or B lymphoma cells (Spector et al. 1992), osteoblasts, promyelocytic leukemia cells (Shakoori et al. 1992; Spector et al. 1993, 1994; Chaufour et al. 1996) normal T cells (Hanash et al. 1993), keratinocytes (Kindas-Mugge and Trautinger 1994), endometrium cells (Devaja et al. 1997), muscle cells (Benjamin et al. 1997), neurons (Mehlen et al. 1999; Loones et al. 2000), and cardiomyocytes (Davidson and Morange 2000). αB-crystallin is expressed during muscle (Benjamin et al. 1997) and epithelial cell differentiation (Boyle and Takemoto 2000). Interestingly, other members of the sHsp family such as MKBP/HspB2 and HspB3 are expressed during myogenic differentiation (Sugiyama et al. 2000).

We have analyzed the biological significance of the transient accumulation of Hsp27 during the early phase of several differentiation processes. An antisense approach allowed us to conclude that a normal expression of Hsp27 is necessary during the differentiation of human promyelocytic HL-60 cells in granulocytes (Chaufour et al. 1996). A similar conclusion was reached concerning the differentiation of the embryonal carcinoma P19 cell line in cardiomyocytes (Davidson and Morange 2000). However, the differentiation of P19 cells in neurons did not appear to be Hsp27-dependent. Other experiments showed that an almost complete inhibition of Hsp27 expression aborted the differentiation process of murine embryonic stem (ES) cells and induced an overall apoptotic process (Mehlen et al. 1997b; see Fig. 3). A similar observation was made for the dopamine-induced differentiation of rat olfactory neurons (Mehlen et al. 1999). Dopamine induced a fraction of these cells to differentiate into olfactory neurons while the remaining cells underwent apoptosis. Hsp27 overexpression was found to drastically decrease the fraction of cells undergoing apoptosis. In contrast, endogenous Hsp27 underexpression led to differentiation abortion and increased the number of apoptotic cells. Furthermore, Hsp27 expression was found to take place only in differentiating cells that were not undergoing apoptosis. In this cell differentiation system, Hsp27 acts as a key protein that controls the decision of olfactory precursor cells to undergo either differentiation or cell death (see Fig. 4). The molecular mechanism by which Hsp27 controls the differentiation program and negatively modulates apoptosis during cell differentiation is unknown but correlates with the formation of large Hsp27 oligomers. One hypothesis could be that the activity of Hsp27 is linked to the above mentioned protection against cytochrome c release-apoptosome formation and/or to the F-actin to mitochondria apoptotic pathway (see Sects. 2.1 and 2.2). This last hypothesis is based on the fact that cell shape and cytoskeleton are drastically changed during cell differentiation. Another interesting possibility relates to the fact that the FADD-dependent pathway is blocked during differentiation of several cell types (i.e. macrophages, lymphocytes, etc.). In this cellular context, the expression of Hsp27 could modulate a putative differentiation-induced DAXX apoptotic pathway (Charette et al. 2000).

4
Hsp27 Modulates Apoptosis in Vivo and Enhances the Tumorigenicity of Transformed Cells

Recent reports have now provided evidence that, in vivo, small stress proteins act as negative regulators of the apoptotic process. For example, it has been shown that the overexpression of Hsp27 by transfection of plasmid constructs in transgenic animals can protect primary cardiac cells from subsequent exposure to severe apoptotic stress (see D.S. Latchman, Chap. 14, this Vol.). A similar protection was also achieved by using herpes simplex virus (HSV) vectors that efficiently delivered the *hsp27* gene in vivo (Brar et al. 1999). Using the same type of HSV-based delivery, it was concluded that Hsp27 is a novel neuroprotective factor (Wagstaff et al. 1999). The potential therapeutic use of HSV-derived vectors that deliver the *hsp27* gene is now actively being investigated.

Another very intriguing finding concerns the fact that elevated titers of serum antibodies recognizing Hsp27 accompany human diseases such as cancer and glaucoma (Conroy et al. 1998). Moreover, antibodies to αB-crystallin have been detected in the blood of patients suffering from multiple sclerosis (Agius et al. 1999). In this regard, novel findings have shown that exogenously applied Hsp27 antibody can enter neuronal cells in the human retina. Once inside the cell, Hsp27 antibody facilitates neuronal apoptosis, probably by decreasing the ability of native Hsp27 to stabilize actin cytoskeleton (Tezel and Wax 2000). This leads to the fascinating conclusion that patients presenting autoantibodies to Hsp27 may impair cell survival in certain human diseases. Since several human tumor cells express high levels of Hsp27 (Têtu et al. 1992; Ciocca et al. 1993; see Ciocca and Vargas-Roig, Chap. 11, this Vol.),

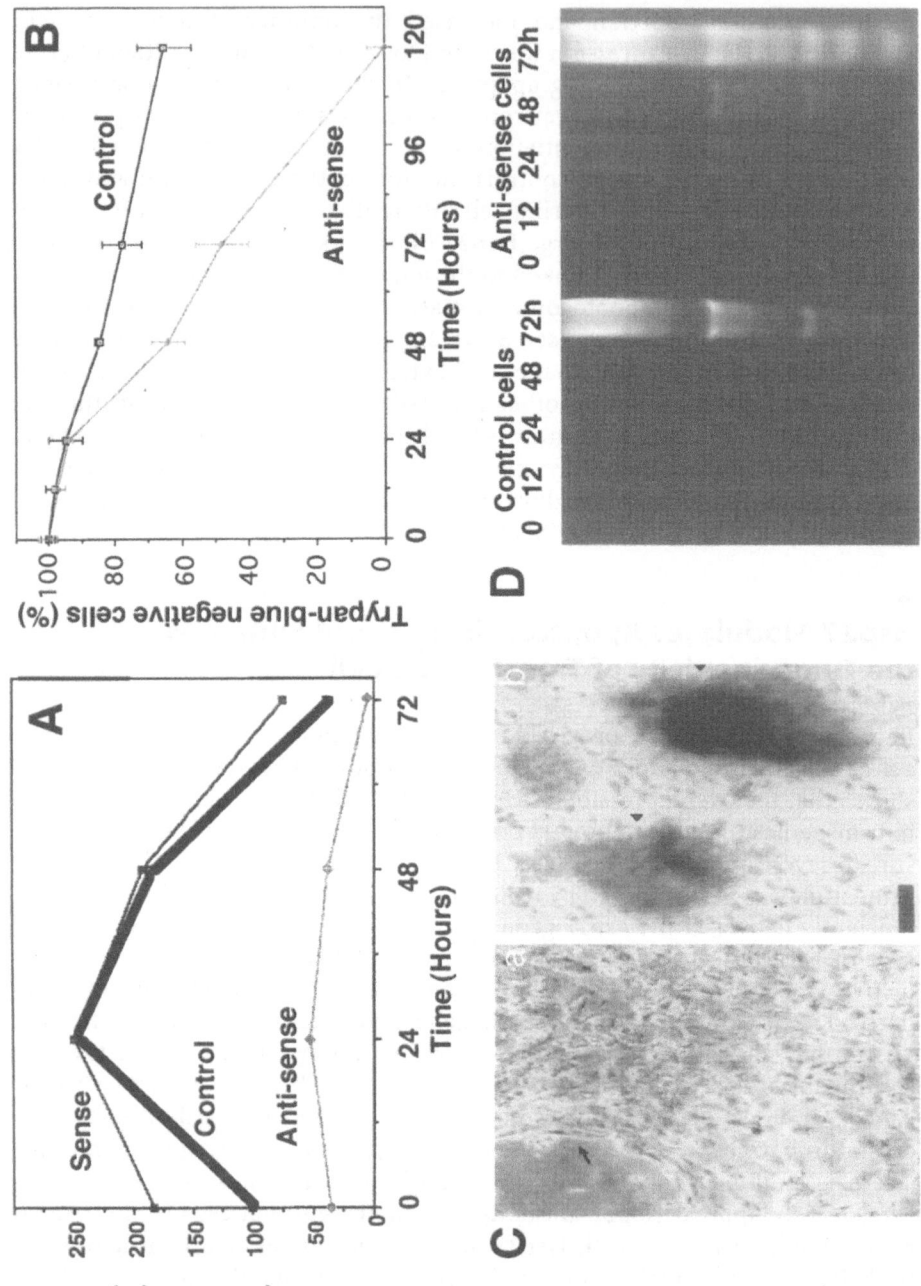

a decrease in the level of this protein may facilitate apoptosis and be beneficial for the patient, as has been observed in the case of some patients with breast cancer (Conroy et al. 1998).

To further explore the role played by Hsp27 once it is highly expressed in tumor cells, nude mice were used to analyze the tumorigenic potential of cell lines expressing Hsp27 (Blackburn et al. 1997). However, these studies did not take into account the fact that Hsp27 can modulate Fas or TNFα-induced apoptosis and can therefore modulate the immune response to tumor cells. Using genetically engineered rat colon carcinoma cells that constitutively express human Hsp27, we observed that these cells form very aggressive tumors in syngenic rats compared to control cells which do not express this protein (Garrido et al. 1998). Fewer apoptotic cells were detected in the tumors expressing Hsp27, hence confirming that this protein has an in vivo anti-apoptotic activity. In contrast, no tumorigenic effect of Hsp27 occurred in nonimmunocompetent animals. Taken together, these observations suggest that Hsp27 allows cancer cells to escape the surveillance mediated by the immune system. Analysis of the structural organization of Hsp27 which counteracts apoptosis in vivo led to surprising results. It was observed that in rat colon carcinoma cells grown in vitro, a wild type and an unphosphorylatable (Ser to Ala) mutant of Hsp27 formed oligomers heterodispersed in size and demonstrated anti-apoptotic activity. In contrast, an Hsp27 phosphomimetic (Ser to Asp) mutant only formed small multimers and was devoid of antiapoptotic activity. However, once the cells were injected into the animal, the aspartate mutant formed large oligomers and protected tumor cells from apoptosis (Bruey et al. 2000b). This led to the conclusion that large oligomers of Hsp27 have an anti-apoptotic activity and that cell to cell contacts induce the formation of large oligomers thereby increasing cell tumorigenicity. This phenomenon could be mimicked in vitro, by showing that when cells reach confluency, the Hsp27 level

◄
_____

Fig. 3. Analysis of the anti-apoptotic effect of Hsp27 transiently expressed in mouse embryonic stem cells. Modulation of the Hsp27 level was obtained using anti-sense or sense gene constructs that were stably transfected in CGR8 cells (ES cells). A Quantification of immunoblot analysis of endogenous Hsp27 accumulation during differentiation of either control cells or cells that under-(anti-sense) or overexpress (sense) murine Hsp27. The level of Hsp27 was analyzed at different time periods in the differentiation process. The differentiation process was induced by leukemia inhibitory factor (LIF) withdrawal. A graph representing the relative accumulation of Hsp27 is shown. B Viability determination of differentiating control and Hsp27 under-expressing (anti-sense) cells by Trypan blue exclusion. C Phase-contrast microscopy analysis of the morphology of control (*a*) and Hsp27 underexpressing (*b*) cells allowed to differentiate for 8 days. Note that normal cells display the typical morphology of differentiating stem cells, while Hsp27 under-expressing cells were still aggregated in early embryoid bodies (*black arrowheads*). D DNA fragmentation analysis. Control and Hsp27 underexpressing (anti-sense) cells were induced to differentiate as above and DNA fragmentation was determined prior to (*0*) and after 12, 24, or 72 h of differentiation. Note that the downregulation of Hsp27 induces an intense DNA fragmentation in ES cells committed to differentiate. (Reproduced from Mehlen et al. 1997b with permission of the American Society for Biochemistry and Molecular Biology)

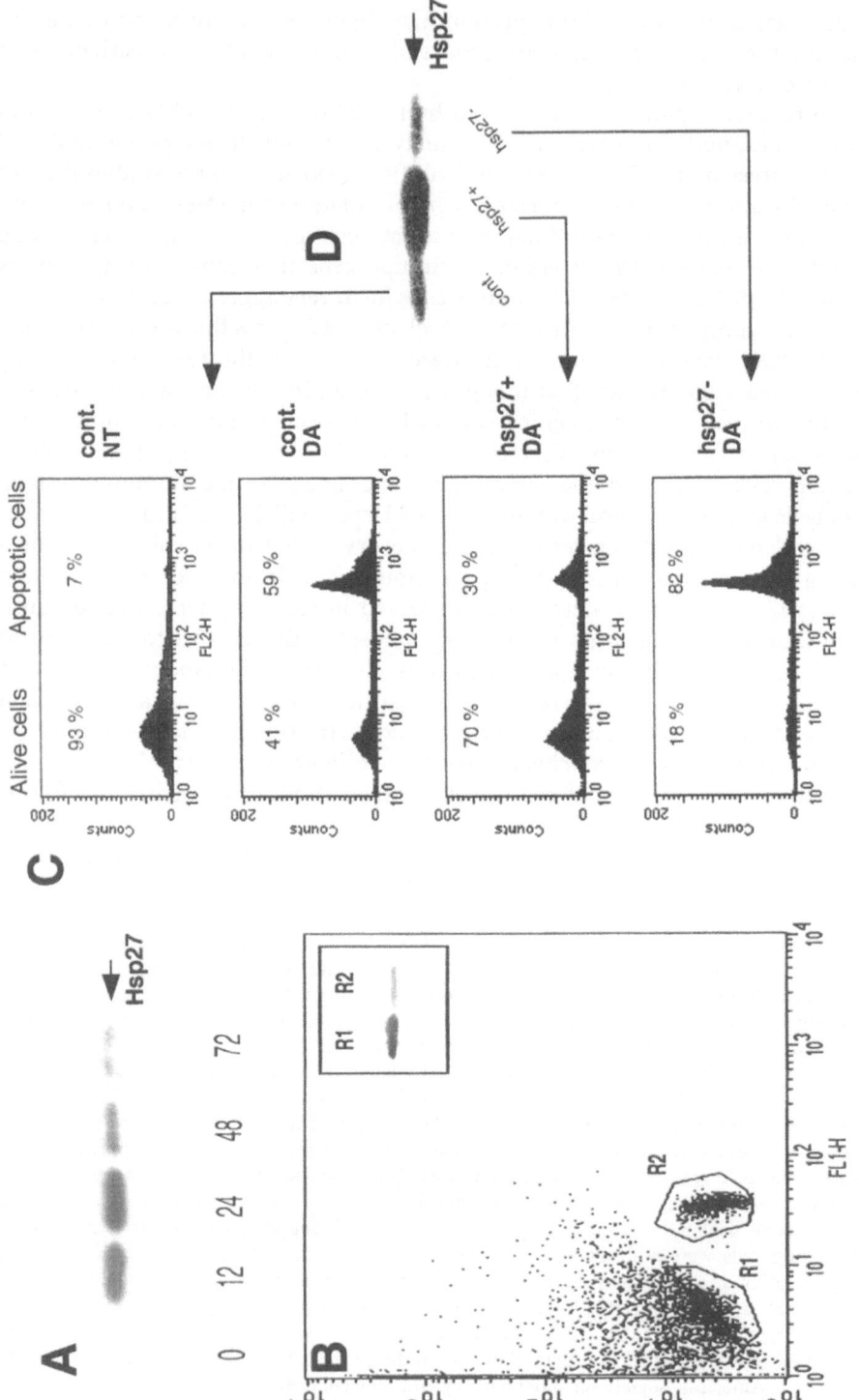

increases (Garrido et al. 1997) and the protein forms large oligomers, whatever the status of phosphorylatable serines could be (Bruey et al. 2000b). Moreover, it was observed that Hsp27 overexpression induced by growing the cells at a high density decreased the level of intracellular ROS (Garrido et al. 1997); a phenomenon which correlated with resistance to TNFα-induced cell death.

5
Conclusions and Future Perspectives

sHsp are new negative regulators of programmed cell death with a broad spectrum of action. In addition to being able to protect cells against oxidative stress, sHsp share the ability to reduce and/or delay apoptotic processes induced independently of ROS overproduction (Arrigo 2000). This leads to the conclusion that, in vivo, sHsp may play a crucial role in the development and integrity of organisms. An interesting example concerns the transient change in Hsp27 expression during early differentiation which appears essential to fulfill a normal differentiation program and to escape cell death. The transient Hsp27 accumulation that takes place during the maturation of B or T cells as well as the protection mediated by these proteins against DAXX-mediated cell death suggest that Hsp27 could also participate in the control of lymphocyte populations. There are also numerous pathological situations (for example, neurodegenerative diseases) where high levels of Hsp27 can be detected (see Krueger-Naus et al., Chap. 13, this Vol.). This expression either induces a protective mechanism that combats the pathological state or could aggravate it. A good example which illustrates the second possibility concerns tumor cells that

---

Fig. 4. Hsp27 as a switch between differentiation and apoptosis in neuronal cells. A Dopamine (*DA*) induces a specific and transient accumulation of Hsp27 in neuronal 13.S.1.24 cells. Cells were either untreated (*0*) or treated with 20 µM of dopamine for the time indicated in growth medium containing 100 µM ascorbic acid. Immunoblot analysis using anti-murine Hsp27 antibody revealed the transient accumulation of murine Hsp27 during the differentiation of these cells. B Hsp27 specifically accumulates in cells escaping from apoptosis. 13.S.1.24 cells were treated for 24 h with 40 µM DA in the presence of ascorbic acid. FACS analysis and cell sorting were performed on either propidium iodide (*PI*) and annexin IV negative cells (living cells, *R1*) or PI negative and annexin V positive (cells undergoing early apoptosis, *R2*) subpopulation. PI fluorescence : FL2-H; annexin V fluorescence : FL1-H. The Hsp27 level, determined by immunoblot analysis, revealed that it is very low in cells committed to apoptosis (see *insert*). C Change in Hsp27 levels modulates dopamine-induced cell death. 13.S.1.24 cells were transfected with control (*cont.*), human Hsp27 sense (*hsp27+*) or anti-sense (*hsp27−*) vectors. 24 h after transfection, cells were either immediately harvested (*NT*) or treated for 48 h with 40 µM DA in the presence of ascorbic acid. Cell death was monitored by PI incorporation and fluorescence. Results are presented as histogram plots showing the number of cells (*counts*) versus PI fluorescence (*FL2-H*). D Immunoblot analysis of the cellular content of Hsp27 in the different cells analyzed in C. Note the relationship between Hsp27 expression and cell survival. (Reproduced from Mehlen et al. 1999 with permission of the Nature Publishing Group)

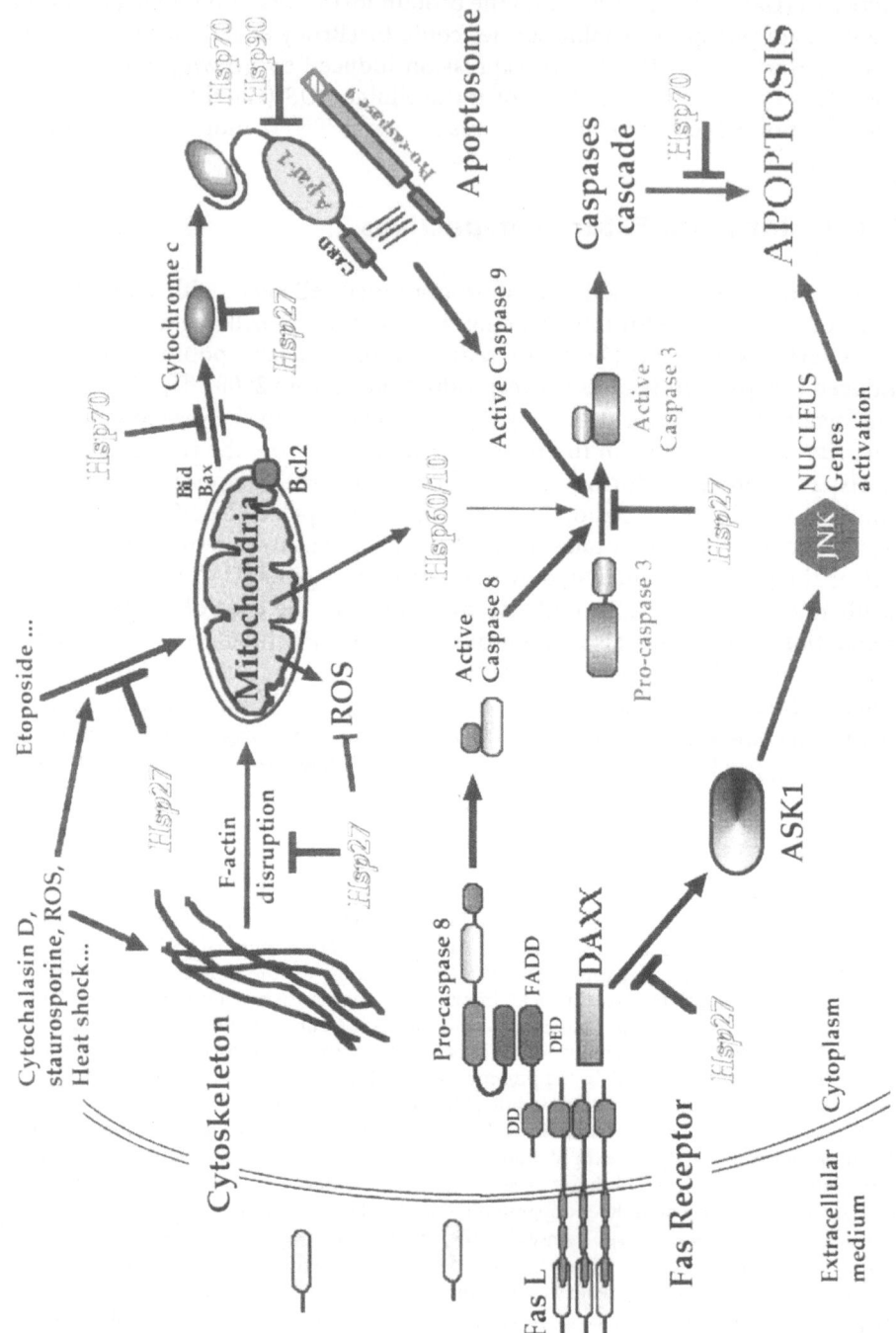

become resistant to apoptosis and more tumorigenic because of the protective effect generated by the expression of survival proteins such as Hsp27.

Studies aimed at understanding the molecular functions of sHsp have enlightened the role they play in intracellular redox potential homeostasis and programmed cell death. The mechanism by which sHsp protect against oxidative stress appears to be related to the ability of these proteins to modulate the redox status of glutathione and to decrease the level of oxidized proteins inside cells (see Arrigo et al., Chap. 10, this Vol.). Concerning the anti-apoptotic function of Hsp27, two essential steps of the mitochondrial apoptotic cascade can be negatively regulated: (1) the release of cytochrome c from the mitochondria and (2) the subsequent activation of the apoptosome complex by cytochrome c (Fig. 5). The first step is a probable consequence of the effects mediated by sHsp upstream of the mitochondria, for example at the cytoskeleton level. The second protective step appears to result from the binding of Hsp27 to cytochrome c, once it is released from the mitochondria. Other results suggest that Hsp27 has a negative effect on the downstream activation of caspases (Pandey et al. 2000b) and can interfere with the DAXX-specific death pathway (Charrette et al. 2000). It is not yet known whether Hsp27 acts solely or in relation to other chaperones, such as Hsp70, which inhibits the formation of the apoptosome and events downstream to caspase activation, or Hsp60/ Hsp10 which stimulate the activation of pro-caspase 3. Additional experiments will also be necessary to better understand the role played by the different structural organization of Hsp27 in the above-mentioned anti-apoptotic activities. We can therefore be confident that future studies will bring us a clear understanding of the molecular mechanisms which regulate the anti-death activity of this family of fascinating chaperone proteins.

Acknowledgements. This work was supported by the Association pour la Recherche sur le Cancer (grant # 5204) and the Région Rhône-Alpes.

◄

Fig. 5. Schematic illustration of the sites of action of Hsp27, Hsp70, Hsp60 and Hsp10 in the apoptotic pathway. Hsp27 interferes with the signal which promotes the release of cytochrome c from the mitochondria. This phenomenon may be due, at least in part, to the modulation of upstream events (i.e. cytoskeleton, ROS, damaged proteins) by Hsp27. A second site is located downstream of cytochrome c release but upstream of caspase 9 activation. It may be a consequence of the binding of Hsp27 to cytochrome c which results in the inhibition of apoptosome formation. This downstream activity necessitated a lower level of Hsp27 expression compared to the upstream activity which interferes with the release of cytochrome c from the mitochondria. Hsp27 also interferes with the DAXX pathway and may also decrease the activation of pro-caspase 3 by direct interaction with this enzyme. Hsp27 may also buffer the release of ROS by apoptotic mitochondria (see Arrigo et al., Chap. 9, this Vol.). Other Hsp, such as Hsp60 and 10, are also released from the mitochondria concomitantly with cytochrome c. These Hsp act as chaperones towards pro-caspase 3 conversion in active caspase 3. The major stress protein Hsp70 is supposed to interfere with cytochrome c release and apoptosome formation and decrease the downstream effects induced by active caspases. Hsp90 interacts with Apaf-1, a phenomenon which inhibits Apaf-1 oligomerization and subsequently caspase 9 activation

References

Agius MA, Kirvan CA, Schafer AL, Gudipati E, Zhu S (1999) High prevalence of anti-alpha-crystallin antibodies in multiple sclerosis: correlation with severity and activity of disease. Acta Neurol Scand 100:139–147

Andley UP, Song Z, Wawrousek EF, Fleming TP, Bassnett S (2000) Differential protective activity of {alpha}A- and {alpha}B-crystallin in lens epithelial cells. J Biol Chem 275:36823–36831

Arrigo A-P (1995) Expression of stress genes during development. Neuropathol Appl Neurobiol 21:488–491

Arrigo A-P, Landry J (1994) Expression and function of the low-molecular-weight heat shock proteins. In: Morimoto RI, Tissieres A, Georgopoulos C (eds) The biology of heat shock proteins and molecular chaperones. Cold Spring Harbor Laboratory Press, Cold Spring Harbor, pp 335–373

Arrigo A-P, Suhan JP, Welch WJ (1988) Dynamic changes in the structure and intracellular locale of the mammalian low-molecular-weight heat shock protein. Mol Cell Biol 8:5059–5071

Arrigo AP (1998) Small stress proteins: chaperones that act as regulators of intracellular redox state and programmed cell death. Biol Chem 379:19–26

Arrigo AP (2000) sHsp as novel regulators of programmed cell death and tumorigenicity. Pathol Biol (Paris) 48:280–288

Beaver JP, Waring P (1995) A decrease in intracellular glutathione concentration precedes the onset of apoptosis in murine thymocytes. Eur J Cell Biol 68:47–54

Beere HM, Wolf BB, Cain K, Mosser DD, Mahboubi A, Kuwana T, Tailor P, Morimoto RI, Cohen GM, Green DR (2000) Heat-shock protein 70 inhibits apoptosis by preventing recruitment of procaspase-9 to the apaf-1 apoptosome. Nat Cell Biol 2:469–475

Benjamin IJ, Shelton J, Garry DJ, Richardson JA (1997) Temporospatial expression of the small HSP/alpha B-crystallin in cardiac and skeletal muscle during mouse development. Dev Dynam 208:75–84

Beresford PJ, Jaju M, Friedman RS, Yoon MJ, Lieberman J (1998) A role for heat shock protein 27 in CTL-mediated cell death. J Immunol 161:161–167

Blackburn RV, Galoforo SS, Berns CM, Armour EP, McEachern D, Corry PM, Lee YJ (1997) comparison of tumor growth between Hsp25- and Hsp27- transfected murine L929 cells in nude mice. Int J Cancer 72:871–877

Bossy-Wetzel E, Newmeyer DD, Green DR (1998) Mitochondrial cytochrome c release in apoptosis occurs upstream of DEVD- specific caspase activation and independently of mitochondrial transmembrane depolarization. EMBO J 17:37–49

Boyle DL, Takemoto L (2000) A possible role for alpha-crystallins in lens epithelial cell differentiation. Mol Vis 6:63–71

Brar BK, Stephanou A, Wagstaff MJ, Coffin RS, Marber MS, Engelmann G, Latchman DS (1999) Heat shock proteins delivered with a virus vector can protect cardiac cells against apoptosis as well as against thermal or hypoxic stress. J Mol Cell Cardiol 31:135–146

Bruey J-M, Ducasse C, Bonniaud P, Ravagnan L, Susin CA, Diaz-Latoud C, Gurbuxani S, Arrigo A-P, Kroemer G, Solary E, Garrido C (2000a) Hsp27 negatively regulates cell death by interacting with cytochrome c. Nat Cell Biol 2:645–652

Bruey JM, Paul C, Fromentin A, Hilpert S, Arrigo AP, Solary E, Garrido C (2000b) Differential regulation of HSP27 oligomerization in tumor cells grown in vitro and in vivo. Oncogene 19:4855–4863

Charette SJ, Lavoie JN, Lambert H, Landry J (2000) Inhibition of daxx-mediated apoptosis by heat shock protein 27. Mol Cell Biol 20:7602–7612

Chaufour S, Mehlen P, Arrigo AP (1996) Transient accumulation, phosphorylation and changes In the oligomerization of Hsp27 during retinoic acid-induced differentiation of HL-60 cells: possible role in the control of cellular growth and differentiation. Cell Stress Chaperones 1:225–235

Chun-Ying Li J-SLY-GK, Jong-Il Kim and Jeong-Sun Seo (2000) Hsp70 inhibits apoptosis downstream of cytochrome c release and upstream of caspase 3 activation. J Biol Chem (in press)

Ciocca DR, Oesterreich S, Chamnes GC, McGuire WL, Fuqua SAW (1993) Biological and clinical implications of heat shock proteins 27000 (Hsp27): a Review. J Natl Cancer Inst 85:1558–1570

Conroy SE, Sasieni PD, Amin V, Wang DY, Smith P, Fentiman IS, Latchman DS (1998) Antibodies to heat-shock protein 27 are associated with improved survival in patients with breast cancer. Br J Cancer 77:1875–1879

Coppola S, Ghibelli L (2000) GSH extrusion and the mitochondrial pathway of apoptotic signalling. Biochem Soc Trans 28:56–61

Davidson SM, Morange M (2000) Hsp25 and the p38 MAPK pathway are involved in differentiation of cardiomyocytes. Dev Biol 218:146–160

Desagher S, Osen-Sand A, Nichols A, Eskes R, Montessuit S, Lauper S, Maundrell K, Antonsson B, Martinou JC (1999) Bid-induced conformational change of bax is responsible for mitochondrial cytochrome c release during apoptosis. J Cell Biol 144:891–901

Devaja O, King RJ, Papadopoulos A, Raju KS (1997) Heat-shock protein 27 (HSP27) and its role in female reproductive organs. Eur J Gynaecol Oncol 18:16–22

Deveraux QL, Roy N, Stennicke HR, Van Arsdale T, Zhou Q, Srinivasula SM, Alnemri ES, Salvesen GS, Reed JC (1998) IAPs block apoptotic events induced by caspase-8 and cytochrome c by direct inhibition of distinct caspases. EMBO J 17:2215–2223

Du C, Fang M, Li Y, Li L, Wang X (2000) Smac, a mitochondrial protein that promotes cytochrome c-dependent caspase activation by eliminating IAP inhibition. Cell 102:33–42

Gaestel M, Gross B, Benndorf R, Strauss M, Schunk W-H, Kraft R, Otto A, Bohm H, Stahl J, Drabsch H, Bielka H (1989) Molecular cloning, sequencing and expression in Escherichia coli of the 25-kDa growth-related protein of Ehrlich ascites tumor and its homology to mammalian stress proteins. Eur J Biochem 179:209–213

Garrido C, Ottavi P, Fromentin A, Hammann A, Arrigo AP, Chauffert B, Mehlen P (1997) HSP27 as a mediator of confluence-dependent resistance to cell death induced by anticancer drugs. Cancer Res 57:2661–2667

Garrido C, Fromentin A, Bonnotte B, Favre N, Moutet M, Arrigo AP, Mehlen P, Solary E (1998) Heat shock protein 27 enhances the tumorigenicity of immunogenic rat colon carcinoma cell clones. Cancer Res 58:5495–5499

Garrido C, Bruey JM, Fromentin A, Hammann A, Arrigo AP, Solary E (1999) HSP27 inhibits cytochrome c-dependent activation of procaspase-9. FASEB J 13:2061–2070

Ghibelli L, Fanelli C, Rotilio G, Lafavia E, Coppola S, Colussi C, Civitareale P, Ciriolo MR (1998) Rescue of cells from apoptosis by inhibition of active GSH extrusion. FASEB J 12:479–486

Green DR, Reed JC (1998) Mitochondria and apoptosis. Science 281:1309–1312

Guenal I, Sidoti-de Fraisse C, Gaumer S, Mignotte B (1997) Bcl-2 and Hsp27 act at different levels to suppress programmed cell death. Oncogene 15:347–360

Hanash S, Strahler J, Chan Y, Kuick R, Teichroew D, Neel J, Hailat N, Keim D, Gratiot-Deans J, Ungar D et al. (1993) Data base analysis of protein expression patterns during T-cell ontogeny and activation. Proc Natl Acad Sci USA 90:3314–3318

Hansen RK, Parra I, Lemieux P, Oesterreich S, Hilsenbeck SG, Fuqua SA (1999) Hsp27 overexpression inhibits doxorubicin-induced apoptosis in human breast cancer cells. Breast Cancer Res Treat 56:187–196

Jäättelä M (1995) Over-expression of Hsp70 confers tumorigenicity to mouse fibrosarcoma cells. Int J Cancer 60:689–693

Jäättelä M, Wissing D, Kokholm K, Kallunki T, Egeblad M (1998) Hsp70 exerts its anti-apoptotic function downstream of caspase-3-like proteases. EMBO J 17:6124–6134

Kindas-Mugge I, Trautinger F (1994) Increased expression of the M(r) 27,000 heat shock protein (hsp27) in vitro differentiated normal human keratinocytes. Cell Growth Differ 5:777–781

Kluck RM, Bossy-Wetzel E, Green DR, Newmeyer DD (1997) The release of cytochrome c from mitochondria: a primary site for Bcl-2 regulation of apoptosis [see comments]. Science 275:1132–1136

Kohler C, Gahm A, Noma T, Nakazawa A, Orrenius S, Zhivotovsky B (1999) Release of adenylate kinase 2 from the mitochondrial intermembrane space during apoptosis. FEBS Lett 447:10–12

Li P, Nijhawan D, Budihardjo I, Srinivasula SM, Ahmad M, Alnemri ES, Wang X (1997) Cytochrome c and dATP-dependent formation of Apaf-1/caspase-9 complex initiates an apoptotic protease cascade. Cell 91:479–489

Loones MT, Chang Y, Morange M (2000) The distribution of heat shock proteins in the nervous system of the unstressed mouse embryo suggests a role in neuronal and non-neuronal differentiation. Cell Stress Chaperones 5:291–305

Mehlen P, Schulze-Osthoff K, Arrigo A-P (1996) Small stress proteins as novel regulators of apoptosis – heat shock protein 27 blocks Fas/APO-1- and staurosporine-induced cell death. J. Biol Chem 271:16510–16514

Mehlen P, Hickey E, Weber L, Arrigo A-P (1997a) Large unphosphorylated aggregates as the active form of hsp27 which controls intracellular reactive oxygen species and glutathione levels and generates a protection against TNFα in NIH-3T3-ras cells. Biochem Biophys Res Commun 241:187–192

Mehlen P, Mehlen A, Godet J, Arrigo A-P (1997b) hsp27 as a switch between differentiation and apoptosis in murine embryonic stem cells. J Biol Chem 272:31657–31665

Mehlen P, Coronas V, Ljubic-Thibal V, Ducasse C, Granger L, Jourdan F, Arrigo AP (1999) Small stress protein Hsp27 accumulation during dopamine-mediated differentiation of rat olfactory neurons counteracts apoptosis. Cell Death Differ 6:227–233

Mosser DD, Caron AW, Bourget L, Meriin AB, Sherman MY, Morimoto RI, Massie B (2000) The chaperone function of hsp70 is required for protection against stress-induced apoptosis. Mol Cell Biol 20:7146–7159

Nicholson DW, Thornberry NA (1997) Caspases: killer proteases. Trends Biochem Sci 22:299–306

Pandey P, Saleh A, Nakazawa A, Kumar S, Srinivasula M, Kumar V, Weichselbaum R, Nalin C, Alnemri ES, Kufe D, Kharbanda S (2000a) Negative regulation of cytochrome c-mediated oligomerization of Apaf-1 and activation of procaspase-9 by heat shock protein 90. EMBO J 19:4310–4322

Pandey P, Farber R, Nakazawa A, Kumar S, Bharti A, Nalin C, Weichselbaum R, Kufe D, Kharbanda S (2000b) Hsp27 functions as a negative regulator of cytochrome c-dependent activation of procaspase-3. Oncogene 19:1975–1981

Pauli D, Tonka C-H, Tissieres A, Arrigo A-P (1990) Tissue-specific expression of the heat shock protein HSP 27 during Drosophila melanogaster development. J Cell Biol 111:817–828

Préville X, Schultz H, Knauf U, Gaestel M, Arrigo AP (1998) Analysis of the role of Hsp25 phosphorylation reveals the importance of the oligomerization state of this small heat shock protein in its protective function against TNFalpha- and hydrogen peroxide-induced cell death. J Cell Biochem 69:436–452

Reed JC (1997) Cytochrome c: can't live with it–can't live without it. Cell 91:559–562

Rogalla T, Ehrnsperger M, Preville X, Kotlyarov A, Lutsch G, Ducasse C, Paul C, Wieske M, Arrigo AP, Buchner J, Gaestel M (1999) Regulation of Hsp27 oligomerization, chaperone function, and protective activity against oxidative stress/tumor necrosis factor alpha by phosphorylation. J Biol Chem 274:18947–18956

Rosse T, Olivier R, Monney L, Rager M, Conus S, Fellay I, Jansen B, Borner C (1998) Bcl-2 prolongs cell survival after Bax-induced release of cytochrome c. Nature 391:496–499

Saleh A, Srinivasula SM, Acharya S, Fishel R, Alnemri ES (1999) Cytochrome c and dATP-mediated oligomerization of Apaf-1 is a prerequisite for procaspase-9 activation. J Biol Chem 274:17941–17945

Saleh A, Srinivasula SM, Balkir L, Robbins PD, Alnemri ES (2000) Negative regulation of the apaf-1 apoptosome by hsp70. Nat Cell Biol 2:476–483

Samali A, Cotter TG (1996) Heat shock proteins increase resistance to apoptosis. Exp Cell Res 223:163–170

Samali A, Cai J, Zhivotovsky B, Jones DP, Orrenius S (1999a) Presence of a pre-apoptotic complex of pro-caspase-3, Hsp60 and Hsp10 in the mitochondrial fraction of Jurkat cells. Embo J 18:2040–2048

Samali A, Holmberg CI, Sistonen L, Orrenius S (1999b) Thermotolerance and cell death are distinct cellular responses to stress: dependence on heat shock proteins. FEBS Lett 461:306–310

Samali A, Orrenius S (1998) Heat shock proteins: regulators of stress response and apoptosis. Cell Stress Chaperones 3:228–236

Samali A, Robertson J, Peterson E, Manero F, van Zeijl L, Paul C, Cotgreave IA, Arrigo A-P, Orrenius S (2000) Small heat shock proteins protect mitochondria of thermotolerant cells. Cell Stress Chaperones 6:49–58

Scaffidi C, Fulda S, Srinivasan A, Friesen C, Li F, Tomaselli KJ, Debatin KM, Krammer PH, Peter ME (1998) Two CD95 (APO-1/Fas) signaling pathways. EMBO J 17:1675–1687

Shakoori AR, Oberdorf AM, Owen TA, Weber LA, Hickey E, Stein JL, Lian JB, Stein GS (1992) Expression of heat shock genes during differentiation of mammalian osteoblasts and promyelocytic leukemia cells. J Cell Biochem 48:277–287

Shimizu S, Narita M, Tsujimoto Y (1999) Bcl-2 family proteins regulate the release of apoptogenic cytochrome c by the mitochondrial channel VDAC. Nature 399:483–487

Slater AF, Stefan C, Nobel I, van den Dobbelsteen DJ, Orrenius S (1995) Signalling mechanisms and oxidative stress in apoptosis. Toxicol Lett 82–83:149–153

Spector NL, Samson W, Ryan C, Gribben J, Urba W, Welch WJ, Nadler LM (1992) Growth arrest of human B lymphocytes is accompanied by induction of the low molecular weight mammalian heat shock protein. J Immunol 148:1668–1673

Spector NL, Ryan C, Samson W, Levine H, Nadler LM, Arrigo A-P (1993) Heat shock protein is a unique marker of growth arrest during macrophage differentiation of HL-60 cells. J Cell Physiol 156:619–625

Spector NL, Mehlen P, Ryan C, Hardy L, Samson W, Levine H, Nadler LM, Fabre N, Arrigo A-P (1994) Regulation of the 28 kDa heat shock protein by retinoic acid during differentiation of human leukemic HL-60 cells. FEBS Lett 337:184–188

Stahl J, Wobus AM, Ihrig S, Lutsch G, Bielka H (1992) The small heat shock protein hsp25 is accumulated in P19 embryonal carcinoma cells and embryonic stem cells of line BLC6 during differentiation. Differentiation 51:33–37

Sugiyama Y, Suzuki A, Kishikawa M, Akutsu R, Hirose T, Waye MM, Tsui SK, Yoshida S, Ohno S (2000) Muscle develops a specific form of small heat shock protein complex composed of MKBP/HSPB2 and HSPB3 during myogenic differentiation. J Biol Chem 275:1095–1104

Susin SA, Lorenzo HK, Zamzami N, Marzo I, Brenner C, Larochette N, Prevost MC, Alzari PM, Kroemer G (1999a) Mitochondrial release of caspase-2 and -9 during the apoptotic process. J Exp Med 189:381–394.

Susin SA, Lorenzo HK, Zamzami N, Marzo I, Snow BE, Brothers GM, Mangion J, Jacotot E, Costantini P, Loeffler M, Larochette N, Goodlett DR, Aebersold R, Siderovski DP, Penninger JM, Kroemer G (1999b) Molecular characterization of mitochondrial apoptosis-inducing factor. Nature 397:441–446

Takayama S, Bimston DN, Matsuzawa S, Freeman BC, Aime-Sempe C, Xie Z, Morimoto RI, Reed JC (1997) BAG-1 modulates the chaperone activity of Hsp70/Hsc70. EMBO J 16:4887–4896

Têtu B, Lacasse B, Bouchard H-L, Lagacé R, Huot J, Landry J (1992) Prognostic influence of HSP-27 expression in malignant fibrous histiocytoma: a clinicopathological and immunohistochemical study. Cancer Res 52:2325–2328

Tezel G, Wax MB (2000) The mechanisms of hsp27 antibody-mediated apoptosis in retinal neuronal cells. J Neurosci 20:3552–3562

Thornberry NA, Lazebnik Y (1998) Caspases: enemies within. Science 281:1312–1316

Van den Dobbelsteen DJ, Nobel CSI, Schlegel J, Cotgreave IA, Orrenius S, Slater AFG (1996) Rapid and specific efflux of reduced glutathione during apoptosis induced by anti-Fas/APO-1 antibody. J Biol Chem 271:15420–15427

Verhagen AM, Ekert PG, Pakusch M, Silke J, Connolly LM, Reid GE, Moritz RL, Simpson RJ, Vaux DL (2000) Identification of DIABLO, a mammalian protein that promotes apoptosis by binding to and antagonizing IAP proteins. Cell 102:43–53

Wagstaff MJ, Collaco-Moraes Y, Smith J, de Belleroche JS, Coffin RS, Latchman DS (1999) Protection of neuronal cells from apoptosis by Hsp27 delivered with a herpes simplex virus-based vector. J Biol Chem 274:5061–5069

Wei MC, Lindsten T, Mootha VK, Weiler S, Gross A, Ashiya M, Thompson CB, Korsmeyer SJ (2000) tBID, a membrane-targeted death ligand, oligomerizes BAK to release cytochrome c. Genes Dev 14:2060–2071

Xanthoudakis S, Roy S, Rasper D, Hennessey T, Aubin Y, Cassady R, Tawa P, Ruel R, Rosen A, Nicholson DW (1999) Hsp60 accelerates the maturation of pro-caspase-3 by upstream activator proteases during apoptosis. EMBO J 18:2049–2056

Yang J, Liu X, Bhalla K, Kim CN, Ibrado AM, Cai J, Peng TI, Jones DP, Wang X (1997) Prevention of apoptosis by Bcl-2: release of cytochrome c from mitochondria blocked. Science 275: 1129–1132

Hsp27 as a Prognostic and Predictive Factor in Cancer

Daniel R. Ciocca and Laura M. Vargas-Roig[1]

1
Introduction

Once the diagnosis of a particular type of cancer has been established, the physicians need to know the aggressiveness of the tumor in order to decide which treatments should be applied. The aggressiveness of the malignant tumor is evaluated by the clinic (e.g., growth rate or given symptoms), by laboratory/image data (e.g., presence of tumor markers in serum, local/distant extension evaluated by CAT scan), and by histopathology (e.g., tumor size, tumor subtype, nuclear grade, lymph node invasion). In the last few years, molecular markers are progressively being incorporated in order to better define the tumor aggressiveness and the treatment options for a particular cancer, and more importantly, for a particular patient. Prognostic factors are those measurements available at the time of diagnosis or surgery that are associated with disease-free survival (DFS) or overall survival (OS) (Clark 1999). In a strict sense, these factors should be evaluated in the absence of systemic adjuvant therapy which is often difficult to achieve. On the other hand, predictive factors are those measurements indicating the response to a particular treatment. Predictive factors may also be useful in the prognosis, e.g., the presence of estrogen receptor (ER) α and progesterone receptor (PR) predicts the responsiveness to tamoxifen in breast cancer patients and these receptors are also associated with a better prognosis. However, sometimes prognostic and predictive factors are not associated, for example, lymph node invasion is useful for disease prognosis but is not associated with the response to a particular therapy.

One serious problem in the search for useful prognostic and predictive factors is the heterogeneity and dynamism of the tumors. The tumors are not static, there are millions of cells changing/evolving biochemically and morphologically they are due to intrinsic genetic/epigenetic variations and responses in the changing conditions of the host (Ciocca and Elledge 2000). The prognostic/predictive factors are usually evaluated at the time of diagnosis or surgery and these represent a unique moment in the tumor evolution.

[1] Laboratory of Reproduction and Lactation (LARLAC), Regional Center for Scientific and Technological Research (CRICYT), Casilla de Correo 855, Parque General San Martín, 5500 Mendoza, Argentina

Progress in Molecular and Subcellular Biology, Vol. 28
A.-P. Arrigo and W.E.G. Müller (Eds.)
© Springer-Verlag Berlin Heidelberg 2002

Another important consideration is that the prognostic/predictive factors should be studied in numerous patients with a long follow-up incorporating homogeneous groups of patients (with the same tumor type/subtypes, similar tumor sizes, same stages, same treatments) in randomized trials. Many of the studies reporting the usefulness of a prognostic/predictive factor have not met these criteria and for these reasons there are many conflicting reports in the literature. Finally, another source of conflicting reports is the marker itself. Many times researchers have not utilized the same methodology to confirm or expand the utility of a prognostic/predictive factor, moreover, the cut-off of the factor is also different in different reports. Thus, we need to reach a consensus in order to obtain more reproducible results.

How did it come about that the small heat shock protein (sHsp) Hsp27 was involved as a prognostic and predictive factor in cancer? Some years ago, one of us (DRC) was working on a project searching for markers to improve the prediction of hormone responsiveness of breast cancers. At that time, an estrogen-responsive protein was found in MCF-7 human breast cancer cells (Edwards et al. 1981), and monoclonal antibodies were generated to study the protein (Ciocca et al. 1983). We then learned that this protein belonged to the heat shock family of proteins and that several other proteins with similar molecular weight, that were studied in different laboratories, were identified as the Hsp27 sHsp (Ciocca et al 1993). Within a tumor, cancer cells are exposed to several stressful conditions (e.g., insufficient blood supply, hypoxia, low pH), which may stimulate the synthesis of Hsps and we know that cells with elevated Hsps have a higher capability to resist damaging conditions. Then, it was of great interest to know whether tumors expressing Hsp27 (and other Hsps as well) were more resistant to cytotoxic drugs and/or if they showed a more aggressive phenotype. In addition, since Hsp27 is under estrogenic regulation (besides the heat shock transcription factors) (Ciocca et al 1993), it was of interest to know whether tumors expressing Hsp27 were more hormone-responsive than tumors lacking this protein. There are another proteins homologous to sHsps (like αB-crystallin) but we have little clinical data about these proteins.

2
Female Genital Tract

2.1
Breast

Several studies have been performed to assess the prognostic value of Hsp27 in breast cancer patients (Table 1). To date, more than 2600 biopsies from patients with breast cancer have been examined. In most studies, the expression of this protein has not correlated with DFS (defined as the time from diagnosis to first recurrence, or last contact with the patient) or OS (defined as the

Table 1. Prognostic significance of Hsp27 in breast cancer patients

| | | | | Correlation with | |
| | | | | --- | --- |
Study	n	Patients	Treatments	DFS[a]	OS
Thor et al. (1991)	300	LN-: 246; LN+: 44 (IHC, Northern, Western) Follow-up: 8 years	Variable	Yes LN-: No LN+: Yes >Hsp27, <DFS	No
Hurlimann et al. (1993)	196	Stages I to IV (IHC) Follow-up: 5 years	Variable (R, C, E)	No	No
Love et al. (1994)	361	Stage I or II (IHC)	Variable	LN-: Yes >Hsp27, <DFS LN+: No	No
Tetu et al. (1995)	890	LN+ (IHC) Follow-up: 2.5–10.5 years	Variable (C, E)	No	No
Oesterreich et al. (1996)	903	LN- (Western, IHC) Follow-up: 4 to 6 years	Variable (C, E)	No	No

[a] Abbreviations used: DFS, disease free survival; OS, overall survival; LN, axillary lymph nodes; IHC, immunohistochemistry; R, radiotherapy; C, chemotherapy; E, endocrine therapy.

time from diagnosis to death from any cause, or last contact with the patient). This lack of correlation has been found both in regional lymph node-negative and lymph node-positive patients. The authors that reported a correlation found a statistical significance of modest value. In multivariate analysis, the expression of Hsp27 did not appear as an independent prognostic marker; there are much more useful factors. For this reason, in the clinic, we are using a combination of factors, ERα, PR, tumor size, tumor grade, lymph node status, HER-2/neu and p53 to make a more objective estimate of the prognosis of individual breast cancer patients (Gago et al. 1998).

Table 2 shows studies that have assessed the predictive value of Hsp27 in breast cancer patients. Hsp27 is like the progesterone receptor and estrogen-regulated protein, its presence correlated with that of ERα (Ciocca et al. 1990). Moreover, in an earlier study, Hsp27 expression was associated with a higher response rate to endocrine therapy in a group of ERα-positive patients (Cano et al. 1986). However, in a relatively large and homogeneous group of patients with a long follow-up period, Hsp27 (and also Hsp70) expression was not significantly associated with response to tamoxifen, the period of time until treatment failure, or survival (Ciocca et al. 1998). It is surprising to find that Hsp27 expression correlates with the presence of ERα but not with the response to tamoxifen. This exemplifies the need to specifically test the usefulness of a molecular marker before its introduction into clinical practice. Perhaps the dynamics of Hsp27 and the wide spectrum of biological functions of the protein in tumor cells restrain its utility as a marker indicating a response to

Table 2. Predictive value of Hsp27 in breast cancer patients

Study	n	Patients	Treatments	Correlation with DFS[a]	Correlation with OS
Andersen et al. (1989) Damstrup et al. (1992)	103	Advanced cancer (IHC) Follow-up: ~10 years	Endocrine (T, Ov)	–	No
Seymour et al. (1990)	51	Metastatic cancer (IHC) Follow-up: 2 years	C (MCV) + T (ER+)	–	Yes >Hsp27, >OS
Hurlimann et al. (1993)	72	Stages I to IV (IHC) Follow-up: 5 years	Variable, 24 only E	No	No
Ciocca et al. (1998)	205	Metastatic cancer (IHC) Follow-up: 9 years	Tamoxifen	–	No
Vargas-Roig et al. (1998)	43	Stage II to III (IHC) Follow-up: 4 years	Induction chemotherapy	Yes >Hsp27, <DFS	No

[a] Abbreviations used: DFS, disease free survival; OS, overall survival; IHC, immunohistochemistry; T, tamoxifen; Ov, ovariectomy; C, chemotherapy; M, mitoxantrone; V, vincristine; C, cyclophosphamide; ER, estrogen receptor; E, endocrine therapy.

tamoxifen. Therefore, at present, ERα, bcl-2 and progesterone receptors are the best molecular markers for predicting response to tamoxifen in breast cancer patients (Ciocca and Elledge 2000).

The basis to study Hsp27 in drug resistance was provided by in vitro studies. Human breast cancer cells overexpressing Hsp27 (constitutively or following a heat shock) showed increased resistance to one of the drugs tested, doxorubicin (Ciocca et al. 1992). Doxorubicin resistance was also noted when a human Hsp27 cDNA was transfected to human breast and colon cancer cells (Garrido et al. 1996). Moreover, human testis cancer cells transfected with the full-length human Hsp27 gene showed increased doxorubicin and cisplatin resistance (Richards et al. 1996). Wachsberger et al. (1997) found cisplatin resistance in Chinese hamster cells and in a human melanoma cell line adapted to grow at low pH (6.7) that caused an elevation in Hsp27 levels. At present, suggested mechanisms involving Hsp27 in drug/cytotoxic resistance are: (1) increased cell growth (Oesterreich et al. 1993); (2) stabilization of actin filaments (Lavoie et al. 1993; Huot et al. 1996); (3) increased resistance to apoptosis (Samali and Cotter 1996); (4) decreased cell growth (Richards et al. 1996); (5) modulation of intracellular redox/reactive oxygen species (Garrido et al. 1997); and (6) altered topoisomerase II expression (Hansen et al. 1999). So far, we do not know with certainty the main mechanism(s) by which Hsp27 confers drug resistance. It is also possible that a relationship exists between Hsp27 (and other Hsps) and the multidrug resistance P170 glycoprotein (encoded by the mdr1 gene) (reviewed in Ciocca and Vargas Roig 1997). Schneider et al. (1998) reported a co-expression of Hsp27 and P170 proteins in biopsy samples from human ovarian cancer. It is of interest to mention here that we have not found multidrug resistance associated with overexpression of Hsp27 in human breast cancer cell lines, and that cultured cells with P170 overexpression were much more resistant to doxorubicin than cells overexpressing Hsp27 (Ciocca et al. 1992). Finally, Hsp27 has not been involved in a decrease in uptake of cytotoxic drugs by tumor cells (Richards et al. 1996). Therefore, it seems that Hsp27 is involved selectively in resistance to certain drugs, that there are more efficient mechanisms that protect tumor cells from cytotoxic drugs, and the protection is for certain tumor cell types and not for others. To illustrate these concepts, we can mention that Hsp27 expression is induced by cisplatin in Ehrlich ascites tumor cells (Bielka et al. 1994; Gotthardt et al. 1996) but not in human breast cancer cells (Ciocca et al. 1992). Colon cancer cells transfected with Hsp27 showed a good correlation between protein content and doxorubicin resistance, however, in colon cancer cells cisplatin treatment increased Hsp27 expression but cell resistance to doxorubicin decreased (Garrido et al. 1996). Tumor cell lines derived from different tissues did not reveal a general relationship between Hsp27 expression levels and cisplatin sensitivity, some of them even showed more cisplatin resistance when the Hsp27 levels were low (Hettinga et al. 1996). It should not be forgotten that Hsp27 phosphorylation changes after heat shock (Ciocca et al. 1992), and seems to play an important role in cell function (Huot et al. 1996; Rogalla et al. 1999). Moreover, Hsp27 phosphorylation changes are generally not detected when the expression levels of this protein are studied.

An especially important question is whether Hsp27 is related to drug resistance in breast cancer patients. We approached this question by studying biopsy samples from locally advanced breast cancer patients ($n = 35$) treated with induction chemotherapy (Vargas-Roig et al. 1998). Comparing the pre- and post-chemotherapy biopsies, we observed that chemotherapy increased Hsp27 and Hsp70 nuclear expression and significantly decreased the cytoplasmic content of Hsp70 and Hsc70. Most of these patients (86%) received relatively high doses of doxorubicin/epirubicin (together with cyclophosphamide and 5-fluorouracil), the other patients received cyclophosphamide, methotrexate and 5-fluorouracil. A high nuclear proportion of Hsp70 in tumor cells (>10%) correlated significantly with drug resistance. We also observed that patients whose tumors expressed nuclear or a high cytoplasmic proportion of Hsp27 (>66%) had a shorter DFS. Therefore, our study suggests that Hsp27 and Hsp70 were involved in drug resistance in breast cancer patients. We need to emphasize here that the predictive value of Hsp27 was useful when the protein levels were measured in the post-chemotherapy biopsies only. In contrast, in another study the expression of cytoplasmic Hsp27 correlated with a high response to cyclophosphamide, mitoxantrone and vincristine and with a longer OS (Seymour et al. 1990). Hurlimann et al. (1993) found no correlation between Hsp27 expression and response to chemotherapy (these authors did not mention the drugs used) (Table 2).Therefore, further studies with homogeneous groups of cancer patients are necessary to clarify the role of Hsp27 and other Hsps in drug resistance.

2.2
Endometrium, Ovary, Uterine Cervix

ERα, PR and Hsp27 have been studied in biopsies from patients with abnormal endometrium (Ciocca et al. 1985, 1989; McGuire et al. 1986). Based on these studies it has been postulated that early endometrial hyperplasia is highly sensitive to estrogenic stimulation, whereas with increasing architectural and cytological atypia, the estrogenic response decreases and cell populations that are more independent of estrogenic influence appear to be growing. Then, the expression of these three proteins was found to be related to the degree of differentiation, histological characteristics of the tumors and clinical stage of disease. We also know that these parameters correlate with the survival of the patients. In fact, a recent study has shown that Hsp27 expression is an independent prognostic factor in patients with endometrial carcinomas (Geisler et al. 1999). It is of interest to mention that Hsp27 is also found in leiomyomas (more than in the surrounding healthy myometrium) (Navarro et al. 1989) and in endometrial stromal sarcomas (more Hsp27 is seen in low-grade than in high-grade stromal sarcomas) (Navarro et al. 1992).

Less clear is the role of Hsp27 in ovarian cancer. Langdon et al. (1995) reported that concentrations of Hsp27 in malignant tumors ($n = 54$) was found to be higher in patients in an advanced stage of the disease; more Hsp27 was found in patients with a shorter time of survival and tumors that had grown progressively after chemotherapy had a significantly higher Hsp27 content. In this study, Hsp27 was measured by an ELISA method in cytosol samples. In contrast, Geisler et al. (1998), using immunohistochemistry, reported that Hsp27 was associated with FIGO (Federation International of Gynecology and Obstetrics) stage of disease ($n = 99$) (>FIGO stage, <Hsp27 content), and that Hsp27 was an independent prognostic factor of survival (>Hsp27, >survival). Clearly, more studies on patients with malignant ovarian carcinomas (including homogeneous groups of patients with longer follow-ups) are necessary to clarify the possible prognostic/predictive value of Hsp27 in ovarian carcinomas.

In the uterine cervix, Hsp27 expression has been explored in normal and abnormal biopsy samples mainly by immunohistochemistry (Ciocca et al. 1986, 1989; Dressler et al. 1986; Puy et al. 1989). These studies have shown that Hsp27 is present in normal squamous cells (in the parabasal cell layer), cervical intraepithelial neoplasias and in squamous cell carcinomas. In squamous cell carcinomas, Hsp27 expression correlates with the degree of differentiation (>Hsp27, >differentiation) but does not correlate with the expression of ERα and PRs. Tamoxifen did not induce modification of Hsp27 in biopsies from patients with invasive cervical carcinomas (Vargas Roig et al. 1993). It is interesting that HPV infection was associated with changes in ERα and Hsp27 expression in cervical tissues (Ciocca et al. 1992). We have recently studied Hsp27 in biopsy samples from patients with locally advanced cervical carcinomas treated with induction chemotherapy and radiation therapy (unpubl. results). These results have implicated Hsp27 expression with resistance to these treatments.

3
Nervous System and Digestive Tract

The prognostic significance of Hsp27 has been evaluated, mainly by immunohistochemistry, in several tumors of the central nervous system including neuroblastomas, pituitary adenomas and astrocytomas (Table 3). It is surprising to observe that in different tumor types of the same origin, the protein can be associated with factors of completely opposite significance; in neuroblastomas Hsp27 is related to limited stage disease and to more differentiated tumors while in astrocytomas it is related to more aggressive and undifferentiated tumors. A similar observation can be made from the studies that evaluated Hsp27 in tumors from the digestive tract (Table 3). In gastric and hepatocel-

Table 3. Prognostic significance of Hsp27 in tumors of the nervous system and digestive tract

Tumors	Characteristics	Reference
Nervous system		
Neuroblastomas (n = 53)	>Hsp27 in limited stage and more differentiated tumors	Ungar et al. (1994)
Pituitary adenomas (n = 20)	>Hsp27 in invasive pituitary adenomas	Gandour-Edwards et al. (1995)
Astrocytomas (n = 66)	>Hsp27 in higher grade and higher growth tumors	Khalid et al. (1995)
Astrocytomas (n = 48)	>Hsp27 in glioblastomas	Hitotsumatsu et al. (1996)
Astrocytomas (n = 95)	>Hsp27 in poorly differentiated tumors	Assimakopoulou et al. (1997)
Digestive tract		
Gastric carcinoma (n = 100)	>Hsp27, <survival	Harrison et al. (1989)
SCC[a], tongue (n = 24)	No correlation with clinical stage, lymph node metastasis and survival	Ito et al. (1998)
SCC, esophagus (n = 102)	>Hsp27, >OS	Kawanishi et al. (1999)
Hepatocellular ca (n = 58)	>Hsp27, <DFS and <OS	King K-L et al. (2000)

[a] Abbreviations used: SCC, squamous cell carcinoma; OS, overall survival; DFS, disease free survival; ca, carcinoma.

lular carcinomas, Hsp27 has been correlated with worse survival while in squa-mous cell carcinomas of the esophagus it has been correlated with better survival.

4
Hematological Malignancies

In childhood acute lymphoblastic leukemia, significantly decreased Hsp27 levels were observed in infants compared to the Hsp27 levels found in older children (who generally have a more favorable prognosis, in other words, >Hsp27, better prognosis) (Hanash et al. 1989; Strahler et al. 1991). Xiao et al. (1996) reported high expression of Hsp27 in some patients with acute lym-phoid leukemia, acute non-lymphoid leukemia and myelodysplastic syndrome compared with that of normal blood cells, but these authors did not correlate the expression of Hsp27 with the outcome of the patients.

Recently, Hsp27 expression has been related to the response to chemother-apy in patients with de novo acute myeloid leukemia. Hsp27 was one of the proteins that could predict treatment failure (associated with other markers); the drugs administered were idarubicin and cytosine arabinoside (Kasimir-Bauer et al. 1998).

Hsp27 was strongly positive in 20% of the Reed-Sternberg cells in Hodgkin's disease, but it was only expressed weakly in the precursor B-lymphoblastic cells in non-Hodgkin lymphoma tissues (Hsu and Hsu 1998). In the same report, Hsp27 was generally found to be absent from lymphoid cells and histiocytes in normal lymphoid cells.

5
Genitourinary Tract

Few studies of small Hsps have been performed in human genitourinary tissues and their neoplasms. αB-crystallin and Hsp27 are present in the normal kidney and in the urinary bladder (Lowe et al. 1992; Storm et al. 1993; Pinder et al. 1994; Khan et al. 1996; Takashi et al. 1998). Hsp27 levels have been found higher in renal carcinomas than in the normal kidney suggesting resistance of this type of tumors to chemotherapy (Takashi et al. 1998). In bladder cancer, the expression of Hsp27 did not correlate with the degree of histological dif-ferentiation, tumor-stage, nodal status, local recurrence, metastases, or survival (Storm et al. 1993). Other researchers have studied the expression of Hsp27 in carcinosarcoma of the urinary bladder, they detected this protein in most epithelial cells and in a small number of tumor cells in the sarcomatous area (Kamishima et al. 1997).

αB-crystallin expression has been detected in prostatic carcinomas (Pinder et al. 1994). Takashi et al. (1998) found that the concentration of αB-crystallin protein in prostatic carcinomas was higher than in benign prostatic hyper-

Table 4. Prognostic value of Hsp27 in others tumors

Tumor	Reference	Cases	Treatment	Methodology	Correlation	
					DFS[a]	OS
Lung carcinoma	Vargas et al. (1998)	84 Men 27 Women	?	IHC	–	None for men 0.046 for women (>Hsp27, <OS)
Squamous cell carcinoma of head and neck	Gandour-Edwards et al. (1998)	Oral cavity 16 Oropharynx 17 Larynx 17	Surgery	IHC	–	No
Osteosarcoma	Uozaki et al. (1997)	60	Chemotherapy/ surgery	IHC	–	Yes (>Hsp27, <OS)
Malignant fibrous histiocytoma	Têtu et al. (1992)	43	Surgery/ chemotherapy	IHC	<0.05	<0.025 (>Hsp27, >OS)

[a] Abbreviations used: DFS, disease free survival; OS, overall survival; IHC, immunohistochemistry.

plasias, whereas in testicular tumors (seminomas and non-seminomas), αB-crystallin levels were lower than in the normal testis. These findings are in agreement with the resistance to treatments of prostatic carcinomas compared with the relative sensitivity of testicular tumors.

In benign prostatic hyperplasia, Hsp27 was detected in glandular epithelium and stroma (Madersbacher et al. 1998), however, in prostate cancer Storm et al. (1993) did not find Hsp27 expression. Thomas et al. (1996) reported that in prostatic carcinoma, stromal Hsp27 staining was similar to that found in benign hyperplasia, and that in the neoplastic epithelial cells there was an apparent loss of Hsp27 staining intensity as the Gleason score and invasiveness increased. Madersbacher et al. (1998) reported an increase of Hsp27 expression after heat treatment in benign and malignant human prostatic cells.

6
Tumors from Others Tissues

Table 4 shows the studies correlating Hsp27 expression with the outcome of patients with cancers of different origins.

7
Conclusion

Hsp27 expression in human tumors and cells has been analyzed in more studies than those presented here, but in this chapter we have emphasized those results showing the evaluation of Hsp27 in the prognosis of the disease or in the response to a specific therapy. From these studies it is not possible to draw a general conclusion, tumors of the same origin but of different cell types have shown disparate associations with disease prognosis or response to a therapy; sometimes there is no association at all. These discrepancies may be, in part, explained by the limits inherent to the studies (heterogeneous and small samples, different methodologies, etc.) however, in each tumor cell type there appears to be a unique role for Hsp27. We should remember that Hsps in general are very promiscuous proteins and the different molecular milieu in each tumor type seems to be affecting the roles and implications of Hsp27 in tumor biology. So far, Hsp27 is emerging as a useful molecular marker in several tumors, but more clinical studies are needed to confirm and expand these observations.

References

Andersen J, Skovbon H, Poulsen H (1989) Immunocytochemical determination of the estrogen-regulated protein M$_r$ 24,000 in primary breast cancer and response to endocrine therapy. Eur J Cancer Clin Oncol 25:641–643

Assimakopoulou M, Sotiropoulo-Bonikou G, Maraziotis T, Varakis I (1997) Prognostic significance of hsp-27 in astrocytic brain tumors: an immunohistochemical study. Anticancer Res 17:2677–2682

Bielka H, Hoinkis G, Oesterreich S, Stahl J, Benndorf R (1994) Induction of the small stress protein, hsp25, in Ehrlich ascites carcinoma cells by anticancer drugs. FEBS Lett 343:165–167

Cano A, Coffer AI, Adatia R, Millis R, Rubens RD, King RJB (1986) Histochemical studies with an estrogen-regulated protein in human breast tumors. Cancer Res 46:6475–6480

Ciocca DR, Elledge R (2000) Molecular markers for predicting response to tamoxifen in breast cancer patients. Endocrine 13:1–10

Ciocca DR, Vargas Roig LM (1997) Heat shock proteins and drug resistance in breast cancer. In: Bernal SD (ed) Drug resistance in oncology. Marcel Dekker, New York, pp 167–190

Ciocca DR, Adams DJ, Edwards DP, Bjercke RJ, McGuire WL (1983) Distribution of an estrogen-induced protein with a molecular weight of 24,000 in normal and malignant human tissues and cells. Cancer Res 43:1204–1210

Ciocca DR, Puy LA, Edwards DP, Adams DJ, McGuire WL (1985) The presence of an estrogen-regulated protein detected by monoclonal antibody in abnormal human endometrium. J Clin Endocrinol Metab 60:137–143

Ciocca DR, Puy LA, Lo Castro G, (1986) Localization of an estrogen-responsive protein in the human cervix during menstrual cycle, pregnancy, and menopause and in abnormal cervical epithelia without atypia. Am J Obst Gynecol 155:1090–1096

Ciocca DR, Stati AO, Fasoli LC (1989) Study of estrogen receptor, progesterone receptor, and the estrogen-regulated Mr 24,000 protein in patients with carcinomas of the endometrium and cervix. Cancer Res 49:4298–4304

Ciocca DR, Stati AO, Amprino de Castro MM (1990) Colocalization of estrogen and progesterone receptors with an estrogen-regulated heat shock protein in paraffin sections of human breast and endometrial cancer tissue. Breast Cancer Res Treat 16:243–251

Ciocca DR, Lo Castro G, Alonio LV, Cobo MF, Lotfi H, Teyssié A (1992) Effect of human papillomavirus infection on estrogen receptor and heat shock protein hsp27 phenotype in human cervix and vagina. Int J Gynecol Pathol 11:113–121

Ciocca DR, Oesterreich S, Chamness GC, McGuire WL, Fuqua SAW (1993) Biological and clinical implications of heat shock protein 27000 (hsp27): a review. J Natl Cancer Inst 85:1558–1570

Ciocca DR, Green S, Elledge RM, Clark GM, Pugh R, Ravdin P, Lew D, Martino S, Osborne CK (1998) Heat shock proteins hsp27 and hsp70: lack of correlation with response to tamoxifen and clinical course of disease in estrogen receptor-positive metastatic breast cancer (a Southwest Oncology Group Study). Clin Cancer Res 5:1263–1266

Clark GM (1999) Prognostic and predictive factors for primary breast cancer. Proc Am Soc Clin Oncol, 35th Annu Meet, 15–18 May 1999, Atlanta, Georgia, pp 205–207

Damstrup L, Andersen J, Kufe DW, Hayes DF, Skovgaard Poulsen H (1992) Immunocytochemical determination of the estrogen-regulated proteins Mr 24,000, Mr 52,000 and DF3 breast cancer associated antigen: clinical value in advanced breast cancer and correlation with estrogen receptor. Ann Oncol 3:71–77

Dressler LG, Ramzy I, Sledge GW, McGuire WL (1986) A new marker of maturation in the cervix: the estrogen-regulated 24K protein. Obst Gynecol 68:825–831

Edwards DP, Adams DJ, McGuire WL (1981) Estradiol stimulates synthesis of a major intracellular protein in human breast cancer cell line (MCF-7). Breast Cancer Res Treat 1:209–223

Gago FE, Tello OM, Diblasi AM, Ciocca DR (1998) Integration of estrogen and progesterone receptors with pathological and molecular prognostic factors in breast cancer patients. J Steroid Biochem Mol Biol 67:431–437

Gandour-Edwards R, Kapadia SB, Janecka IP, Martinez AJ, Barnes L (1995) Biologic markers of invasive pituitary adenomas involving the spenoid sinus. Modern Pathol 8:160–164

Garrido C, Mehlen P, Fromentin A, Hammann A, Assem M, Arrigo A-P, Chauffert B (1996) Inconstant association between 27-kDa heat-shock protein (Hsp27) content and doxorubicin resistance in human colon cancer cells. The doxorubicin-protecting effect of Hsp27. Eur J Biochem 237:653–659

Garrido C, Ottavi P, Fromentin A, Hammann A, Arrigo A-P, Chauffert B, Mehlen P (1997) Hsp27 as a mediator of confluence-dependent resistance to cell death induced by anticancer drugs. Cancer Res 57:2661–2667

Geisler JP, Geisler HE, Tammela J, Wiemann MC, Zhou Z, Miller GA, Crabtree W (1998) Heat shock protein 27: an independent prognostic indicator of survival in patients with epithelial ovarian carcinoma. Gynecol Oncol 69:14-16.

Geisler JP, Geisler HE, Tammela J, Miller GA, Weimann MC, Zhou Z (1999) A study of heat shock protein 27 in endometrial carcinoma. Gynecol Oncol 72:347-350

Gotthardt R, Neininger A, Gaestel M (1996) The anti-cancer drug cisplatin induces hsp25 in Ehrlich ascites tumor cells by a mechanism different from transcriptional stimulation influencing predominantly hsp25 translation. Int J Cancer 66:790-795

Hanash SM, Kuick R, Strahler J, Richardson B, Reaman G, Stoolman L, Hanson C, Nichols D, Tueche HJ (1989) Identification of a cellular polypeptide that distinguishes between acute lymphoblastic leukemia in infants and in older children. Blood 73:527-532

Hansen RK, Parra I, Lemieux P, Oesterreich S, Hilsenbeck SG, Fuqua SAW (1999) Hsp27 overexpression inhibits doxorubicin-induced apoptosis in human breast cancer cells. Breast Cancer Res Treat 56:187-196

Harrison JD, Morris DL, Ellis IO, Jones JA, Jackson I (1989) The effect of tamoxifen and estrogen receptor status on survival in gastric carcinoma. Cancer 64:1007-1010

Hettinga JVE, Lemstra W, Meijer C, Los G, de Vries EGE, Konings AWT, Kampinga HH (1996) Heat-shock protein expression in cisplatin-sensitive and -resistant human tumor cells. Int J Cancer 67:800-807

Hitotsumatsu T, Iwaki T, Fukui M, Tateishi J (1996) Distinctive immunohistochemical profiles of small heat shock proteins (heat shock protein 27 and αb-crystallin) in human brain tumors. Cancer 77:352-361

Hsu P-L, Hsu S-M (1998) Abundance of heat-shock proteins (hsp89, hsp60, and hsp27) in malignant cells of Hodgkin's disease. Cancer Res 58:5507-5513

Huot J, Houle F, Spitz DR, Landry J (1996) Hsp27 phosphorylation-mediated resistance against actin fragmentation and cell death induced by oxidative stress. Cancer Res 56:273-279

Hurlimann J, Gebhard S, Gomez F (1993) Oestrogen receptor, progesterone receptor, pS2, ERD5, HSP27 and cathepsin D in invasive ductal breast carcinomas. Histopathology 23:239-248

Ito T, Kawabe R, Kurasono Y, Hara M, Kitamura H, Fujita K, Kanisawa M (1998) Expression of heat shock proteins in squamous cell carcinoma of the tongue: an immunohistochemical study. J Oral Pathol Med 27:18-22

Kamishima T, Fukuda T, Usuda H, Takato H, Iwamoto H, Kaneko H (1997) Carcinosarcoma of the urinary bladder: expression of epithelial markers and different expression of heat shock proteins between epithelial and sarcomatous elements. Pathol Int 47:166-173

Kasimir-Bauer S, Ottinger H, Meusers P, Beelen DW, Brittinger G, Seeber S, Scheulen ME (1998) In acute myeloid leukemia, coexpression of at least two proteins, including P-glycoprotein, the multidrug resistance-related protein, bcl-2, mutant p53, and heat-shock protein 27, is predictive of the response to induction chemotherapy. Exp Hematol 26:1111-1117

Kawanishi K, Shiozaki H, Doki Y, Sakita I, Inoue M, Yano M, Tsujinaka T, Shamma A, Monden M (1999) Prognostic significance of heat shock proteins 27 and 70 in patients with squamous cell carcinoma of the esophagus. Cancer 85:1649-1657

Khalid H, Tsutsumi K, Yamashita H, Kishikawa M, Yasunaga A, Shibata S (1995) Expression of the small heat shock protein (hsp) 27 in human astrocytomas correlates with histologic grades and tumor growth fractions. Cell Mol Neurobiol 15:257-268

Khan W, McGuirt JP, Sens MA, Sens DA, Todd JH (1996) Expression of heat shock protein 27 in developing and adult human kidney. Toxicol Lett 84:69-79

King K-L, Li AF-Y, Chau G-Y, Chi C-W, Wu C-W, Huang C-L, Lui W-Y (2000) Prognostic significance of heat shock protein-27 expression in hepatocellular carcinoma and its relation to histologic grading and survival. Cancer 88:2464-2470

Langdon SP, Rabiasz GJ, Hirst GL, King RJB, Hawkins RA, Smyth JF, Miller WR (1995) Expression of the heat shock protein hsp27 in human ovarian cancer. Clin Cancer Res 1:1603-1609

Lavoie JN, Gingras-Breton G, Tanguay RM, Landry J (1993) Induction of Chinese hamster hsp27 gene expression in mouse cells confers resistance to heat shock. Hsp27 stabilization of the microfilament organization. J Biol Chem 268:3420-3429

Love S, King RJB (1994) A 27 kDa heat shock protein that has anomalous prognostic powers in early and advanced breast cancer. Br J Cancer 69:743–748

Lowe J, McDermott H, Pike I, Spendlove I, Landon M, Mayer RJ (1992) Alpha B crystallin expression in non-lenticular tissues and selective presence in ubiquitinated inclusion bodies in human disease. J Pathol 166:61–68

Madersbacher S, Gröbl M, Kramer G, Dirnhofer S, Steiner GE, Marberger M (1998) Regulation of heat shock protein 27 expression of prostatic cells in response to heat treatment. Prostate 37:174–181

McGuire WL, Dressler LG, Sledge GW, Ramzy I, Ciocca DR (1986) An estrogen-regulated protein in normal and malignant human endometrium. J Steroid Biochem 24:155–159

Navarro D, Cabrera JJ, Falcón O, Jiménez P, Ruiz A, Chirino R, López A, Rivero JF, Díaz-Chico JC, Díaz-Chico BN (1989) Monoclonal antibody characterization of progesterone receptors, estrogen receptors and the stress-responsive protein of 27 kDa (srp27) in human uterine leiomyoma. J Steroid Biochem 34:491–498

Navarro D, Cabrera JJ, León L, Chirino R. Fernández L, López A, Rivero JF, Fernández P, Falcón O, Jiménez P, Pestano J, Díaz-Chico JC, Díaz-Chico BN (1992) Endometrial stromal sarcoma expression of estrogen receptors, progesterone receptors and estrogen-induced srp27 (24K) suggests hormone responsiveness. J Steroid Biochem Mol Biol 41:589–596

Oesterreich S, Weng C-H, Qiu M, Hilsenbeck SG, Osborne CK, Fuqua SAW (1993) The small heat shock protein hsp27 is correlated with growth and drug resistance in human breast cancer cell lines. Cancer Res 53:4443–4448

Oesterreich S, Hilsenbeck SG, Ciocca DR, Allred DC, Clark GM, Chamness GC, Osborne CK, Fuqua SAW (1996) The small heat shock protein Hsp27 is not an independent prognostic marker in axillary lymph node-negative breast cancer. Clin Cancer Res 2:1199–1206

Pinder SE, Balsitis M, Ellis IO, Landon M, Mayer RJ, Lowe J (1994) The expression of alpha B-crystallin in epithelial tumours: a useful tumour marker? J Pathol 174:209–215

Puy LA, Lo Castro G, Olcese JE, Lotfi HO, Brandi HR, Ciocca DR (1989) Analysis of a 24-kiladalton (KD) protein in the human uterine cervix during abnormal growth. Cancer 64:1067–1073

Richards EH, Hickey E, Weber L, Masters JRW (1996) Effect of overexpression of the small heat shock protein hsp27 on the heat and drug sensitivities of human testis tumor cells. Cancer Res 56:2446–2451

Rogalla T, Ehrnsperger M, Preville X, Kotlyarov A, Lutsch G, Ducasse C, Paul C, Wieske M, Arrigo AP, Buchner J, Gaestel M (1999) Regulation of Hsp27 oligomerization, chaperone function, and protective activity against oxidative stress/tumor necrosis factor alpha by phosphorylation. J Biol Chem 274:18947–18956

Salami A, Cotter TG (1996) Heat shock proteins increase resistance to apoptosis. Exp Cell Res 223:163–170

Schneider J, Jimenez E, Marenbach K, Marx D, Meden H (1998) Co-expression of the mdr1 gene and hsp27 in human ovarian cancer. Anticancer Res 18:2967–2972

Seymour L, Bezwoda WR, Meyer K (1990) Tumor factors predicting for prognosis in metastatic breast cancer. Cancer 66:2390–2394

Storm FK, Mahvi DM, Gilchrist KW (1993) Hsp-27 has no diagnostic or prognostic significance in prostate or bladder cancers. Urology 42:379–382

Strahler JR, Kuick R, Hanash SM (1991) Diminished phosphorylation of a heat shock protein (Hsp 27) in infant acute lymphoblastic leukemia. Biochem Biophys Res Commun 175:134–142

Takashi M, Katsuno S, Sakata T, Ohshima S, Kato K (1998) Different concentrations of two small stress proteins, αB crystallin and HSP27 in human urological tumor tissues. Urol Res 26:395–399

Têtu B, Lacasse B, Bouchard H-L, Lagacé R, Huot J, Landry J (1992) Prognostic influence of HSP-27 expression in malignant fibrous histiocytoma: a clinicopathological and immunohistochemical study. Cancer Res 52:2325–2328

Têtu B, Brisson J, Landry J, Huot J (1995) Prognostic significance of heat-shock protein-27 in node-positive breast carcinoma: an immunohistochemical study. Breast Cancer Res Treat 36:93–97

Thomas SA, Brown IL, Hollins GW, Hocken A, Kirk D, King RJB, Leake RE (1996) Detection and distribution of heat shock proteins 27 and 90 in human benign and malignant prostatic tissue. Br J Urol 77:367–372

Thor A, Benz C, Moore II D, Goldman E, Edgerton S, Landry J, Schwartz L, Mayall B, Hickey E, Weber LA (1991) Stress response protein (srp-27) determination in primary human breast carcinomas: clinical, histologic, and prognostic correlations. J Natl Cancer Inst 83:170–178

Ungar DR, Hailat N, Strahler JR, Kuick RD, Brodeur GM, Seeger RC, Reynolds CP, Hanash SM (1994) Hsp27 expression in neuroblastoma: correlation with disease stage. J Natl Cancer Inst 86:780–784

Uozaki H, Horiuchi H, Ishida T, Iijima T, Imamura T, Machinami R (1997) Overexpression of resistance-related proteins (metallothioneins, glutathione-S-transferase pi, heat shock protein 27, and lung resistance-related protein) in osteosarcoma. Relationship with poor prognosis. Cancer 79:2336–2344

Vargas SO, Leslie KO, Vacek PM, Socinski MA, Weaver DL (1998) Estrogen-receptor-related protein p29 in primary nonsmall cell lung carcinoma. Cancer 82:1495–1500

Vargas-Roig LM, Lotfi H, Olcese JE, Lo Castro G, Ciocca DR (1993) Effects of short-term tamoxifen administration in patients with invasive cervical carcinoma. Anticancer Res 13:2457–2464

Vargas-Roig LM, Gago FE, Tello O, Aznar JC, Ciocca DR (1998) Heat shock protein expression and drug resistance in breast cancer patients treated with induction chemotherapy. Int J Cancer (Pred Oncol) 79:468–475

Wachsberger PR, Landry J, Storck C, Davis K, O'Hara MD, Owen CS, Leeper DB, Coss RA (1997) Mammalian cells adapted to growth at pH 6.7 have elevated HSP27 levels and are resistant to cisplatin. Int J Hyperthermia 13:251–255

Xiao K, Liu W, Qu S, Sun H, Tang J (1996) Study of heat shock protein HSP90 alpha, HSP70 mRNA expression in human acute leukemia cells. J Tongji Med Univ 16:212–216

Cytoskeletal Competence Requires Protein Chaperones

Roy Quinlan[1]

1
Summary

The cytoskeleton is the internal structure of the cell that makes diverse cellular functions possible. These structures are extremely dynamic, undergoing continual remodelling within the cytoplasm and requiring continual change in the protein-protein interactions as part of their function. It is therefore not surprising that cytoskeletal proteins require the attention of protein chaperones at all stages of their life. Initially, chaperonins ensure the nascent chains of actin and tubulin fold correctly as they emerge from the ribosome and in this review, the role of the small heat shock proteins (sHSPs) in further cytoskeletal function will be discussed. Chaperones are most certainly important components in the birth and life of the cytoskeleton.

This review will discuss the importance of protein chaperones, and specifically sHSPs, to the maintenance and control of the cytoskeleton. A key viewpoint within the article is the consideration of the cytoskeleton as a whole, removing the convenient, but artificial, boundaries of actin, tubulin and intermediate filaments to discuss sHSP function. Importantly, viewing chaperones and the cytoskeleton as a functional unit suggests exciting new possibilities that should cause us to rethink the commonly perceived role of chaperones as just "quality control" proteins. At the very least sHSPs improve the efficiency of cytoskeletal function and prevent potentially damaging liaisons, but there is greater potential in sHSP-cytoskeleton complex. This review will discuss what is currently known about the interaction of sHSPs with the cytoskeleton and will end with some ideas to inspire future research based on this emerging holistic view of the cell!

[1] Department of Biological Sciences, South Road, Durham DH1 3LE, UK

Progress in Molecular and Subcellular Biology, Vol. 28
A.-P. Arrigo and W.E.G. Müller (Eds.)
© Springer-Verlag Berlin Heidelberg 2002

2
Human Diseases Caused by sHSP Mutations

2.1
Protein Inclusions Involving the Cytoskeleton and sHSPs

The first critical observations linking the cytoskeleton with sHSPs came in the late 1980s. Firstly, αB-crystallin mRNA was upregulated in scrapie-infected brain together with GFAP, the glial specific intermediate filament protein (Duguid et al. 1988). Then, in 1989, αB-crystallin was identified as a major component of Rosenthal Fibres (Iwaki et al. 1989), which also contained GFAP (Goldman and Corbin 1988). From this point onwards, sHSPs were recognised as characteristic components of these protein inclusions (Iwaki et al. 1993) and the link between intermediate filaments and sHSPs was established. This led to the identification of αB-crystallin as a component of Mallory bodies (Lowe et al. 1992), neurofilament tangles (Kato et al. 1992; Lowe et al. 1992) and granulo-filamentous material containing desmin in myopathy patients (Goebel and Bornemann 1993), all within the space of a couple of years. It was also realised that HSP27, another sHSP was often a component of these intermediate filament inclusions (e.g. Iwaki et al. 1993). From these collective data, it was clear that sHSPs were associated with intermediate filaments in pathological situations. Did this reflect a specific requirement of intermediate filaments during cell stress or disease? Was this the only cytoskeletal component that associated with sHSPs?

Over the next 5 years these questions were answered. Evidence was presented to suggest the sHSP-intermediate filament interaction was direct and independent of pathological situations (Longoni et al. 1990; Nicholl and Quinlan 1994). The first association of sHSPs with the actin cytoskeleton was discovered at about the same time (Chiesi et al. 1990; Miron et al. 1991; Bennardini et al. 1992; Gopalakrishnan and Takemoto 1992). These data clearly suggested a far greater potential for sHSPs in attending the needs of the whole cytoskeleton in both healthy and diseased cells, but before I explore these in more detail, it is useful to discuss those human diseases and their associated pathologies that are caused by sHSP mutations. After all, it is from this source that the most unequivocal and strongest evidence has emerged that sHSPs are absolutely required for the efficient function of the cytoskeleton.

2.2
Cardiomyopathies and Eye Lens Cataract

In the last 3 years there have been two mutations in sHSPs reported to cause autosomal dominant cataract (Litt et al. 1998; Vicart et al. 1998). The mutations affected two different sHSPs, namely αA- and αB-crystallin respectively. The R120G mutation in αB-crystallin was identified primarily as the genetic basis

```
              100       110       120       130       140       150
αA-Crys   HNERQDDHGYISREFHRRYRLPSNVDQSALSCSLSADGMLTFCGPKIQTGL
αB-Crys   HEERQDEHGFISREFHRKYRIPADVDPLTITSSLSSDGVLTVNGPRKQVSG
HSP27     HEERQDEHGYISRCFTRKYTLPPGVDPTQVSSSLSPEGTLTVEAPMPKLAT
HSPb2     HPQRLDRHGFVSREFCRTYVLPADVDPWRVRAALSHDGILNLEAPRGGRHL
HSPb3     HGTRMDEHGFISRSFTRQYKLPDGVEIKDLSAVLCHDGILVVEVKDPVGTK
HSP20     HEERPDEHGFVAREFHRRYRLPPGVDPAAVTSALSPEGVLSIQAAPASAQA
CvHSP     ---KLAADGTVMNTFAHKCQLPEDVDPTSVTSALREDGSLTIRARRHPHTE
            :    .* : . * :      :* .*:    : . *  :* * .
```

Fig. 1. Alignment of the α-crystallin domain region in the mammalian sHSPs spanning the two putative actin binding domains (108–119; 136–143) and including the arginine 116 (*bold*) mutated in the cataract and cardiomyopathy diseases in αA- and αB-crystallin. Sequences are αA crystallin (accession no. P02489), αB-crystallin (accession no. CYHUAB), HSP27 (accession no. BAB17232), HSB2/MKBP (accession no. BAA24004), HSPB3 (accession no. NP_006299), HSP20 (accession no. AX01376), cvHSP (accession no. NP_055239)

of cardiomyopathy, although cataract was also an autosomal dominant trait in the family (Vicart et al. 1998). The mutations in αA- and αB-crystallin involve the same residue which is very conserved amongst sHSPs (de Jong et al. 1998). Both proteins are part of the same sHSP complex i.e. α-crystallin in the eye lens. As seen in Fig. 1, the sequence alignment identifies these residues as R116 and R120 in αA- and αB-crystallin respectively. The model based upon the crystal structure of Mj16.5 shows this residue is important for protein-protein interactions in the α-crystallin complex (Kim et al. 1998; van den IJssel et al. 1999). The mutations affected the secondary and tertiary structure of αA- and αB-crystallin (Bova et al. 1999; Perng et al. 1999b), which in turn altered the quaternary structure of the protein and reduced the in vitro chaperone activity (Bova et al. 1999; Shroff et al. 2000).

At first these mutations appear to tell us little about the possible physiological function of sHSPs. The key lies within the pathology of the disease. Cardiomyopathy caused by the αB-crystallin mutation displayed the same pathology as that seen in inherited cardiomyopathies caused by desmin mutations (Goldfarb et al. 1998; Li et al. 1999). Collapse of the desmin intermediate filaments into characteristic cytoplasmic bodies are used histopathogically to define the cardiomyopathy type and both the desmin and αB-crystallin mutations resulted in similar pathologies. These data suggest that αB-crystallin is intimately involved in the management of the intermediate filament network in the normal cardiomyocyte. This conclusion is the more feasible when it was subsequently shown that the R120G mutation in αB-crystallin not only destroyed the ability of this protein to chaperone intermediate filaments, but also appeared to drive the accumulation of intermediate filaments in vitro (Perng et al. 1999b). A similar course of events could underlie cataract caused by sHSPs (Vicart et al. 1998; Jakobs et al. 2000) and intermediate filament mutations, but the nature of the clinical treatment of the disease prevents such histopathological comparisons. The questions that now arise are whether this is the only function of αB-crystallin and what might be the contribution of the other sHSPs?

3
sHSP Interaction with the Cytoskeleton

These questions are addressed in this section where the interaction of sHSPs with the different components of the cytoskeleton will be discussed.

3.1
Association with Actin and Regulation of the Interaction

The first element of the cytoskeleton identified as a partner for HSP27/25 by biochemical assays was actin. This important discovery was made by serendipity when it was determined that a small molecular weight component of turkey gizzard extract had barbed end capping activity in actin assembly assays (Miron et al. 1988). This component was in fact the sHSP, HSP25 (the mouse equivalent of human HSP27; Miron et al. 1991). This discovery was the more exciting when it was realised that HSP27/25 was a physiological substrate for the stress kinase pathway and was specifically phosphorylated by the kinase, MAPKAP-K2 (Stokoe et al. 1992). Here for the first time was a direct link between a signal transduction pathway, sHSPs and the cytoskeleton, or more specifically at that point in time, the actin cytoskeleton.

The association of HSP27/25 with actin is in response to dramatic changes in the cytoskeleton as a result of stress, drug treatment, growth factor stimulation etc. (Landry and Huot 1999). It has been shown that these signals trigger sHSP phosphorylation, which in turn drives actin assembly. In vitro (Benndorf et al. 1994) and in vivo (Lavoie et al. 1993b, 1995; Guay et al. 1997; Huot et al. 1998; Loktionova and Kabakov 1998) data show HSP27/25 phosphorylation is an important modulator of actin dynamics by HSP27/25. Similar arguments can be made for HSP20 association with actin in vitro (Brophy et al. 1999a); (see Rembold et al. 2000 for conflicting data) and in vivo (Brophy et al. 1999a). Nevertheless, the overexpression of HSP27/25 can also have similar effects (Lavoie et al. 1993b), suggesting that free sHSP concentration is also a factor in regulating actin assembly. So sHSPs are important mediators of the signal transduction pathways and are directly involved in translating external stimuli into cytoskeletal remodelling.

In mammalian cells there are now seven different sHSPs identified (Krief et al. 1999) and actin binding is a feature of many sHSPs. In addition to HSP25/27, the lens-specific complex of αA- and αB-crystallin also binds actin filaments (Gopalakrishnan and Takemoto 1992) as does HSP20 (Brophy et al. 1999b; Rembold et al. 2000). These proteins are more closely related by sequence analysis (Krief et al. 1999) within the sHSP family. Myotonic dystrophy protein kinase binding protein (MKBP) is part of a second subgroup (Krief et al. 1999), but there is some evidence to suggest it also binds actin (Shama et al. 1999). From these data, it is quite apparent that actin association is not a feature easily assigned to an individual sHSP. To make sense of these observations, we need more information on the control and regulation of **all** the sHSP-actin interactions.

One explanation for this apparent overlap in sHSP activity could be a common sequence motif. A putative actin binding region spanning residues 110–121 in HSP20 has been identified (Rembold et al. 2000). This region is reasonably well conserved amongst the other sHSPs (see Fig. 1; Rembold et al. 2000). Others have suggested that the motif G(V/I)LT(X3)P, located at 140–147 in HSP20, competes with actin for incorporation into the filament. This motif is also found in an assembly-critical domain in actin itself (Kaukinen et al. 1996) and overlaps with a putative chaperone site for sHSPs (Muchowski et al. 1999a). Crystal structures for mammalian sHSPs will determine the relative importance of these putative actin-binding sequences. It is, however, quite clear that the different sHSPs do not all behave in the same way. For instance, although HSP25 had apparent barbed end capping activity that prevents filament growth, αB-crystallin appeared to increase actin assembly rather than inhibit it (Wang and Spector 1996). HSP20 binding to actin filaments appears to be dependent upon tropomyosin (Rembold et al. 2000). In conclusion, most sHSPs interact preferentially with polymerised actin (F-actin), but HSP20 can interact with both polymerised and soluble (G-) actin (Brophy et al. 1999b). These examples indicate two different mechanisms to regulate cytoskeletal dynamics in cells. One is via a direct interaction with the filament itself whilst the other is with the soluble filament subunits. A third is discussed in the next section.

3.2
Interaction with Cytoskeleton Associated Proteins

The two-hybrid technique has revealed some important partners for sHSPs. Notably, the sHSP MKBP (formerly (HSPB2) was found to specifically bind to the Rho-kinase family member, myotonic protein kinase (Suzuki et al. 1998). MKBP enhanced the activity of this kinase as well as protecting it against heat denaturation, but it is pertinent to the sometimes degenerate function of sHSPs, that HSP27 and αB-crystallin were also able to increase the activity of the kinase, but to a lesser extent. It should be noted, however, that their binding to MDPK could not be detected by two hybrid assay (Sugiyama et al. 2000). The Rho-kinase protein family are important modulators of the cytoskeleton (Hirose et al. 1998; Madaule et al. 1998; Kosako et al. 1999). In fact, HSP27 was also reported to bind protein kinase B under stress conditions (Konishi et al. 1997). These data suggest that some sHSPs could modulate the cytoskeleton by regulating kinases that induce the changes in the cytoskeleton and for instance the ability of HSP27 to inhibit Src family protein kinases is another example (Kasi and Kuppuswamy 1999).

The most recent addition to the family, cvHSP was used to Z-hybridscreen a mouse embryo library and from this emerged a very interesting binding partner, filamin (Krief et al. 1999). It is an actin binding protein that promotes actin polymerisation as well as the attachment of membrane proteins to actin filaments. Immunoprecipitation experiments verified that the filamin-cvHSP interaction was a *bone fide* physiological interaction (Krief et al. 1999). Other sHSPs did not appear to bind filamin and so this is a rare example of the binding of a sHSP to a specific partner. It was also shown that cvHSP bound to a region overlapping with the MKK-4 (mitogen-activated protein kinase kinase kinase kinase) binding site (Krief et al. 1999), once again suggesting that sHSPs modulate the cytoskeleton via signal transduction pathways. In conclusion, the third way of regulating the actin cytoskeleton is via interactions with actin binding proteins or proteins that regulate these actin binding proteins.

3.3
Intermediate Filament and Microtubule Association

Immunoprecipitation has been very effective in showing that the sHSPs, αB-crystallin and HSP27 are associated with soluble complexes of vimentin and glial fibrillary acid protein (GFAP) (Nicholl and Quinlan, 1994; Perng et al. 1999a). Both sHSPs, including HSP27, also associate with intact intermediate filaments (Nicholl and Quinlan 1994; Perng et al. 1999a; see Fig. 2). So in stark contrast to the situation with actin, neither the assembly status of the intermediate filament proteins nor the phosphorylation status of αB-crystallin affected binding in vitro to GFAP filaments (Nicholl and Quinlan 1994). So in these respects the association of sHSPs with intermediate filaments is different to their association with actin filaments.

In terms of the sHSP association with intermediate filaments, a key function is to maintain the individuality of intermediate filaments (Perng et al. 1999a) by managing the network in cells. This idea is supported by evidence that in cell models of pathological situations, the over-expression of αB-crystallin can dissolve intermediate filament aggregates (Koyama and Goldman 1999). Stress also increases the association of sHSPs with intermediate filaments (Djabali et al. 1997; Muchowski et al. 1999b) and in the light of the previous data, it would appear this is specifically to prevent inappropriate interactions by intermediate filaments and thus maintain an efficient cytoskeleton.

Binding to intermediate filaments is not the only way to manage their networks. After all, sHSPs also bind soluble intermediate filament subunits in vivo (Nicholl and Quinlan 1994; Perng et al. 1999a). This would provide a very direct mechanism to regulate assembly by either controlling the availability of assembly-competent subunits or by driving filament disassembly through the sequestration of subunits (Nicholl and Quinlan 1994). Indeed, heat shock increases the soluble pool of intermediate filament proteins (Liao et al. 1995) and there is extensive rearrangement of the filament network so this is an attractive idea.

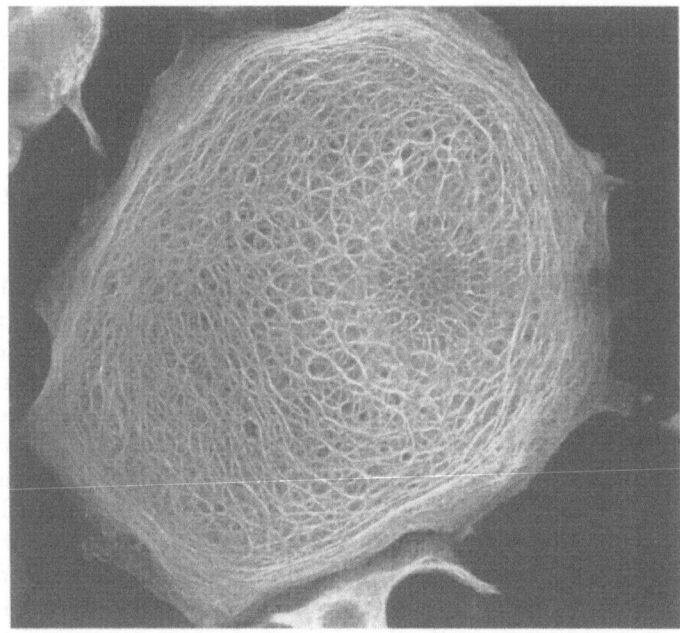

Fig. 2. A human epithelial MCF7 cell stained with anti-HSP27 antibodies showing a filamentous network for this sHSP. This pattern corresponds to the keratin intermediate filaments within this cell and illustrates the tight association seen between HSP27 and intermediate filaments. Note that HSP27 is apparently not located at the actin cytoskeleton as seen with these HSP27 antibodies. A colour version of this image can be viewed at my web site

Details of sHSP binding to microtubules and tubulin are scarce. It has been reported that αB-crystallin can be co-immunoprecipitated with soluble tubulin dimers (Arai and Atomi 1997) and it is also capable of protecting microtubule networks in overexpression studies (Bluhm et al. 1998). HSP27 was unable to give the same protection to microtubules (Bluhm et al. 1998) arguing against an indirect stabilising effect via intermediate filaments. Collectively though, the data show sHSPs (HSP27 and αB-crystallin) are well placed to help in the regulation of the whole cytoskeleton

4
Functional Implications of the Interactions

The previous sections have given an overview of the wide range of links between sHSPs and the cytoskeleton. At present, it is difficult to draw all these observations into a single cohesive picture because the field is establishing the nature and meaning of these many different interactions. Nevertheless, it is this wider picture that is missing from our current thinking. It is important to understand the mechanisms that allow one sHSP, like αB-crystallin, to interact with actin, intermediate filaments, microtubules and unfolding cellular proteins in the same cell. So in this section, particular emphasis will be given to developing this broader picture of sHSP function to the cytoskeleton. By default this tends to concentrate the discussion onto HSP27/25 and αB-crystallin as these are the best studied.

4.1
Cell Stress Identifies an Important Link Between Chaperones and the Cytoskeleton

The first evidence of a tight association between sHSPs and the cytoskeleton came from studies on cell stress. The cytoskeleton is very sensitive to stresses such as heat and osmotic shock as well as oxidative and chemical stresses (Thomas et al. 1982; Welch and Feramisco 1985; Welch et al. 1985). Indeed, alterations to the distribution of the cytoskeleton are some of the early events in the stress response along with the induction of protein chaperones (Mizzen and Welch 1988; Welch and Mizzen 1988). The overexpression of HSP27/25 and αB-crystallin was able to prevent these early changes in the cytoskeleton (Landry et al. 1989; Huot et al. 1991, 1997; Lavoie et al. 1993a,b), clearly indicating the inter-dependence of sHSPs and the cytoskeleton.

Recently, ischemia has become a focus of stress studies representing a physiological situation of cell and tissue damage. One consistent message is that αB-crystallin becomes associated with actin microfilaments during ischemia. Stress induces αB-crystallin to translocate to the Z-lines and intercalated discs in the myocardium sarcomere (Chiesi et al. 1990; Barbato et al. 1996; van de Klundert et al. 1998; Golenhofen et al. 1999) and eventually the I-band under extreme ischemia (Golenhofen et al. 1999). Phosphorylation of αB-crystallin is involved (Golenhofen et al. 1998), but the specific phosphorylation events and kinases involved in the translocation have yet to be determined. The p38 pathway is a good candidate (Ito et al. 1997; Kato et al. 1998; Hoover et al. 2000). An increased interaction between αB-crystallin and desmin was not apparent in this ischemic tissue (Golenhofen et al. 1999), although this was anticipated because of the cardiomyopathy studies (Goldfarb et al. 1998; Vicart et al. 1998). Perhaps this is not too surprising as other studies have shown αB-crystallin is associated with intermediate filaments in normal and stressed cells, and although thermal stress can increase the association (Djabali et al. 1997; Muchowski et al. 1999b), it is more difficult to see a relatively small increase immunohistochemically. Similarly, microtubules are also protected by sHSPs during ischemia (Bluhm et al. 1998). What feature(s) in actin, intermediate filament and tubulin filament allow the same sHSP to act on all three at the same time?

4.2
sHSP Regulation of the Cytoskeleton in Unstressed Cells

As sHSPs associate with all the major elements of the cytoskeleton, it is tempting to speculate a co-ordinating role for these chaperones. After all, the cytoskeleton functions as a whole in cells (e.g. Waterman-Storer et al. 2000). Changes in one aspect of the cytoskeleton induces changes in the rest. For instance, the depolymerisation or assembly of tubulin can induce the reorganisation of actin (Bershadsky et al. 1996; Waterman-Storer et al. 1999) as well as the collapse of the intermediate filament network (Perng et al. 1999a). Microtubules are also stabilised by focal adhesions (Kaverina et al. 1998). The transport of intermediate filament subunits is dependent upon microtubules (Prahlad et al. 1998) and intermediate filament and actin assembly is linked through mixed complexes (Correia et al. 1999). There are numerous proteins that link the different systems (e.g. Herrmann and Aebi 2000) to make sure the cytoskeleton functions as a whole, but sHSPs provide one exciting common thread. Perhaps sHSPs contribute to the regulated redistribution of the whole network in both stressed and unstressed cells.

This proposal requires an active rather than a passive involvement of sHSPs chaperones in cytoskeletal organisation. This has been proven for both the actin and intermediate filament networks and there are indications that other chaperones will contribute. For instance, co-operation between different chaperone families (HSP70 and 90 classes) during cytoskeletal change is an emerging new theme (Rousseau et al. 2000). It is also worth remembering that sHSPs can also bind to the plasma membrane (Piotrowicz and Levin 1997), regulate kinase activities (Suzuki et al. 1998), enter the nucleus (Bhat et al. 1999) and bind to specific DNA sequences (Pietrowski et al. 1994). Perhaps we have still to realise the full cell biological potential of stress proteins and particularly sHSPs in assisting many cellular functions.

5
The Chaperone–Cytoskeleton Complex. The Guardian of the Cytoplasm?

Cells are very much committed to efficiency in all their cellular processes and although we have discussed the cytoskeleton as the major beneficiary of chaperone attention, there might also be a net gain to the chaperones by their association with the cytoskeleton. After all, intermediate filaments have sHSPs, the HSP70 (Perng et al. 1999a) and the HSP90 (Czar et al. 1996) class of protein chaperones associated with them. HSP70 and sHSPs function as a chaperone machine (Ehrnsperger et al. 1997) and chaperones are more efficient when bound to a support (Altamirano et al. 1997). Although stress causes the perinuclear collapse of the cytoskeleton, this serves to concentrate damaged proteins and the chaperones into the same space (Fig. 3). Intermediate filaments also present binding sites for proteosomes too (Olinkcoux et al. 1994), providing the necessary degradation machinery to deal with those proteins that are beyond repair by the bound chaperone machines. This provides the most efficient environment to maximise protein refolding and therefore enhance the chances of cell survival. Removal of non-recoverable misfolded proteins is also

The Chaperone - Cytoskeleton
Complex. Guardian of the cytoplasm?

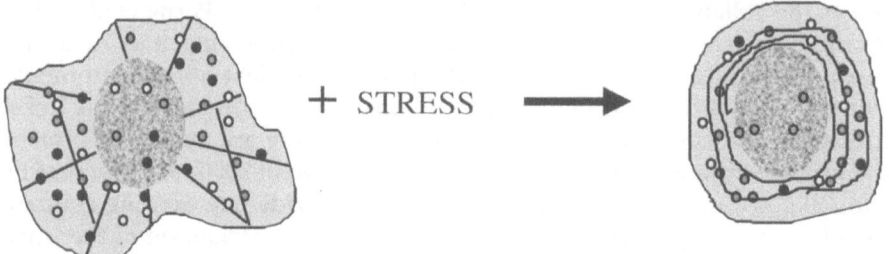

Normal Cell:	Stressed Cell:
Chaperone - Cytoskeleton Complex	Chaperone - Cytoskeleton Complex
• Supports Chaperone Machines	• Collapse of the complex to a perinuclear location
• Improves efficiency of the Cytoskeleton	• Concentrated with damaged cell proteins
• Improves efficiency of Chaperones	• Prevents irreversible damage to the cytoskeleton
• Part of the Signal Transduction Machinery	• Improves cell survival

Fig. 3. The Guardian of Cytoplasm – the chaperone–cytoskeleton complex. In this diagram, the cytoskeleton has been represented by the *straight black lines*. The chaperones are represented as *spots*, shaded differently to represent the different types of chaperones that associate with the cytoskeleton e.g. sHSP, HSP70 and 90 classes. In normal unstressed cells, the chaperone–cytoskeleton complex (CCC) is important for maintaining an efficient cytoskeleton. This includes transducing signals, co-ordinating change in the cytoskeleton and ensuring that the cytoskeleton and chaperones work efficiently. In the stressed cell, the main function of the CCC is to ensure cell survival. This is achieved by making sure that damaged proteins and the machinery needed to repair or destroy (proteosome) damaged proteins are in the best possible working environment

important to survival and intermediate filaments also provide docking sites for proteosomes (Olinkcoux et al. 1994). It might be by just such an association that the cytoskeleton and protein chaperones have become the guardian of the cytoplasm!!

Acknowledgements. The financial support of the Wellcome Trust is gratefully acknowledged. Thanks to all the members of the lab past and present who did the research to make this article possible – particularly Iain Nicholl, Lindsay Cairns, Paul van den IJssel, Ming Der Perng, Aileen Hutcheson and Heather Long. Thanks also to Alan Prescott for helping with Fig. 2.

References

Altamirano MM, Golbik R, Zahn R, Buckle AM, Fersht AR (1997) Refolding chromatography with immobilized mini-chaperones. Proc Natl Acad Sci USA 94:3576–3578
Arai H, Atomi Y (1997) Chaperone activity of alpha B-crystallin suppresses tubulin aggregation through complex formation. Cell Struct Funct 22:539–544

Barbato R, Menabo R, Dainese P, Carafoli E, Schiaffino S, Di Lisa F (1996) Binding of cytosolic proteins to myofibrils in ischemic rat hearts. Circ Res 78:821–828

Bennardini F, Wrzosek A, Chiesi M (1992) Alpha B-crystallin in cardiac tissue. Association with actin and desmin filaments. Circ Res 71:288–294

Benndorf R, Hayess K, Ryazantsev S, Wieske M, Behl J, Lutsch G (1994) Phosphorylation and supramolecular organization of murine small heat shock protein HSP25 abolish its actin polymerization-inhibiting activity. J Biol Chem 269:20780–20784

Bershadsky A, Chausovsky A, Becker E, Lyubimova A, Geiger B (1996) Involvement of microtubules in the control of adhesion-dependent signal-transduction. Curr Biol 6:1279–1289

Bhat SP, Hale IL, Matsumoto B, Elghanayan D (1999) Ectopic expression of alpha B-crystallin in Chinese hamster ovary cells suggests a nuclear role for this protein. Eur J Cell Biol 78:143–150

Bluhm WF, Martin JL, Mestril R, Dillmann WH (1998) Specific heat shock proteins protect microtubules during simulated ischemia in cardiac myocytes. Am J Physiol 275:H2243–H2249

Bova MP, Yaron O, Huang Q, Ding L, Haley DA, Stewart PL, Horwitz J (1999) Mutation R120G in alphaB-crystallin, which is linked to a desmin- related myopathy, results in an irregular structure and defective chaperone-like function [In Process Citation]. Proc Natl Acad Sci USA 96:6137–6142

Brophy CM, Dickinson M, Woodrum D (1999a) Phosphorylation of the small heat shock-related protein, HSP20, in vascular smooth muscles is associated with changes in the macromolecular associations of HSP20. J Biol Chem 274:6324–6329

Brophy CM, Lamb S, Graham A (1999b) The small heat shock-related protein-20 is an actin-associated protein. J Vasc Surg 29:326–333

Chiesi M, Longoni S, Limbruno U (1990) Cardiac alpha-crystallin. III. Involvement during heart ischemia. Mol Cell Biochem 97:129–136

Correia I, Chu D, Chou YH, Goldman RD, Matsudaira P (1999) Integrating the actin and vimentin cytoskeletons. adhesion-dependent formation of fibrin-vimentin complexes in macrophages. J Cell Biol 146:831–842

Czar MJ, Welsh MJ, Pratt WB (1996) Immunofluorescence localization of the 90-kDa heat-shock protein to cytoskeleton. Eur J Cell Biol 70:322–330

De Jong WW, Caspers GJ, Leunissen JA (1998) Genealogy of the alpha-crystallin–small heat-shock protein superfamily. Int J Biol Macromol 22:151–162

Djabali K, deNechaud B, Landon F, Portier MM (1997) Alpha B-crystallin interacts with intermediate filaments in response to stress. J Cell Sci 110:2759–2769

Duguid JR, Rohwer RG, Seed B (1988) Isolation of cDNAs of scrapie-modulated RNAs by subtractive hybridization of a cDNA library. Proc Natl Acad Sci USA 85:5738–5742

Ehrnsperger M, Graber S, Gaestel M, Buchner J (1997) Binding of non-native protein to Hsp25 during heat shock creates a reservoir of folding intermediates for reactivation. EMBO J 16:221–229

Goebel HH, Bornemann A (1993) Desmin pathology in neuromuscular diseases. Virchows Arch B Cell Pathol Incl Mol Pathol 64:127–135

Goldfarb LG, Park KY, Cervenakova L, Gorokhova S, Lee HS, Vasconcelos O, Nagle JW, Semino-Mora C, Sivakumar K, Dalakas MC (1998) Missense mutations in desmin associated with familial cardiac and skeletal myopathy. Nat Genet 19:402–403

Goldman JE, Corbin E (1988) Isolation of a major protein component of Rosenthal fibers. Am J Pathol 130:569–578

Golenhofen N, Ness W, Koob R, Htun P, Schaper W, Drenckhahn D (1998) Ischemia-induced phosphorylation and translocation of stress protein alpha B-crystallin to Z lines of myocardium. Am J Physiol 274:H1457–H1464

Golenhofen N, Htun P, Ness W, Koob R, Schaper W, Drenckhahn D (1999) Binding of the stress protein alpha B-crystallin to cardiac myofibrils correlates with the degree of myocardial damage during ischemia/reperfusion in vivo. J Mol Cell Cardiol 31:569–580

Gopalakrishnan S, Takemoto L (1992) Binding of actin to lens alpha crystallins. Curr Eye Res 11:929–933

Guay J, Lambert H, GingrasBreton G, Lavoie JN, Huot J, Landry J (1997) Regulation of actin filament dynamics by p38 map kinase-mediated phosphorylation of heat shock protein 27. J Cell Sci 110:357–368

Herrmann H, Aebi U (2000) Intermediate filaments and their associates: multi-talented structural elements specifying cytoarchitecture and cytodynamics. Curr Opin Cell Biol 12:79–90

Hirose M, Ishizaki T, Watanabe N, Uehata M, Kranenburg O, Moolenaar WH, Matsumura F, Maekawa M, Bito H, Narumiya S (1998) Molecular dissection of the Rho-associated protein kinase (p160ROCK)-regulated neurite remodeling in neuroblastoma N1E-115 cells. J Cell Biol 141:1625–1636

Hoover HE, Thuerauf DJ, Martindale JJ, Glembotski CC (2000) Alpha B-crystallin gene induction and phosphorylation by MKK6-activated p38. A potential role for alpha B-crystallin as a target of the p38 branch of the cardiac stress response. J Biol Chem 275:23825–23833

Huot J, Roy G, Lambert H, Chretien P, Landry J (1991) Increased survival after treatments with anticancer agents of Chinese-hamster cells expressing the human mr 27,000 heat-shock protein. Cancer Res 51:5245–5252

Huot J, Houle F, Marceau F, Landry J (1997) Oxidative stress-induced actin reorganization mediated by the p38 mitogen-activated protein kinase heat shock protein 27 pathway in vascular endothelial cells. Circ Res 80:383–392

Huot J, Houle F, Rousseau S, Deschesnes RG, Shah GM, Landry J (1998) SAPK2/p38-dependent F-actin reorganization regulates early membrane blebbing during stress-induced apoptosis. J Cell Biol 143:1361–1373

Ito H, Okamoto K, Nakayama H, Isobe T, Kato K (1997) Phosphorylation of alpha B-crystallin in response to various types of stress. J Biol Chem 272:29934–29941

Iwaki T, Kume-Iwaki A, Liem RKH, Goldman JE (1989) αB-crystallin is expressed in non-lenticular tissues and accumulates in Alexander's disease brain. Cell 57:71–78

Iwaki T, Iwaki A, Tateishi J, Sakaki Y, Goldman JE (1993) Alpha-b-crystallin and 27-kd heat-shock protein are regulated by stress conditions in the central-nervous-system and accumulate in Rosenthal fibers. Am J Pathol 143:487–495

Jakobs PM, Hess JF, FitzGerald PG, Kramer P, Weleber RG, Litt M (2000) Autosomal-dominant congenital cataract associated with a deletio n mutation in the human beaded filament protein gene bFSP2. Am J Hum Genet 66:1432–1436

Kasi VS, Kuppuswamy D (1999) Inhibition of src family kinases by a combinatorial action of 5′-AMP and small heat shock proteins, identified from the adult heart. Mol Cell Biol 19:6858–6871

Kato K, Ito H, Kamei K, Inaguma Y, Iwamoto I, Saga S (1998) Phosphorylation of alphaB-crystallin in mitotic cells and identification of enzymatic activities responsible for phosphorylation. J Biol Chem 273:28346–28354

Kato S, Hirano A, Umahara T, Llena JF, Herz F, Ohama E (1992) Ultrastructural and immunohistochemical studies on ballooned cortical neurons in Creutzfeldt-Jakob disease: expression of alpha B-crystallin, ubiquitin and stress-response protein 27. Acta Neuropathol Berl 84:443–448

Kaukinen KH, Tranbarger TJ, Misra S (1996) Post-termination-induced and hormonally dependent expression of low- molecular-weight heat shock protein genes in Douglas fir. Plant Mol Biol 30:1115–1128

Kaverina I, Rottner K, Small JV (1998) Targeting, capture, and stabilization of microtubules at early focal adhesions. J Cell Biol 142:181–190

Kim KK, Kim R, Kim SH (1998) Crystal structure of a small heat-shock protein. Nature 394:595–599

Konishi H, Matsuzaki H, Tanaka M, Takemura Y, Kuroda S, Ono Y, Kikkawa U (1997) Activation of protein kinase B (Akt/RAC-protein kinase) by cellular stress and its association with heat shock protein Hsp27. FEBS Lett 410:493–498

Kosako H, Goto H, Yanagida M, Matsuzawa K, Fujita M, Tomono Y, Okigaki T, Odai H, Kaibuchi K, Inagaki M (1999) Specific accumulation of Rho-associated kinase at the cleavage furrow during cytokinesis: cleavage furrow-specific phosphorylation of intermediate filaments [In Process Citation]. Oncogene 18:2783–2788

Koyama Y, Goldman JE (1999) Formation of GFAP cytoplasmic inclusions in astrocytes and their disaggregation by alphaB-crystallin [In Process Citation]. Am J Pathol 154:1563–1572

Krief S, Faivre JF, Robert P, Le Douarin B, Brument-Larignon N, Lefrere I, Bouzyk MM, Anderson KM, Greller LD, Tobin FL et al. (1999) Identification and characterization of cvHsp. A novel human small stress protein selectively expressed in cardiovascular and insulin- sensitive tissues. J Biol Chem 274:36592–36600

Landry J, Huot J (1999) Regulation of actin dynamics by stress-activated protein kinase 2 (SAPK2)-dependent phosphorylation of heat-shock protein of 27 kDa (Hsp27). Biochem Soc Symp 64:79–89

Landry J, Chretien P, Lambert H, Hickey E, Weber LA (1989) Heat shock resistance conferred by expression of the human HSP27 gene in rodent cells. J Cell Biol 109:7–15

Lavoie JN, Gingras Breton G, Tanguay RM, Landry J (1993a) Induction of Chinese hamster HSP27 gene expression in mouse cells confers resistance to heat shock. HSP27 stabilization of the microfilament organization J Biol Chem 268:3420–3429

Lavoie JN, Hickey E, Weber LA, Landry J (1993b) Modulation of actin microfilament dynamics and fluid phase pinocytosis by phosphorylation of heat shock protein 27. J Biol Chem 268: 24210–24214

Lavoie JN, Lambert H, Hickey E, Weber LA, Landry J (1995) Modulation of cellular thermoresistance and actin filament stability accompanies phosphorylation-induced changes in the oligomeric structure of heat shock protein 27. Mol Cell Biol 15:505–516

Li D, Tapscoft T, Gonzalez O, Burch PE, Quinones MA, Zoghbi WA, Hill R, Bachinski LL, Mann DL, Roberts R (1999) Desmin mutation responsible for idiopathic dilated cardiomyopathy. Circulation 100:461–464

Liao J, Lowthert LA, Ghori N, Omary MB (1995) The 70-kDa heat shock proteins associate with glandular intermediate filaments in an ATP-dependent manner. J Biol Chem 270:915–922

Litt M, Kramer P, LaMorticella DM, Murphey W, Lovrien EW, Weleber RG (1998) Autosomal dominant congenital cataract associated with a missense mutation in the human alpha crystallin gene CRYAA. Hum Mol Genet 7:471–474

Loktionova SA, Kabakov AE (1998) Protein phosphatase inhibitors and heat preconditioning prevent Hsp27 dephosphorylation, F-actin disruption and deterioration of morphology in ATP-depleted endothelial cells. FEBS Lett 433:294–300

Longoni S, Lattonen S, Bullock G, Chiesi M (1990) Cardiac alpha-crystallin. II. Intracellular localization. Mol Cell Biochem 97:121–128

Lowe J, Errington DR, Lennox G, Pike I, Spendlove I, Landon M, Mayer RJ (1992) Ballooned neurons in several neurodegenerative diseases and stroke contain alpha B crystallin. Neuropathol Appl Neurobiol 18:341–350

Madaule P, Eda M, Watanabe N, Fujisawa K, Matsuoka T, Bito H, Ishizaki T, Narumiya S (1998) Role of citron kinase as a target of the small GTPase Rho in cytokinesis. Nature 394:491–494

Miron T, Wilchek M, Geiger B (1988) Characterization of an inhibitor of actin polymerization in vinculin- rich fraction of turkey gizzard smooth-muscle. Eur J Biochem 178:543–553

Miron T, Vancompernolle K, Vandekerckhove J, Wilchek M, Geiger B (1991) A 25-kd inhibitor of actin polymerization is a low-molecular mass heat-shock protein. J Cell Biol 114:255–261

Mizzen LA, Welch WJ (1988) Characterization of the thermotolerant cell. I. Effects on protein synthesis activity and the regulation of heat-shock protein 70 expression. J Cell Biol 106: 1105–1116

Muchowski PJ, Hays LG, Yates III JR, Clark JI (1999a) ATP and the core 'alpha-crystallin' domain of the small heat shock protein alphaB-crystallin. J Biol Chem (submitted)

Muchowski PJ, Valdez MM, Clark JI (1999b) αB-crystallin selectively targets intermediate filament proteins during thermal stress. Invest Ophthalmol Vis Sci 40:951–958

Nicholl ID, Quinlan RA (1994) Chaperone activity of α-crystallins modulates intermediate filament assembly. EMBO J 13:945–953

Olinkcoux M, Arcangeletti C, Pinardi F, Minisini R, Huesca M, Chezzi C,Scherrer K (1994) Cytolocation of prosome antigens on intermediate filament subnetworks of cytokeratin, vimentin and desmin type. J Cell Sci 107:353–366

Perng MD, Cairns L, van den IJssel P, Prescott A, Hutcheson AM, Quinlan RA (1999a) Intermediate filament interactions can be altered by HSP27 and αB- crystallin. J Cell Sci 112:2099–2112

Perng MD, Muchowski PJ, van den IJssel P, Wu GJS, Clark JI, Quinlan RA (1999b) The cardiomyopathy and lens cataract mutation in αB-crystallin compromises secondary, tertiary and quaternary protein structure and reduces in vitro chaperone activity. J Biol Chem 274:33235–33243

Pietrowski D, Durante MJ, Liebstein A, Schmitt-John T, Werner T, Graw J (1994) Alpha-crystallins are involved in specific interactions with the murine gamma D/E/F-crystallin-encoding gene. Gene 144:171–178

Piotrowicz RS, Levin EG (1997) Basolateral membrane-associated 27-kDa heat shock protein and microfilament polymerization. J Biol Chem 272:25920–25927

Prahlad V, Yoon M, Moir RD, Vale RD, Goldman RD (1998) Rapid movements of vimentin on microtubule tracks: kinesin-dependent assembly of intermediate filament networks. J Cell Biol 143:159–170

Rembold CM, Foster DB, Strauss JD, Wingard CJ, Eyk JE (2000) cGMP-mediated phosphorylation of heat shock protein 20 may cause smooth muscle relaxation without myosin light chain dephosphorylation in swine carotid artery. J Physiol (Lond) 524 Pt 3:865–878

Rousseau S, Houle F, Kotanides H, Witte L, Waltenberger J, Landry J, Huot J (2000) Vascular endothelial growth factor (VEGF)-driven actin-based motility is mediated by VEGFR2 and requires concerted activation of stress- activated protein kinase 2 (SAPK2/p38) and geldanamycin-sensitive phosphorylation of focal adhesion kinase. J Biol Chem 275:10661–10672

Shama KM, Suzuki A, Harada K, Fujitani N, Kimura H, Ohno S, Yoshida K (1999) Transient up-regulation of myotonic dystrophy protein kinase-binding protein, MKBP, and HSP27 in the neonatal myocardium. Cell Struct Funct 24:1–4

Shroff NP, Cherian-Shaw M, Bera S, Abraham EC (2000) Mutation of R116 C results in highly oligomerized alpha A-crystallin with modified structure and defective chaperone-like function. Biochemistry 39:1420–1426

Stokoe D, Engel K, Campbell DG, Cohen P, Gaestel M (1992) Identification of MAPKAP kinase 2 as a major enzyme responsible for the phosphorylation of the small mammalian heat shock proteins. FEBS Lett 313:307–313

Sugiyama Y, Suzuki A, Kishikawa M, Akutsu R, Hirose T, Waye MM, Tsui SK, Yoshida S, Ohno S (2000) Muscle develops a specific form of small heat shock protein complex composed of MKBP/HSPB2 and HSPB3 during myogenic differentiation. J Biol Chem 275:1095–1104

Suzuki A, Sugiyama Y, Hayashi Y, Nyu-i N, Yoshida M, Nonaka I, Ishiura S, Arahata K, Ohno S (1998) MKBP, a novel member of the small heat shock protein family, binds and activates the myotonic dystrophy protein kinase. J Cell Biol 140:1113–1124

Thomas GP, Welch WJ, Matthews MB, Feramisco JR (1982) Molecular and cellular effects of heat-shock and related treatments on mammalian tissue-culture cells. Cold Spring Harbor Symp Quant Biol 46:985–996

Van de Klundert FAJ M, Gijsen MLJ, van den IJssel PRLA, Snoeckx LHEH, de Jong WW (1998) Alpha B-crystallin and hsp25 in neonatal cardiac cells – differences in cellular localization under stress conditions. Eur J Cell Biol 75:38–45

Van den IJssel P, Norman DG, Quinlan RA (1999) Small heat shock proteins in the limelight [In Process Citation]. Curr Biol 9:R103–R105

Vicart P, Caron A, Guicheney P, Li Z, Prevost MC, Faure A, Chateau D, Chapon F, Tome F, Dupret JM et al. (1998) A missense mutation in the alphaB-crystallin chaperone gene causes a desmin-related myopathy. Nat Genet 20:92–95

Wang K, Spector A (1996) α-crystallin stabilises actin filaments and prevents cytochalasin-induced depolymerisation in a phosphorylation-dependent manner. Eur J Biochem 242: 56–66

Waterman-Storer CM, Worthylake RA, Liu BP, Burridge K, Salmon ED (1999) Microtubule growth activates Rac1 to promote lamellipodial protrusion in fibroblasts [see comments]. Nat Cell Biol 1:45–50

Waterman-Storer CM, Salmon WC, Salmon ED (2000) Feedback interactions between cell-cell adherens junctions and cytoskeletal dynamics in newt lung epithelial cells. Mol Biol Cell 11:2471–2483

Welch WJ, Feramisco JR (1985) Disruption of the three cytoskeletal networks in mammalian cells does not affect transcription, translation, or protein translocation changes induced by heat shock. Mol Cell Biol 5:1571–1581

Welch WJ, Mizzen LA (1988) Characterization of the thermotolerant cell. II. Effects on the intracellular distribution of heat-shock protein 70, intermediate filaments, and small nuclear ribonucleoprotein complexes. J Cell Biol 106:974–981

Welch WJ, Feramisco JR, Blose SH (1985) The mammalian stress response and the cytoskeleton: alterations in intermediate filaments. Ann NY Acad Sci 455:57–67

Hsp27 in the Nervous System: Expression in Pathophysiology and in the Aging Brain

A.M.R. Krueger-Naug, J-C.L. Plumier, D.A. Hopkins, and R.W. Currie[1]

1
Introduction

Heat shock proteins have been detected in many mammalian tissues, including the nervous system. In recent years, the family of small molecular weight heat shock proteins, Hsp27, Hsp32 (heme oxygenase), and αB-crystallin, have been shown to be similarly expressed and regulated in response to a variety of challenges. While Hsp27 has been detected in several organs in mammals (Klemenz et al. 1993; Tanguay et al. 1993), its expression and distribution in the central nervous system are particularly striking in both development and in the adult, normally and after pathophysiological challenge. Therefore, in this chapter, we will focus on the expression of Hsp27 in the nervous system. In our work we have focused on the constitutive and pathophysiological expression of Hsp27, particularly in comparison to Hsp70. In the central and peripheral nervous systems, Hsp27 is expressed constitutively in well-defined subsets of neurons but only occasionally in neuroglia. In contrast, expression of Hsp27 in neurons and neuroglia is markedly increased in response to physiological challenges and in various models of nervous system injury, leading to the idea of specific cell-type, stress-dependent expression of Hsp27. While little is known about the role of Hsp27 in neurodegenerative diseases, their detection in these diseases leads one to consider whether such expression is beneficial and slows the degenerative process or is detrimental and indicative of the severity of the disease process. The present chapter will emphasize results on the expression of Hsp27 in the nervous system after various challenges.

2
Constitutive Expression of Hsp27 in the Nervous System

2.1
Central Nervous System: Hsp27 in the Brain and Spinal Cord

During development, Hsp27 is detected around the 17th week of gestation in human brains (Aquino et al. 1996) and in the cerebellum of rodents (Gernold

[1] Laboratory of Molecular Neurobiology, Department of Anatomy and Neurobiology, Dalhousie University, Halifax, Nova Scotia, Canada B3H 4H7

Progress in Molecular and Subcellular Biology, Vol. 28
A.-P. Arrigo and W.E.G. Müller (Eds.)
© Springer-Verlag Berlin Heidelberg 2002

et al. 1993). In humans, the developmental and adult expression of Hsp27 has not been thoroughly examined. However, in the mouse cerebellum, Hsp27 is developmentally regulated (Armstrong et al. 2001) and is expressed in a select population of parasagittal bands of Purkinje neurons in the adult (Armstrong et al. 2000). In the adult rat, the distribution of Hsp27 has been described in detail. It is constitutively expressed in most sensory and motor neurons of the adult rat brainstem and spinal cord (Plumier et al. 1997c; Fig. 1A). While Hsp27 is not normally expressed in neurons of the forebrain, including the cerebral cortex, a small number of neurons in the arcuate nucleus do express the protein. Hsp27 is detected normally in many motor neurons of the oculomotor nuclei (oculomotor, trochlear, and abducens) of the brain stem. Even more striking expression of Hsp27 is observed in the motor neurons in the trigeminal, facial, vagal, and hypoglossal cranial nerve nuclei. In contrast, autonomic preganglionic motor neurons of the vagus nerve have low levels of Hsp27. Hsp27 is also normally expressed at high levels in anterior horn motor neurons in the spinal cord. In addition, the central afferent processes of many peripheral sensory neurons also contain Hsp27.

2.2
Peripheral Nervous System

2.2.1
Dorsal Root Ganglia and Nodose Ganglion

Many, but not all, first-order somatosensory neurons located in dorsal root ganglia and cranial nerve ganglia (Plumier et al. 1997c; Costigan et al. 1998) express varying levels of Hsp27 normally. Similarly, visceral sensory neurons

---

Fig. 1A–E. Cell type- and stress-specific expression of Hsp27 in the adult rat central nervous system in control and following various stressors. **A** Hsp27 expression in a sagittal section of a normal adult rat brain. Little or no expression of Hsp27 is detected in the forebrain of control animals with the exception of occasional astrocytes. Glial cells in the fimbria (*f*) are Hsp27 positive. Note the constitutive expression of Hsp27 in the mesencephalic nucleus (*MeV*) and motor nucleus (*MoV*) of the trigeminal nerve, facial nucleus (*VIIn*), facial nerve (*VII*) and the nucleus ambiguus (*Amb*). **B** Hilar region (*h*) of the dentate gyrus showing Hsp27 expression 24h following hyperthermic treatment. The majority of Hsp27-positive cells are astrocytes. Note that some neurons in the dentate granule cell layer (*GC*) are Hsp27 positive (*arrows*). **C** Hilar region (*h*) of the dentate gyrus showing Hsp27 expression 24h following kainic acid treatment. Note the strong astrocytic induction of Hsp27. **D** Inducible expression of Hsp27 4 days following middle cerebral artery occlusion. Note the inducible expression of Hsp27 in astrocytes in all the cortical layers ipsilateral to the injury. *LV* Lateral ventricle. **E** Inducible expression of Hsp27 in the dorsal vagal complex 90 days following vagotomy. Note the increased expression of Hsp27 in the left dorsal motor nucleus of the vagus (*DMV*) and the left nucleus of tractus solitarius (*NTS*). Hypoglossal nucleus (*XII*) showing constitutive expression of Hsp27. *AP* Area postrema. *Scale bars:* **A** 1 mm, **B** 50 μm, **C** 100 μm, **D** 200 μm, **E** 2 mm. A was originally published in Plumier et al. 1997c, Wiley-Liss, Inc., a subsidiary of John Wiley & Sons, Inc.

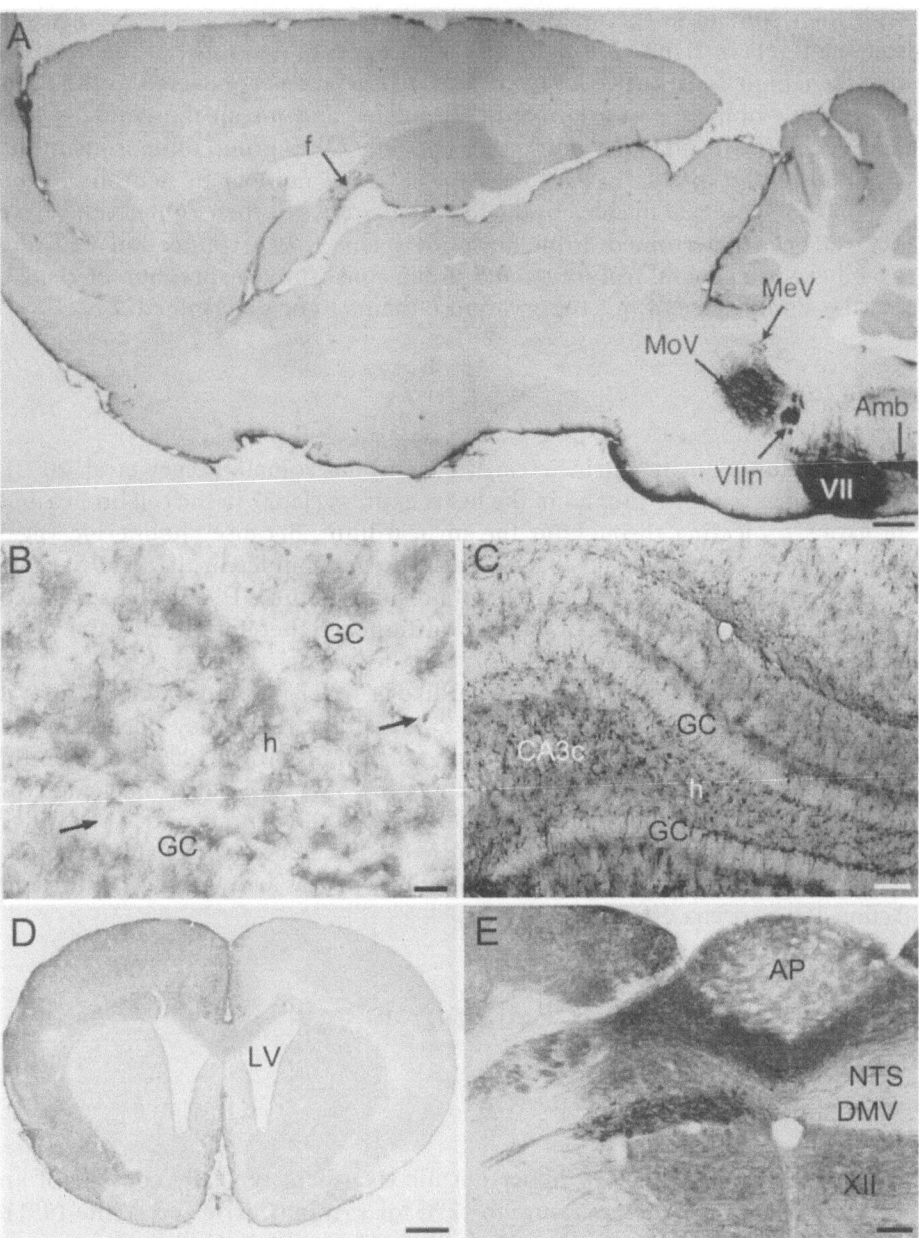

with their somata in the nodose ganglion (Hopkins et al. 1998) also express varying levels of Hsp27 normally. In fact, it appears that most of the neurons that have high constitutive levels of Hsp27 have axons connected to the periphery. This holds true for spinal nerve nuclei, and for all the cranial nerve nuclei, except the olfactory and optic. The only other group of neurons in the rat brain that express Hsp27 consists of a small number of neurons in the region of the arcuate nucleus of the hypothalamus, but their connections have not yet been determined. Nonetheless, it seems that a connection with the periphery is a general feature related to the constitutive expression of Hsp27, but the significance of this observation remains to be determined.

2.2.2
Cardiac Neurons

Hsp27 is abundant in intrinsic cardiac neuronal somata (Leger et al. 2000). Peripheral autonomic ganglia in the heart express Hsp27 in the cell bodies and in axons projecting throughout the myocardium. Because cholinergic vagal preganglionic neurons, especially cardiac preganglionics in the external formation of the nucleus ambiguus do not normally express Hsp27 unless injured (Hopkins et al. 1998), axons in the myocardium are, therefore, likely to be postganglionic or sensory ones.

The precise function of Hsp27 in cardiac or other neurons and their axonal projections is presently unclear. Hsp27, like other Hsps, may play a role in axonal transport or in synaptic function (Clark and Brown 1985; Karunanithi et al. 1999). In the heart, constitutive expression of Hsp27 in neuronal elements may facilitate responses to physiological stress in the neural control of cardiac function. Interestingly, Hsp27 in intrinsic cardiac neurons may stabilize neuronal function during ischemia and possibly suppress arrhythmias (Steare and Yellon 1993; Joyeux et al. 1997).

3
Expression of Hsp27 in Pathophysiology

3.1
Heat Shock and Fever

Experimental heat shock or hyperthermic treatment typically consists of an increase in core body temperature to 42 °C for 15 min (Currie and White 1981). Whether the temperature increase is brought about extrinsically by heating an anesthetized animal (Krueger et al. 1999; Krueger-Naug et al. 2000) or intrinsically by administration of drugs such as methamphetamine (Goto et al. 1993), morphological evidence of cell loss or tissue damage in the central nervous system has not been reported. Interestingly, there is considerable evidence suggesting that prior hyperthermic treatment may protect cells from subsequent more stressful insults (Barbe et al. 1988; Chopp et al. 1989; Kitagawa et al. 1991; Samali and Cotter 1996; Karunanithi et al. 1999). However, prolonged exposure to high temperature has been shown to have detrimental effects on cell division and differentiation during embryonic development (Edwards et al. 1997).

In the brain, the heat-inducible expression of the small heat shock protein Hsp27 was first studied by Northern analysis to detect increased levels of Hsp27 mRNA in rat brain following hyperthermic treatment (Blake et al. 1990). Western analysis indicated increased levels of Hsp27 protein in the cerebral cortex, hippocampus and cerebellum of rats from 8–16h following hyperthermic treatment (Inaguma et al. 1995). Using immunohistochemistry, in the cerebellum after hyperthermia, Hsp27 was localized to Bergmann and other types of glia and appeared to be transported distally along the radial Bergmann processes (Bechtold and Brown 2000; Krueger-Naug et al. 2000). Bechtold and Brown (2000) proposed that the expression of Hsp27 along with another small heat shock protein, Hsp32, in radial processes of Bergmann glia may be protective to synapses in the molecular layer of the cerebellum. In our study of the heat-inducible expression of Hsp27 in the rat central nervous system, we observed a generalized, time-dependent, increase in Hsp27 in astrocytes (Fig. 1B), Bergmann glia, the ependyma and the choroid plexus (Krueger-Naug et al. 2000). Hyperthermic treatment also induced Hsp27 in neurons of the hippocampal formation (Fig. 1B), the hypothalamus, the dorsal vagal complex and circumventricular organs including subfornical organ and area postrema. The neuronal response was apparently more specific than the glial response, being prominent in neurons and neural systems that are involved in regulation of physiological homeostasis, including body fluid regulation (subfornical organ and area postrema) and neuroendocrine responses to stress (hypothalamus). Thus, Hsp27 expression following hyperthermia appears to occur in physiologically relevant neuronal pathways for regulation of homeostasis. Interestingly, Liang et al. (1997) have shown that the expression of a Hsp27-like protein, p26, in encysted (desiccated) Artremia (brine shrimp) embryos, has the potential to protect macromolecular components of these embryos as they encyst or upon exposure to extreme environmental stress (Liang et al. 1997). This expression of p26 in Artremia could be interpreted as being analogous to that of Hsp27 in the adult rat nervous system and may represent a highly conserved response for the maintenance of fluid homeostasis.

In addition to the glial and neuronal expression of Hsp27 following hyperthermic treatment, unusual Hsp27-positive cells with few processes resembling immature neurons were identified adjacent to the ependyma of the third ventricle (Krueger-Naug et al. 2000). These cells appear to be associated with or have migrated away from the ependyma or subependymal region of the third ventricle into the hypothalamus. These Hsp27-positive cells did not react with antibodies to OX-42, a marker for microglia, but did react with antibodies to vimentin, a marker found in differentiating neurons and glia (Bignami et al.

1982; Zerlin et al. 1995). These Hsp27-positive cells have features that resemble stem cells with one process found in close proximity to the ventricular wall (Jankovski and Sotelo 1996; Jankovski et al. 1996). Therefore, these Hsp27-positive cells may be a type of multipotent stem cell similar to the neural stem cells that have been isolated from the striatal subventricular zone of adult mammalian brain (Reynolds and Weiss 1992; Vescovi et al. 1993; Shimazaki et al. 1999).

3.2
Kainic Acid-Induced Seizures

Kainic acid is a powerful excitatory neurotoxic analogue of glutamate (Olney et al. 1974) that is known to cause cell death in various regions of the rat central nervous system including the hippocampus, amygdala, entorhinal and piriform cortex. The damage is considered similar to that seen with temporal lobe epilepsy (Ben-Ari 1985). Heat shock proteins, specifically Hsp70, have been reported in injured neurons following kainic acid administration. Consequently, it was proposed that Hsp70 is a marker of stressful excitation (Gonzalez et al. 1989; Sloviter and Lowenstein 1992; Armstrong et al. 1996) and its expression could be used to define circuitry activated by this neurotoxin. We have studied the expression of Hsp27 following kainic acid-induced status epilepticus (Plumier et al. 1996) and compared it with the expression of Hsp70 (Armstrong et al. 1996). These studies demonstrated that these two heat shock proteins are differentially expressed in a cell-type specific manner. After kainic acid-induced seizures, Hsp70 is expressed in neurons in the hippocampus, piriform cortex and parietal cortex (Armstrong et al. 1996; Krueger et al. 1999). In the same model, Hsp27 is expressed in astrocytes (Fig. 1C) in those areas where neurons express Hsp70 (Plumier et al. 1996), demonstrating a cell type-specific induction of Hsp27 and Hsp70. Kato et al. (1999) have detected Hsp27 in astrocytes as well as in neurons of the hippocampus, piriform cortex and entorhinal cortex indicating that some neurons in the affected regions express Hsp27. It is not yet known whether these Hsp27-positive neurons also express Hsp70. It has been suggested that the glial expression of Hsp27 may be indicative of a role for these cells in protection or recovery of adjacent neurons following induced status epilepticus (Plumier et al. 1996).

3.3
Cortical Spreading Depression

Cortical spreading depression is characterized by a transient disruption of neural activity in the brain that spreads from a central location (Leao 1944). The wave of neural activity spreads to adjacent neurons causing sodium, calcium, and chloride ions to enter into the neurons, resulting in a burst of action potentials followed by electrical silence. Experimentally, spreading depression occurs as a consequence of ischemic injury in the brain or can be triggered by direct application of potassium chloride to the cortex (Matsushima et al. 1996). After induction of cortical spreading depression by cortical application of potassium chloride, Hsp27 is expressed in astrocytes throughout most of the ipsilateral cortex (Plumier et al. 1997b). In this experimental paradigm, brains are resistant to ischemic injury as indicated by reduced infarct size (Matsushima et al. 1998). Maximal accumulation of Hsp27 occurs 1–4 days after the application of potassium chloride. Blocking of cortical spreading depression with MK-801, an N-methyl-D-aspartate receptor antagonist, suppresses the expression of Hsp27 throughout the cortex, suggesting that glutamate release plays a role in the expression of Hsp27 in astrocytes (Plumier et al. 1997b). Thus, potassium chloride-induced cortical spreading depression up-regulates Hsp27 in astrocytes, raising the possibility that increased Hsp27 has a protective effect in reducing stroke-like injury.

3.4
Ischemia, Preconditioning and Cell Survival

The first studies of the effect of ischemic injury to the brain on the expression of heat shock proteins revealed the increased expression of Hsp70 (Currie and White 1981; Nowak 1985; Dienel et al. 1986). More recent studies have demonstrated the localization of Hsp gene products to injured and apparently uninjured brain regions. For example, following cortical photothrombotic injury, a non-reversible model of cerebral infarction, Hsp27 is expressed in astrocytes in the penumbra, as well as in the ipsilateral apparently non-ischemic cortex while mRNA for Hsp70 is expressed in the penumbra area between the necrotic area and the non-ischemic area (Plumier et al. 1997a). Following 10-min occlusion of the middle cerebral artery, a reversible model of ischemic injury, Hsp27 is expressed in astrocytes not only in the ischemic area but also in the non-ischemic cortex, while Hsp70 is expressed only in the ischemic area (Currie et al. 2000).

Brief ischemia preconditions the brain to be more resistant to subsequent episodes of severe ischemic injury (Kato et al. 1994; Kitagawa et al. 1997; Puisieux et al. 2000). For example, in a model of ischemic preconditioning, a 10 min occlusion of the middle cerebral artery produces significant tolerance to subsequent permanent middle cerebral artery occlusion (Barone et al. 1998). From 1 to 7 days after the preconditioning stimulus there are significant reductions in infarct size and neurological deficits following permanent middle cerebral artery occlusion. In this model, Hsp27 is expressed throughout the ipsilateral cortex (Fig. 1D), mainly in astrocytes for up to 4 weeks after preconditioning, while Hsp70 is expressed mainly in neurons in the ischemic area for only 1–4 days after preconditioning (Currie et al. 2000). As mentioned previously, preconditioning with potassium chloride-induced cortical spreading depression also results in a significant resistance to subsequent ischemic injury (Kobayashi et al. 1995; Matsushima et al. 1996). Taken together, these results suggest the intriguing possibility of an astrocyte-neuron interaction, such that

Hsp27 may protect astrocyte function during and after ischemia and improve neuronal survival through astrocyte-neuron interactions. In support of this idea, a protective effect of astrocytes on the survival of neurons after oxidative injury has been demonstrated in vitro in mixed astrocyte-neuron cultures (Desagher et al. 1996; Xu et al. 1999). Thus, Hsp27 may confer tolerance to astrocytes against ischemic injury by preserving functions such as antioxidant activity or glutamate uptake that may, in turn, protect neurons in the cortex from more severe injury.

3.5
Peripheral Nerve Injury

Hsp27 is specifically upregulated in neurons after peripheral nerve injury. Following cervical resection of the vagus nerve, Hsp27 is up-regulated in both sensory and motor neurons (Hopkins et al. 1998; Fig. 1E). Hsp27 is detected at high levels in most injured neuronal cell bodies from 2 to 90 days following vagotomy. After transection of the sciatic nerve and dorsal roots (Costigan et al. 1998; Lewis et al. 1999), similar increased expression of Hsp27 was detected in sensory and motor neurons. Comparable mechanisms may be operating in the central nervous system because lesioning of axons in the fimbria fornix also up-regulates the expression of Hsp27 in some neurons of the medial septum by 10 days post-lesion (Anguelova and Smirnova 2000). Thus, both peripheral and central neurons are capable of expressing high levels of Hsp27 for sustained periods following direct injury. This is in contrast to the expression of Hsp27 in neurons after hyperthermia (Krueger-Naug et al. 2000) and in neuroglia after ischemic injury, seizure activity or cortical spreading depression. Together these results indicate that there is a cell type-specific expression of Hsp27 that is stress dependent, i.e., there are different, specific signaling pathways for upregulation of Hsp27 in neuroglia and neurons.

It has been suggested that increased expression of Hsp27 in neurons after peripheral nerve axotomy or central fimbria fornix lesion may contribute to the prevention of neuronal cell death (Costigan et al. 1998; Anguelova and Smirnova 2000). In addition, the observation that there are elevated levels of Hsp27 in injured sensory and motor neurons suggests a possible role for Hsp27 in neuron survival or in axon regeneration. For example, prior crush or cut injury to a nerve results in a more rapid and a longer regeneration of the axons following a second such injury (Jacob and McQuarrie 1993; Jacob and Croes 1998; Lankford et al. 1998). Hsp27 may also play a role in regeneration of axons. For example, axotomy not only induces Hsp27 expression in axons of injured neurons but also appears to induce sprouting of Hsp27-positive afferent axons. In the vagotomy model, many Hsp27-positive axons are found bilaterally in the rostral nucleus of the tractus solitarius and in the contralateral dorsal motor nucleus of the vagus nerve, regions of the dorsal vagal complex that do not normally receive vagal afferents (Hopkins et al. 1998). These results have been interpreted as evidence for sprouting of vagal afferent axons.

A nerve preconditioning injury is a complex stimulus. It causes major ion fluxes, changes in gene expression, altered axonal transport, and induction of local inflammatory responses. Actin and tubulin concentrations in neurons also increase after injury and may play a role in axon regrowth and axonal transport (Miller et al. 1989; Tetzlaff et al. 1991). Related to this, it has been shown that inhibition of Hsp27 interferes with the cytoskeleton and normal cellular growth (Mairesse et al. 1996). When regenerating axons reinnervate their targets, actin and tubulin levels return to normal. In such models of nerve injury and regeneration, Hsp27 may serve as a molecular chaperone, interacting with actin, regulating antioxidative activity, or protecting cells from apoptosis. Improved regeneration of axons and possibly neuronal survival is consistent with the functions of Hsp27 (Costigan et al. 1998). The long-term neuronal expression of Hsp27 after nerve injury (Hopkins et al. 1998) strengthens this idea that Hsp27 may have other injury-related or survival roles.

4
Expression of Hsp27 in Diseases of the Aging Brain

Because Hsp27 is expressed at increased levels following a variety of injuries and physiological challenges, it seems reasonable to expect that in diseases of the aging brain, such as stroke or Alzheimer's disease, Hsp27 would be expressed at increased levels because of the injury caused by the disease. While stroke is one of the most common afflictions of the human aging brain, little or no data is available on the expression of Hsp27 after stroke. However, as reviewed above, in animal models of stroke, Hsp27 is expressed at elevated levels mostly in neuroglia, which is where changes might be expected in the human brain.

Neurodegenerative diseases represent the other most common affliction of the aging brain. Hsp27 is expressed at elevated levels in the human brain afflicted with neurodegenerative diseases such as Alzheimer's disease (Renkawek et al. 1994), Parkinson's disease with dementia (Renkawek et al. 1999), and multiple sclerosis (Aquino et al. 1997). Primary tumors of the brain such as astrocytomas, meningiomas, and glioblastomas, also express elevated levels of Hsp27 (Kato et al. 1992, 1993; Khalid et al. 1995; Hitotsumatsu et al. 1996).

4.1
Neurodegenerative Diseases

The related small heat shock proteins, Hsp27 and αB-crystallin, have been detected in Alzheimer's diseased brain. Alzheimer's disease is a progressive neurodegenerative disorder that is characterized neuroanatomically by the presence of amyloid beta peptides that assemble into hyperphosphorylated tau neurofibrillary tangles that are deposited as plaques during neuritic degeneration, with neuron loss in the affected cortex (Bornemann and Straufenbiel 2000). Hsp27-immunoreactive degenerating neurons and αB-crystallin-immunoreactive astrocytes are more frequent in Alzheimer's disease brains (Shinohara et al. 1993). In addition, Hsp27 is found in plaques in Alzheimer's disease brains (Shinohara et al. 1993; Stege et al. 1999). The increased accumulations of Hsp27 and αB-crystallin appear to be part of reactive processes of glial cells and neurons under pathologic conditions (Shinohara et al. 1993). The expression of Hsp27 has also been detected in large numbers of reactive and degenerative glial cells, in areas rich in plaques and in neurofibrillary tangles (Renkawek et al. 1994). The expression of Hsp27 appears to increase with the severity of the Alzheimer's disease-related morphological changes and with the duration of the associated dementia. Thus, Renkawek et al. (1994) propose that the induction and increased expression of Hsp27 in astrocytes is a response to the cellular stress associated with sustained neurodegenerative processes. Reactive astrocytes may also play a role in the pathogenesis of Alzheimer's disease (Frederickson 1992).

There are suggestions that chronic oxidative stress may play a role in the development of Alzheimer's disease (Pappolla et al. 1996). Oxidative stress clearly increases the expression of antioxidant enzymes and heat shock proteins (Omar and Pappolla 1993), suggesting that the Hsps are markers of oxidative stress.

In a recent study, Hsp27 and αB-crystallin have been colocalized in the plaques of Alzheimer's diseased brains (Stege et al. 1999). Keeping in mind that the small Hsps may act as molecular chaperones that prevent aggregation of unfolded peptides, increased levels of Hsp27 and αB-crystallin may reflect a response to prevent amyloid fibril formation and toxicity. While αB-crystallin does prevent in vitro amyloid beta aggregation into fibrils, it actually increases the toxicity of amyloid beta. The interaction of αB-crystallin with amyloid beta may keep the latter in a nonfibrillar, highly toxic form (Stege et al. 1999).

Hsp27 is also expressed at increased levels in the brains of Parkinson's disease patients with dementia (Renkawek et al. 1999). Parkinson's disease is the selective degeneration of dopamine-producing neurons in the substantia nigra resulting in a progressive loss in the ability to initiate or sustain coordinated movement. Symptoms include tremor, muscle rigidity and bradykinesia (slowing of movement and a loss of spontaneous movement) (reviewed in Sian et al. 1999). In the later stages of Parkinson's disease, Alzheimer's disease-like dementia becomes a prominent feature. In the brains of these patients, there is reactive gliosis and increased expression of Hsp27 in the cortex similar to that observed in Alzheimer's disease. However, there is little or no detectable expression (above that of controls) of Hsp27 in substantia nigra neurons of Parkinson's disease patients without dementia (Renkawek et al. 1994, 1999). While Hsp27 has been shown to be induced in various cell types including neurons and glial cells following oxidative stress, in the early stages of

Parkinson's disease, this may not be the case even though degeneration of the dopaminergic substantia nigra neurons may be due to increased free radical formation in these cells (Han et al. 1999; Zhang et al. 1999). Renkawek et al. (1999) have proposed that the pathologies of dementia in Parkinson's disease and Alzheimer's disease are related although the function of Hsp27 in reactive astrocytes in dementia is not yet clear.

In multiple sclerosis, increased expression of Hsp27 has been found in and on the margins of sclerotic lesions (Aquino et al. 1997). Multiple sclerosis is a chronic, often progressive, inflammatory demyelinating disease of the central nervous system (Birnbaum 1995; van Noort 1996). Hsp27 is found in fibrous astrocytes in the sclerotic lesion region and in oligodendrocytes on the margin of the lesion region. Hsp27 was also detected in multiple sclerosis myelin but not in control myelin (Aquino et al. 1997). This led to the suggestion that the increased myelin-associated expression of Hsp27 (and other Hsps) may initially prevent destruction of the myelin sheath, but the prolonged expression of Hsps may exacerbate the disease. Other studies have implicated heat shock proteins in the pathology of the multiple sclerosis lesion (van Noort et al. 1995, 1998). One suggestion has been that an immune response to the Hsps of an infectious agent could result in a cross-reactive immune response to central nervous system myelin associated Hsps (reviewed in Birnbaum 1995). It is interesting to note that expression of αB-crystallin, another member of the small heat shock protein family, is elevated in oligo-dendrocytes and astrocytes in the region of multiple sclerosis lesions (van Noort et al. 1995). This protein has been shown to elicit an immune response from T cells isolated from multiple sclerosis and control patients, lending further support to the hypothesis that expression of Hsps may be involved in the progression of this inflammatory disease. For example, prolonged up-regulated expression of Hsps may provide additional targets for the immune response and may contribute to the progression of multiple sclerosis (Aquino et al. 1997).

In Alexander's disease Hsp27 has been detected in Rosenthal fibers. Alexander's disease is characterized by progressive psychomotor retardation and is frequently accompanied by seizures and a variable degree of mega-loencephaly (Borrett and Becker 1985). Histologically, there is a considerable degree of gliosis and lack of myelination, but the most significant diagnostic feature is the presence of Rosenthal fibers found associated with skeins of intermediate filaments within reactive and neoplastic astrocytes (Head et al. 1993; Iwaki et al. 1993). The presence of Rosenthal fibers in astroglia and the occurrence of these fibers in astrocytomas and glial scars have led to the hypothesis that Alexander's disease represents a dysfunction of astrocytes (Borrett and Becker 1985; Head et al. 1993). Hsp27 in brain tissue and various glioma cell lines raises the possibility that Hsp27 is involved in Rosenthal fiber formation (Head et al. 1993; Iwaki et al. 1993). Head et al. (1993) proposed that the conditions resulting in Rosenthal fiber formation are chronic and appear to be irreversible as opposed to more transitory stress such as heat shock. A possible explanation for Rosenthal fiber formation is the chronic over-expression of the small heat shock proteins, Hsp27 and αB-crystallin. Rosenthal fibers are present in large numbers in Alexander's disease but it is unclear as to whether these inclusions are beneficial to the astrocytes, or whether Rosenthal fibers promote the development of Alexander's disease. Further study is needed to determine the function or dysfunction of the small heat shock proteins in this fatal neurological disorder.

The possibility that the expression of Hsp27 is altered in Creutzfeldt-Jakob disease has not been addressed directly. Creutzfeldt-Jakob disease is a lethal prion disease in humans and is a member of the family of transmissible spongiform encephalophathies like scrapies in sheep. Neuropathological features include late onset dementia, in some cases ataxia, and spongiform degeneration of brain tissue and gliosis (reviewed in Prusiner and Hsiao 1994). Interestingly, Hsp27 and Hsp70 (both highly inducible under stress conditions) are not induced by heat shock in scrapie-infected mouse neuroblastoma cells. In these cells, the constitutively expressed Hsc70 does not translocate to the nucleus with heat shock and remains localized to various intensely labeled regions of the cytoplasm (Tatzelt et al. 1995, 1998). The role that this apparent unresponsiveness, after heat shock in scrapie infected cells, may play in prion disease pathology remains to be determined. Although no study has looked specifically at the expression and induction of Hsp27 in prion diseased brains, Hsps with chaperone function could be implicated in this lethal degenerative disease.

In conclusion, Hsp27 is found at increased levels mostly in reactive astrocytes or neuroglia in neurodegenerative diseases of the central nervous system, suggesting that the expression of this stress-induced protein is disease-related (Brzyska et al. 1998).

4.2
Brain Tumors

Primary and metastatic brain tumors have been shown to express various heat shock proteins, including Hsp27 (Kato et al. 1993). In general, Hsp27 has been shown to be expressed at higher levels in anaplastic and metastatic brain tumors and is relatively absent from low grade or well differentiated tumors (Hitotsumatsu et al. 1996). In astrocytomas, oligodendrogliomas, ependymomas and meningiomas the expression of Hsp27 and the grade of malignancy is correlated (Hitotsumatsu et al. 1996). Interestingly, high levels of Hsps in tumors provide resistance to tumor cells against the toxic effects of chemotherapeutic regimens. For example, malignant glioma cells with high levels of Hsps, including Hsp27, are resistant to both chemical and radiation-induced cell death (Hermisson et al. 2000). While Hsps generally appear to have a beneficial role enhancing cell function and survival, in cancer biology this effect is undesirable.

5
Conclusions

Hsp27 is a uniquely interesting heat shock protein. Several lines of evidence indicate that Hsp27 has fundamental roles in neural development, the support of sensory and motor neurons, and in cellular responses to physiological challenges or injury. In the nervous system Hsp27 is found normally in neurons that have a direct connection with the periphery. However, Hsp27 is found at varying levels in such neurons and is not found normally in cortical neurons. Hsp27 is generally and highly inducible in cortical astrocytes after seizure activity, ischemia, cortical spreading depression and fever-like temperatures. On the other hand, Hsp27 is specifically inducible in neurons after axotomy, suggesting that in the nervous system the expression of Hsp27 is cell-specific and stress- or injury-dependent. Functionally, Hsp27 may play several roles in the normal and injured nervous system, acting as a chaperone, as an inhibitor of apoptosis, promoting axon regeneration, and possibly as a regulator of fluid homeostasis. Hsp27 is also expressed in numerous degenerative diseases of the nervous system. While the expression of Hsp27 in neurodegenerative diseases is likely to be in response to the cellular injury, it is still an open question as to whether the chronic expression of Hsp27 is protective and buffers the effects of the disease or is detrimental and exacerbates the disease. Clearly much more research is required to delineate the roles of Hsp27 in the nervous system.

Acknowledgements. Anne Marie Krueger-Naug received a Faculty of Graduate Studies Studentship from Dalhousie University. Dr. Plumier was a graduate student with Dr. Currie and is currently a research associate with Dr. Michael A. Moskowitz at Harvard Medical School, Massachusetts General Hospital, Charlestown, MA 02129. Dr. Hopkins' research is funded by the Heart and Stroke Foundation of Nova Scotia. Dr. Currie's research is funded by the Heart and Stroke Foundation of New Brunswick and by the Canadian Stroke Network.

References

Anguelova E, Smirnova T (2000) Differential expression of small heat shock protein 27 in the rat hippocampus and septum after fimbria-fornix lesion. Neurosci Lett 280:99–102

Aquino DA, Padin C, Perez JM, Peng D, Lyman WD, Chiu FC (1996) Analysis of glial fibrillary acidic protein, neurofilament protein, actin and heat shock proteins in human fetal brain during the second trimester. Dev Brain Res 91:1–10

Aquino DA, Capello E, Weisstein J, Sanders V, Lopez C, Tourtellotte W, Brosnan CF, Raine CS, Norton WT (1997) Multiple sclerosis: Altered expression of 70- and 27-kDa heat shock proteins in lesion and myelin. J Neuropath Exp Neurol 56:664–672

Armstrong CL, Krueger-Naug AM, Currie RW, Hawkes R (2000) Constitutive expression of the 25-kDa heat shock protein Hsp25 reveals novel parasagittal bands of Purkinje cells in the adult mouse cerebellar cortex. J Comp Neurol 416:383–397

Armstrong CL, Krueger-Naug AM, Currie RW, Hawkes R (2001) Expression of heat-shock protein hsp25 in mouse Purkinje cells during development reveals novel features of cerebellar compartmentation. J Comp Neurol 429:7–21

Armstrong JN, Plumier JCL, Robertson HA, Currie RW (1996) The inducible 70,000 molecular/weight heat shock protein is expressed in the degenerating dentate hilus and piriform cortex after systemic administration of kainic acid in the rat. Neuroscience 74:685–693

Barbe MF, Tytell M, Gower DJ, Welch WJ (1988) Hyperthermia protects against light damage in the rat retina. Science 241:1817–1820

Barone FC, White RF, Spera PA, Ellison J, Currie RW, Wang X, Feuerstein GZ (1998) Ischemic preconditioning and brain tolerance: temporal histological and functional outcomes, protein synthesis requirement, and interleukin-1 receptor antagonist and early gene expression. Stroke 29:1937–1950

Bechtold DA, Brown IR (2000) Heat shock proteins Hsp27 and Hsp32 localize to synaptic sites in the rat cerebellum following hyperthermia. Mol Brain Res 75:309–320

Ben-Ari Y (1985) Limbic seizure and brain damage produced by kainic acid: mechanisms and relevance to human temporal lobe epilepsy. Neuroscience 14:375–403

Bignami A, Raju T, Dahl D (1982) Localization of vimentin, the nonspecific intermediate filament protein, in embryonal glia and in early differentiating neurons. In vivo and in vitro immunofluorescence study of the rat embryo with vimentin and neurofilament antisera. Dev Biol 91:286–295

Birnbaum G (1995) Stress proteins: their role in the normal central nervous system and in disease states, especially multiple sclerosis. Springer Semin Immunopathol 17:107–118

Blake MJ, Gershon D, Fargnole J, Holbrook NJ (1990) Discordant expression of heat shock protein mRNAs in tissues of heat-stressed rats. J Biol Chem 265:15275–15279

Bornemann KD, Staufenbiel M (2000) Transgenic mouse models of Alzheimer's disease. Ann N Y Acad Sci 908:260–266

Borrett D, Becker LE (1985) Alexander's disease. A disease of astrocytes. Brain 108:367–385

Brzyska M, Stege GJJ, Renkawek K, Bosman GJ (1998) Heat shock, but not the reactive state per se, induces increased expression of the small stress proteins hsp25 and αB-crystallin in glial cells in vitro. NeuroReport 9:1549–1552

Chopp M, Chen H, Ho KL, Dereski MO, Brown E, Hetzel FW, Welch KM (1989) Transient hyperthermia protects against subsequent forebrain ischemic cell damage in the rat. Neurology 39:1396–1398

Clark BD, Brown IR (1985) Axonal transport of a heat shock protein in the rabbit visual system. Proc Natl Acad Sci USA 82:1281–1285

Costigan M, Mannion RJ, Kendall G, Lewis SE, Campagna JA, Coggeshall RE, Meridith-Middleton J, Tate S, Woolf CJ (1998) Heat shock protein 27: developmental regulation and expression after peripheral nerve injury. J Neurosci 18:5891–5900

Currie RW, White FP (1981) Trauma-induced protein in rat tissues: a physiological role for a "heat shock" protein? Science 214:72–73

Currie RW, Ellison JA, White RF, Feuerstein GZ, Wang X, Barone FC (2000) Benign focal ischemic preconditioning induces neuronal Hsp 70 and prolonged astrogliosis with expression of Hsp27. Brain Res. 863:169–181

Desagher S, Glowinski J, Premont J (1996) Astrocytes protect neurons from hydrogen peroxide toxicity. J Neurosci 16:2553–2562

Dienel GA, Kiessling M, Jacewicz M, Pulsinelli WA (1986) Synthesis of heat shock proteins in rat brain cortex after transient ischemia. J Cereb Blood Flow Metab 6:505–510

Edwards MJ, Walsh DA, Li Z (1997) Hyperthermia, teratogenesis and the heat shock response in mammalian embryos in culture. Int J Dev Biol 41:345–358

Frederickson RC (1992) Astroglia in Alzheimer's disease. Neurobiol Aging 13:239–253

Gernold M, Knauf U, Gaestel M, Stahl J, Kloetzel PM (1993) Development and tissue-specific distribution of mouse small heat shock protein hsp25. Dev Genet 14:103–111

Gonzalez MF, Shiraishi K, Hisanaga K, Sagar SM, Mandabach M, Sharp FR (1989) Heat shock proteins as markers of neural injury. Mol Brain Res 6:93–100

Goto S, Korematsu K, Oyama T, Yamada K, Hamada J, Inoue N, Nagahiro S, Ushio Y (1993) Neuronal induction of 72-kDa heat shock protein following methamphetamine-induced hyperthermia in the mouse hippocampus. Brain Res 626:351–356

Han J, Cheng FC, Yang Z, Dryhurst G (1999) Inhibitors of mitochondrial respiration, iron (II), and hydroxyl radical evoke release and extracellular hydrolysis of glutathione in rat striatum and substantia nigra: Potential implications to Parkinson's disease. J Neurochem 73:1683–1695

Head MW, Corbin E, Goldman JE (1993) Overexpression and abnormal modification of the stress proteins αB-crystallin and Hsp27 in Alexander disease. Am J Pathol 143:1743–1753

Hermisson M, Strik H, Rieger J, Dichgans J, Meyermann R, Weller M (2000) Expression and functional activity of heat shock proteins in human glioblastoma multiforme. Neurology 54: 1357–1365

Hitotsumatsu T, Iwaki T, Fukui M, Tateishi J (1996) Distinctive immunohistochemical profiles of small heat shock proteins (heat shock protein 27 and alpha B-crystallin) in human brain tumors. Cancer 77:352–361

Hopkins DA, Plumier J-CL, Currie RW (1998) Induction of the 27-kDa heat shock protein (Hsp27) in the rat medulla oblongata after vagus nerve injury. Exp Neurol 153:173–183

Inaguma Y, Hasegawa K, Goto S, Ito H, Kato K (1995) Induction of the synthesis of hsp27 and αB crystalline in tissues of heat-stressed rats and its suppression by ethanol or an α_1- adrenergic antagonist. J Biochem 117:1238–1243

Iwaki T, Iwaki A, Tateishi J, Sakaki Y, Goldman JE (1993) αB-crystallin and 27-kd heat shock protein are regulated by stress conditions in the central nervous system and accumulate in Rosenthal fibers. Am J Pathol 143:487–495

Jacob JM, Croes SA (1998) Acceleration of axonal outgrowth in motor axons from mature and old F344 rats after a conditioning lesion. Exp Neurol 152:231–237

Jacob JM, McQuarrie IG (1993) Acceleration of axonal outgrowth in rat sciatic nerve at one week after axotomy. J Neurobiol 24:356–367

Jankovski A, Sotelo C (1996) Subventricular zone-olfactory bulb migratory pathway in the adult mouse: cellular composition and specificity as determined by heterochronic and heterotopic transplantation. J Comp Neurol 371:376–396

Jankovski A, Rossi F, Sotelo C (1996) Neuronal precursors in the postnatal mouse cerebellum are fully committed cells: evidence from heterochronic transplantations. Eur J Neurosci 8:2308–2319

Joyeux M, Ribuot C, Bourlier V, Verdetti J, Durand A, Richard MJ, Godin-Ribuot D, Demenge P (1997) In vitro antiarrhythmic effect of prior whole body hyperthermia: implication of catalase. J Mol Cell Cardiol 29:3285–3292

Karunanithi S, Barclay JW, Robertson RM, Brown IR, Atwood HL (1999) Neuroprotection at Drosophila synapses conferred by prior heat shock. J Neurosci 19:4360–4369

Kato H, Liu Y, Kogure K, Kato K (1994) Induction of 27-kDa heat shock protein following cerebral ischemia in a rat model of ischemic tolerance. Brain Res 634:235–244

Kato K, Katoh-Semba R, Takeuchi IK, Ito H, Kamei K (1999) Responses of heat shock proteins hsp27, alphaB-crystallin, and hsp70 in rat brain after kainic acid-induced seizure activity. J Neurochem 73:229–236

Kato S, Hirano A, Umahara T, Kato M, Herz F, Ohama E (1992) Comparative immunohistochemical study in the expression of alpha B crystalline, ubiquitin and stress-response protein 27 in ballooned neurons in various disorders. Neuropathol Appl Neurobiol 18:335–340

Kato S, Hirano A, Kato M, Herz F, Ohama E (1993) Comparative study on the expression of stress-response protein (srp) 72, srp 27, alpha B-crystallin and ubiquitin in brain tumours. An immunohistochemical investigation. Neuropathol Appl Neurobiol 19:436–442

Khalid H, Tsutsumi K, Yamashita H, Kishikawa M, Yasunaga A, Shibata S (1995) Expression of the small heat shock protein (hsp) 27 in human astrocytomas correlates with histologic grades and tumor growth fractions. Cell Mol Neurobiol 15:257–268

Kitagawa K, Matsumoto M, Kuwabara K, Tagaya M, Ohtsuki T, Hata R, Ueda H, Handa N, Kimura K, Kamada T (1991) 'Ischemic tolerance' phenomenon detected in various brain regions. Brain Res 561:203–211

Kitagawa K, Matsumoto M, Mabuchi T, Yagita Y, Mandai K, Matsushita K, Hori M, Yanagihara T (1997) Ischemic tolerance in hippocampal CA1 neurons studied using contralateral controls. Neuroscience 81:989–998

Klemenz R, Andres AC, Frohli E, Schafer R, Aoyama A (1993) Expression of the murine small heat shock proteins hsp 25 and alpha B crystallin in the absence of stress. J Cell Biol 120:639–645

Kobayashi S, Harris VA, Welsh FA (1995) Spreading depression induces tolerance of cortical neurons to ischemia in rat brain. J Cereb Blood Flow Metab 15:721–727

Krueger AM, Armstrong JN, Plumier JC, Robertson HA, Currie RW (1999) Cell specific expression of Hsp70 in neurons and glia of the rat hippocampus after hyperthermia and kainic acid-induced seizure activity. Mol Brain Res 71:265–278

Krueger-Naug AM, Hopkins DA, Armstrong JN, Plumier JC, Currie RW (2000) Hyperthermic induction of the 27-kDa heat shock protein (Hsp27) in neuroglia and neurons of the rat central nervous system. J Comp Neurol 428:495–510

Lankford KL, Waxman SG, Kocsis JD (1998) Mechanisms of enhancement of neurite regeneration in vitro following a conditioning sciatic nerve lesion. J Comp Neurol 391:11–29

Leao AAP (1944) Spreading depression of activity in the cerebral cortex. J Neurophysiol 7:359–390

Leger JP, Smith FM, Currie RW (2000) Confocal microscopic localization of constitutive and heat shock-induced proteins HSP70 and HSP27 in the rat heart. Circulation 102:1703–1709

Lewis SE, Mannion RJ, White FA, Coggeshall RE, Beggs S, Costigan M, Martin JL, Dillmann WH, Woolf CJ (1999) A role for HSP27 in sensory neuron survival. J Neurosci 19:8945–8953

Liang P, Amons R, Clegg JS, MacRae TH (1997) Molecular characterization of a small heat shock/alpha-crystallin protein in encysted Artemia embryos. J Biol Chem 272:19051–19058

Mairesse N, Horman S, Mosselmans R, Galand P (1996) Antisense inhibition of the 27 kDa heat shock protein production affects growth rate and cytoskeletal organization in MCF-7 cells. Cell Biol Int 20:205–212

Matsushima K, Hogan MJ, Hakim AM (1996) Cortical spreading depression protects against subsequent focal cerebral ischemia in rats. J Cereb Blood Flow Metab 16:221–226

Matsushima K, Schmidt-Kastner R, Hogan MJ, Hakim AM (1998) Cortical spreading depression activates trophic factor expression in neurons and astrocytes and protects against subsequent focal brain ischemia. Brain Res 807:47–60

Miller FD, Tetzlaff W, Bisby MA, Fawcett JW, Milner RJ (1989) Rapid induction of the major embryonic alpha-tubulin mRNA, T alpha 1, during nerve regeneration in adult rats. J Neurosci 9:1452–1463

Nowak TS Jr (1985) Synthesis of a stress protein following transient ischemia in the gerbil. J Neurochem 45:1635–1641

Olney JW, Rhee V, Ho OL (1974) Kainic acid: a powerful neurotoxic analogue of glutamate. Brain Res 77:507–512

Omar R, Pappolla M (1993) Oxygen free radicals as inducers of heat shock protein synthesis in cultured human neuroblastoma cells: Relevance to neurodegenerative diseases. Eur Arch Psychiatry Clin Neurosci 242:262–267

Pappolla MA, Sos M, Omar RA, Sambamurti K (1996) The heat shock/oxidative stress connection. Relevance to Alzheimer disease. Mol Chem Neuropathol 28:21–34

Plumier JC, Armstrong JN, Landry J, Babity JM, Robertson HA, Currie RW (1996) Expression of the 27,000 mol. wt heat shock protein following kainic acid-induced status epilepticus in the rat. Neuroscience 75:849–856

Plumier JC, Armstrong JN, Wood NI, Babity JM, Hamilton TC, Hunter AJ, Robertson HA, Currie RW (1997a) Differential expression of c-fos, Hsp70 and Hsp27 after photothrombotic injury in the rat brain. Mol Brain Res 45:239–246

Plumier JC, David JC, Robertson HA, Currie RW (1997b) Cortical application of potassium chloride induces the low-molecular weight heat shock protein (Hsp27) in astrocytes. J Cereb Blood Flow Metab 17:781–790

Plumier JC, Hopkins DA, Robertson HA, Currie RW (1997c) Constitutive expression of the 27-kDa heat shock protein (Hsp27) in sensory and motor neurons of the rat nervous system. J Comp Neurol 384:409–428

Prusiner SB, Hsiao KK (1994) Human prion diseases. Ann Neurol 35:385–395

Puisieux F, Deplanque D, Pu Q, Souil E, Bastide M, Bordet R (2000) Differential role of nitric oxide pathway and heat shock protein in preconditioning and lipopolysaccharide-induced brain ischemic tolerance. Eur J Pharmacol 389:71–78

Renkawek K, Bosman GJ, de Jong WW (1994) Expression of small heat-shock protein hsp27 in reactive gliosis in Alzheimer disease and other types of dementia. Acta Neuropathol 87: 511–519

Renkawek K, Stege GJ, Bosman GJ (1999) Dementia, gliosis and expression of the small heat shock proteins Hsp27 and αB-crystallin in Parkinson's disease. NeuroReport 10:2273–2276

Reynolds BA, Weiss S (1992) Generation of neurons and astrocytes from isolated cells of the adult mammalian central nervous system. Science 255:1707–1710

Samali A, Cotter TG (1996) Heat shock proteins increase resistance to apoptosis. Exp Cell Res 223:163–170

Shimazaki T, Arsenijevic Y, Ryan AK, Rosenfeld MG, Weiss S (1999) A role for the POU-III transcription factor Brn-4 in the regulation of striatal neuron precursor differentiation. EMBO J 18:444–456

Shinohara H, Inaguma Y, Goto S, Inagaki T, Kato K (1993) Alpha B crystallin and HSP28 are enhanced in the cerebral cortex of patients with Alzheimer's disease. J Neurol Sci 119:203–208

Sian J, Gerlach M, Youdim MB, Riederer P (1999) Parkinson's disease: a major hypokinetic basal ganglia disorder. J Neural Transm 106:443–476

Sloviter RS, Lowenstein DH (1992) Heat shock protein expression in vulnerable cells of the rat hippocampus as an indicator of excitation-induced neuronal stress. J Neurosci 12:3004–3009

Steare SE, Yellon DM (1993) The protective effect of heat stress against reperfusion arrhythmias in the rat. J Mol Cell Cardiol 25:1471–1481

Stege GJ, Renkawek K, Overkamp PS, Verschuure P, van Rijk AF, Reijnen-Aalbers A, Boelens WC, Bosman GJ, de Jong WW (1999) The molecular chaperone alphaB-crystallin enhances amyloid beta neurotoxicity. Biochem Biophys Res Commun 262:152–156

Tanguay RM, Wu Y, Khandjian EW (1993) Tissue-specific expression of heat shock proteins of the mouse in the absence of stress. Dev Gen 14:112–118

Tatzelt J, Zuo J, Voellmy R, Scott M, Hartl U, Prusiner SB, Welch WJ (1995) Scrapie prions selectively modify the stress response in neuroblastoma cells. Proc Natl Acad Sci USA 92:2944–2948

Tatzelt J, Voellmy R, Welch WJ (1998) Abnormalities in stress proteins in prion diseases. Cell Mol Neurobiol 18:721–729

Tetzlaff W, Alexander SW, Miller FD, Bisby MA (1991) Response of facial and rubrospinal neurons to axotomy: changes in mRNA expression for cytoskeletal proteins and GAP-43. J Neurosci 11:2528–2544

Van Noort JM (1996) Multiple sclerosis: An altered immune response or an altered stress response? J Mol Med 74:285–296

Van Noort JM, van Sechel AC, Bajramovic JJ, el Ouagmiri M, Polman CH, Lassmann H, Ravid R (1995) The small heat-shock protein alpha B-crystallin as candidate autoantigen in multiple sclerosis. Nature 375:798–801

Van Noort JM, van Sechel AC, van Stipdonk MJ, Bajramovic JJ (1998) The small heat shock protein alpha B-crystallin as key autoantigen in multiple sclerosis. Prog Brain Res 117:435–452

Vescovi AL, Reynolds BA, Fraser DD, Weiss S (1993) bFGF regulates the proliferative fate of unipotent (neuronal) and bipotent (neuronal/astroglial) EGF-generated CNS progenitor cells. Neuron 11:951–966

Xu L, Lee JE, Giffard RG (1999) Overexpression of bcl-2, bcl-XL or hsp70 in murine cortical astrocytes reduces injury of co-cultured neurons. Neurosci Lett 277:193–197

Zerlin M, Levison SW, Goldman JE (1995) Early patterns of migration, morphogenesis, and intermediate filament expression of subventricular zone cells in the postnatal rat forebrain. J Neurosci 15:7238–7249

Zhang J, Perry G, Smith MA, Robertson D, Olson SJ, Graham DG, Montine TJ (1999) Parkinson's disease is associated with oxidative damage to cytoplasmic DNA and RNA in substantia nigra neurons. Am J Pathol 154:1423–1429

Protection of Neuronal and Cardiac Cells by HSP27

David S. Latchman[1]

1
Protective Effect of the Heat Shock Proteins

One of the aspects of heat shock protein biology, which has attracted the most attention, is their protective effect when over-expressed (for reviews see Latchman 1991; Yellon and Latchman 1992; Parsell and Lindquist 1993; Benjamin and McMillan 1998; Gray, et al. 1999). As well as being of interest in itself, this phenomenon is evidently of importance in medical terms, since it may be possible to achieve a therapeutic benefit in diseases such as stroke or cardiac ischaemia by inducing HSP over-expression either pharmacologically or using gene delivery procedures.

The initial evidence that heat shock proteins might have a protective effect came from a wide variety of studies, which showed that an initial mild stress, sufficient to induce HSP expression, produced a protective effect against subsequent more severe stress. This effect has now been demonstrated in a very wide variety of different cell types both when cultured in vitro and within the intact animal in vivo. Thus, for example, primary neuronal cells cultured in vitro are protected by prior exposure to mild heat or ischaemic stress from a subsequent more severe heat or ischaemic stress, or exposure to the excitotoxin glutamate (Lowenstein et al. 1991; Rordorf et al. 1991; Amin et al. 1995). Similar effects have also been demonstrated, for example, in cultured cardiac cells exposed to HSP-inducing stimuli before exposure to a more severe stress (Cumming et al. 1996b). In these experiments, a specific mild stress such as a heat shock could protect against several different subsequent severe stresses such as thermal or ischaemic stress, and the protective effect correlated with the amount of heat shock protein produced, suggesting that it was HSP synthesis rather than any other effect of the mild stress which was protective (see for example, Amin et al. 1995).

These experiments with cultured cells in vitro are paralleled by experiments using either isolated organs or the intact animal. Thus, Marber et al. (1994), showed a correlation between the amount of the 72-kDa heat shock protein (HSP70) produced by heat stress of papillary muscle and the muscle's ability to recover function following a period of hypoxia, whereas Hutter et al. (1994)

[1] Institute of Child Health, University College London, 30 Guilford Street, London WC1N 1EH, UK

Progress in Molecular and Subcellular Biology, Vol. 28
A.-P. Arrigo and W.E.G. Müller (Eds.)
© Springer-Verlag Berlin Heidelberg 2002

demonstrated a similar correlation between the amount of HSP70 and the ability to limit infarct size following exposure of the heart to ischaemia and subsequent reperfusion. These studies in the heart are paralleled by similar studies in the brain where it has been shown that a mild HSP-inducing stress can protect neuronal cells in the intact brain in vivo against subsequent exposure to light (Barbe et al. 1988), or ischaemia (Chopp et al. 1989; Kitagawa et al. 1990).

These experiments suggested therefore that HSP over-expression induced by a mild stress was able to protect cells both in vitro and in vivo against a subsequent more severe stress. Evidently however, these experiments produced only indirect evidence in favour of this hypothesis and did not identify the HSP(s) involved. A number of subsequent studies therefore involved over-expressing individual HSP(s) in cultured cells or in vivo and assessed their protective effect.

The majority of these studies concentrated on HSP70 since it is a major HSP, which is absent in unstressed cells and is induced to high levels following stress. In general, these experiments demonstrated that HSP70 could have a protective effect against different stresses. This was demonstrated in experiments in which HSP70 was over-expressed using a plasmid construct and the transfected cells exposed to a subsequent stress. Thus both the ND7 immortalised neuronal cell line and primary cultures of dorsal root ganglion neurones can be protected against exposure to subsequent thermal or ischaemic stress by over-expression of HSP70 (Uney et al. 1993; Mailhos et al. 1994; Amin et al. 1996; Wyatt et al. 1996). Moreover, a similar protective effect against both thermal and ischaemic stress could also be observed by over-expressing the 90-kDa HSP (HSP90) in the ND7 cell line and primary neurones (Mailhos et al. 1994; Wyatt et al. 1996).

Similar experiments have also been carried out in cell lines of cardiac origin or primary cultures of cardiac cells. In this case, as in neuronal cells, over-expression of HSP70 was able to protect the cells against subsequent thermal or ischaemic stress (Mestril et al. 1994; Heads et al. 1995; Cumming et al. 1996a). Interestingly however, over-expression of HSP90 produced a protective effect only against thermal stress in these experiments and the 60kDa heat shock protein (HSP60) had no protective effect (Heads et al. 1995; Cumming et al. 1996a). These in vitro experiments therefore confirm that over-expression of individual HSPs can have a protective effect against a subsequent stress, although they raise the interesting possibility that different HSPs can have different protective abilities against different stresses.

These in vitro experiments on the protective effect of the HSPs have been supplemented more recently by studies in the intact animal. Thus, several groups have produced transgenic mice over-expressing HSP70 and have shown that these mice are protected from the damaging effects of cardiac ischaemia, relative to control mice (Marber et al. 1995; Plumier et al. 1995; Radford et al. 1996). Similarly, directed delivery of the HSP70 gene by intra-coronary perfusion of the gene packaged into liposomal particles has been shown to protect the heart from endotoxin-induced myocardial disfunction (Meldrum et al. 1999) whilst Suzuki et al. (1997) observed a protective effect against the damaging effects of ischaemia/reperfusion when the hsp70 gene was delivered to the heart using liposomes continuing inactivated haemaglutinating virus of Japan.

These in vivo studies in the heart are paralleled also in the brain. Thus, a similar but less dramatic protective effect to that observed in the heart is also observed when transgenic mice over-expressing HSP70 are exposed to cerebral ischaemia (Plumier et al. 1997). Similarly, a defective herpes simplex virus vector over-expressing HSP70 produces a protective effect in rat models of transient focal cerebral ischaemia and excitotoxin-induced seizures (Yenari et al. 1998).

These in vitro and in vivo studies therefore clearly establish that over-expression of HSP70 and in some instances HSP90 can have a protective effect against subsequent stress. However, they do indicate that different HSPs can have different abilities to protect cells against individual stresses. This is supported by the work of Fink et al. (1997), who were able to protect cultured hippocampal neurones against subsequent heat shock by over-expressing HSP70 but did not observe any protective effect against glutamate toxicity. This represents a paradox since heat shock itself is known to protect such cells against the toxic effects of glutamate (Lowenstein et al. 1991; Rordorf et al. 1991). A similar paradox, in which heat shock itself has a protective effect whereas over-expression of HSP70 does not, has also been observed in other neuronal cell types. Thus, in both ND7 neuronal cells and primary neurones, a prior heat shock is highly effective in conferring protection against the induction of programmed cell death (apoptosis) by withdrawal of serum or nerve growth factor (Mailhos et al. 1993, 1994). However, over-expression of either HSP70 or HSP90 has no protective effect in this situation even though the same cells are protected against thermal or ischaemic stress (Mailhos et al. 1994; Wyatt et al. 1996). A similar lack of protection against apoptosis is also observed when HSP90 was over-expressed in these cells.

These studies suggest that whilst heat shock proteins have a clear protective effect in a number of different situations, the focus of attention on HSP70 needs to be reassessed. Thus, although HSP70 protects in many situations, the different abilities of different HSPs to protect against different stresses suggest that other HSPs may be of particular importance in specific situations. These studies therefore focus attention on to the protective effect of other HSPs and, in particular, that of HSP27.

2
Protective Effect of HSP27

Until recently, the great majority of protective studies on the HSPs have concentrated on HSP70 and neglected HSP27. As well as reflecting the very strong induction of HSP70 by heat shock, this neglect reflected the finding that HSP27 becomes multiply phosphorylated on serine residues following exposure to stressful stimuli such as heat shock. This may well have led investigators to consider that post-translational modification of HSP27 might be necessary for its protective effect and that therefore over-expression of unmodified HSP27 would not produce a protective effect.

Nonetheless, more recently, a number of studies have indeed demonstrated a protective effect of HSP27 in a wide variety of situations. Thus, for example, it has been shown that over-expression of HSP27 can protect cells against heat shock itself (Carper et al. 1997), treatment with anti-cancer agents (Huot et al. 1991), and cell death induced by oxidative stress or treatment with tumour necrosis factor α (Mehlen et al. 1996a; Rogalla et al. 1999). Most importantly, HSP27 acts by reducing the level of programmed cell death (apoptosis) induced by particular agents. Thus, the extent of apoptosis is reduced by over-expression of HSP27 in monocytic cells exposed to DNA damaging agents such as camtothecin and etoposide (Samali and Cotter et al. 1996) and in fibrosarcoma cells exposed to Fas-induced apoptosis or to staurosporine (Mehlen et al. 1996b).

These results therefore do indicate that over-expression of HSP27 can have a protective effect against a variety of different stimuli and are of particular interest since they may indicate a key role for this protein in protecting cells from apoptosis. Interestingly however, it is still unclear what role the phosphorylation of HSP27 plays in its protective effect. Thus, initial studies suggested that an artificial mutant of HSP27 in which the three serines targeted for phosphorylation had been replaced with non-phosphorylatable glycine residues, was unable to protect cells against heat shock (Lavoie et al. 1995; Huot et al. 1996). In contrast however, another study indicated that non-phosphorylatable HSP25 in which the serine residues have been replaced with non-phosphorylatable alanine residues showed no difference in its protective effect compared to wild-type HSP25 (Knauf et al. 1994). Moreover, more recent studies, in which the serine residues have been replaced with charged aspartate residues mimicking constitutive phosphorylation, have suggested that phosphorylation prevents HSP27 from conferring resistance against oxidative stress, or tumour necrosis factor α (Preville et al. 1998; Rogalla et al. 1999). This suggests that the failure of HSP27, containing glycine rather than serine residues, to protect cells may represent an abnormal structure of this protein rather than being due to the inability to be phosphorylated.

Indeed, whatever the precise role of its phosphorylation, the available data certainly do suggest that in several different cell types over-expression of unmodified HSP27 can have a protective effect. This therefore indicated that it was appropriate to investigate whether such a protective effect could also be observed in neuronal and cardiac cells.

3
Protective Effect of HSP27 in Neuronal and Cardiac Cells

3.1
Neuronal Cells

As noted above, we have previously observed that a mild heat shock can protect neuronal cells from stimuli which would otherwise induce apoptosis (Mailhos et al. 1993), whereas over-expression of HSP70 alone does not produce this effect (Mailhos et al. 1994; Wyatt et al. 1996). In view of the protective effect of HSP27 against apoptosis in other cell types described above, we wished to investigate whether HSP27 could protect neuronal cells against apoptosis. To do this, we constructed a herpes simplex virus (HSV)-based vector capable of over-expressing HSP27 in neuronal cells with high efficiency (for review of HSV vectors see Latchman 1999; Latchman and Coffin 2000). Similar HSV vectors capable of over-expressing HSP70 or HSP56 were also constructed for comparison.

In these experiments (Wagstaff et al. 1999), over-expression of either HSP70 or HSP27 was able to protect the immortalised ND7 neuronal cell line or dorsal root ganglion neurones from subsequent exposure to heat shock or severe ischaemia, compared to control cells infected with an HSV vector expressing only the marker β-galactosidase protein. In contrast, over-expression of HSP56 using a virus vector had no protective effect against either stimulus. These results agree with our previous data demonstrating a protective effect for HSP70 in these cells (Mailhos et al. 1994; Wyatt et al. 1996), but also extend them to HSP27 and show that it can have a similar protective effect to HSP70 against thermal or ischaemic stress.

Most importantly, in assays of apoptosis where HSP70 had previously been shown to have no protective effect (Mailhos et al. 1994; Wyatt et al. 1996), over-expression of HSP27 had a clear protective effect. This effect was observed in ND7 cells, both in assays of total cell death following serum withdrawal and when apoptosis was specifically measured using a TUNEL assay (Fig. 1). Similarly, when apoptosis was induced in primary dorsal root ganglion neurones by withdrawal of nerve growth factor, cells infected with the HSP27 virus showed a clear reduction of apoptosis (Fig. 2) compared to cells infected with the control vector or left uninfected. As expected, HSP70 had no protective effect against apoptosis in ND7 cells or primary neurones in accordance with our previous results (Mailhos et al. 1994; Wyatt et al. 1996).

These experiments thus show that HSP27 has a protective effect against specific apoptotic stimuli in neuronal cells whereas this effect is not observed by over-expressing HSP70. They therefore indicate that the protective effect of a prior heat shock against apoptosis in neuronal cells, which we previously observed (Mailhos et al. 1993, 1994), is likely to be dependent, at least in part, on the enhanced synthesis of HSP27 induced by heat shock. Clearly, it would

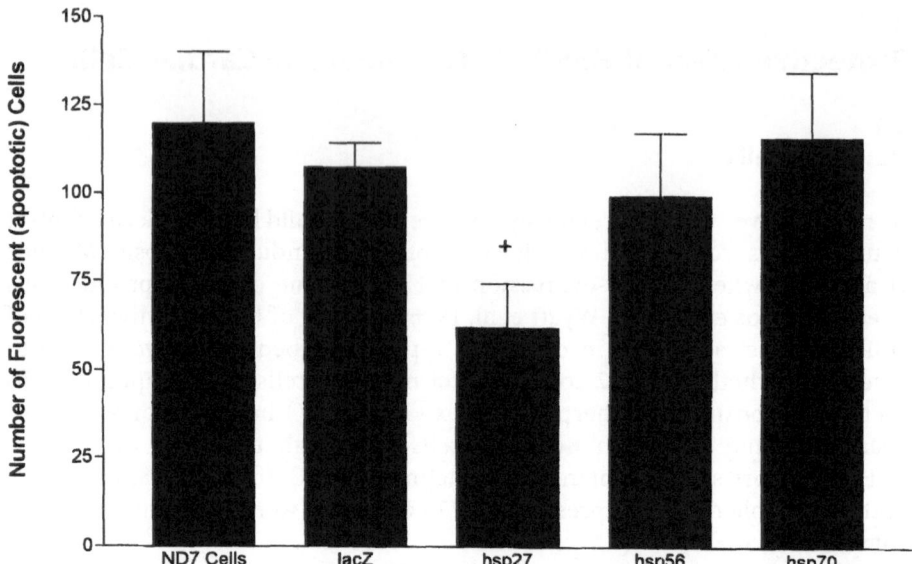

Fig. 1. Number of ND7 cells undergoing programmed cell death following 48 h of serum withdrawal in the presence of all-TRANS-retinoic acid as assayed by the number of TUNEL-labelled fluorescent cells. The cells were pre-treated by mock infection or by infection with *LacZ*, *HSP27*, *HSP56* and *HSP70*-expressing viruses. Values are the mean of four experiments whose standard error is shown by the *bar*. + indicates a significant difference in survival compared with Lac Z virus-infected ND7 cells ($p<0.05$)

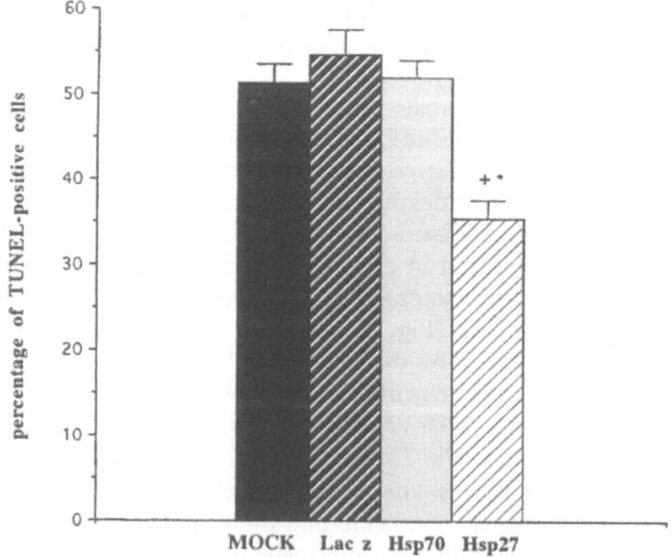

Fig. 2. Number of DRG neurones undergoing apoptosis after NGF withdrawal following prior infection with the indicated virus. Values are the mean of three determinations whose standard error is shown by the *bars*. Significant enhancement of survival ($p<0.05$) was observed only with HSP27-expressing virus

also be of interest to determine whether over-expression of HSP27 protects neuronal cells against glutamate toxicity, in order to determine whether the protective effect of heat shock against such toxicity (Lowenstein et al. 1991; Rordorf et al. 1991), which cannot be reproduced by over-expressing HSP70 (Fink et al. 1997), is also due to the over-expression of HSP27.

3.2
Cardiac Cells

As in neuronal cells, a number of studies have demonstrated that stimuli which protect either cultured cardiac cells or the intact heart from the effects of subsequent stress result in the enhanced synthesis of heat shock proteins and, in particular, these stimuli have been shown to induce HSP27 as well as HSP70 and other shock proteins (see for example, Das et al. 1993; Maulik et al. 1995).

It is therefore of interest to investigate whether over-expression of HSP27 can protect cardiac cells against different stressful stimuli and against apoptosis. Martin et al. (1997) prepared adenovirus vectors over-expressing HSP27 or the related protein αB-crystallin. They were able to demonstrate that both these proteins were able to protect cardiac myocytes from the effects of simulated ischaemia and that decreasing the level of endogenous HSP27 using an anti-sense approach enhanced the damaging effect of a subsequent ischaemic stimulus. These data therefore indicate that the endogenous level of HSP27 plays a key role in protecting cardiac cells from the effects of ischaemia. The damaging effect can be enhanced by stimuli, which reduced HSP27 levels and minimised by stimuli, which increase it. Interestingly, the protective effect of HSP27 in cardiac cells was not affected by mutation of the phosphorylatable serine residues (see above) to either glycine or alanine indicating that phosphorylation is unlikely to be required for the protective effect (Martin et al. 1999). In another study (Bluhm et al. 1998), the same group was able to demonstrate that the protective effect of αB-crystallin in cardiac cells is associated with its ability to protect microtubules from the effects of simulated ischaemia, whereas the protective effect of HSP27 is not associated with enhanced microtubule integrity in cells exposed to simulated ischaemia.

Although these studies demonstrated a clear protective effect against death induced by ischaemia in cardiac cells, they utilised assays based on the release of soluble enzymes such as lactate dehydrogenase and creatine phosphokinase, which measure total cell death. To extend these results to investigate the ability of HSP27 to protect cardiac cells against apoptosis, we used our various HSP-expressing HSV vectors. When these vectors were used to infect cultures of neonatal cardiac cells, we observed that both HSP27 and HSP70 were able to protect the cells against a subsequent heat shock or ischaemia compared to cells infected with control virus, whereas over-expression of HSP56 had no effect (Brar et al. 1999). This effect was observed both when the cells were assayed for total cell death on the basis of their ability to exclude trypan blue, or in assays of apoptosis such as TUNEL labelling. These findings suggested

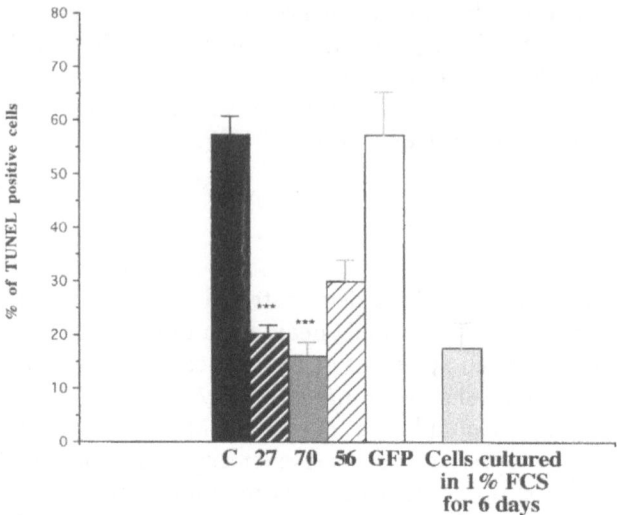

Fig. 3. Percentage of primary cardiac cells undergoing apoptosis following mock infection (*C*) or infection with viruses expressing HSP27, HSP70, HSP56 or green fluorescent protein (*GFP*). Following viral infection, cells were subjected to serum removal and assayed 24h later for the number of TUNEL-positive cells. Values are the mean of eight determinations whose standard error is shown by the *bars*. Both HSP27 and HSP70 expressing-viruses protected the cardiac cells from apoptosis (*** *p*<0.001)

therefore that in cardiac cells, over-expression of HSP27 and HSP70 not only protect cells again thermal or ischaemic stress but also act, at least in part, by minimising the extent of apoptosis induced by these stimuli.

The protective effect of both these proteins against apoptosis in cardiac cells was confirmed by investigating their effect against stimuli which specifically induce apoptosis in primary cardiac cells, such as serum removal or treatment with ceramide. In these experiments (Fig. 3) a clear protective effect of HSP27 and HSP70 against serum removal was observed using a TUNEL assay for apoptosis. Moreover, a strong protective effect of HSP27 and HSP70 was also observed when apoptosis was induced in the cardiac cells by exposure to ceramide and apoptosis was assayed by an alternative method, surface-staining using Annexin V (Fig. 4).

These results therefore indicate that HSP27 can protect cardiac cells against different apoptotic stimuli with these results being confirmed using assays which measure two distinct aspects of apoptosis, namely, DNA fragmentation (TUNEL labelling) and translocation of phosphotidyl-serine to the outer surface of the plasma membrane (Annexin V). When taken together with our results in neuronal cells, they suggest that HSP27 has a very broad protective effect against a variety of stressful and apoptotic stimuli. Interestingly, in our experiments we also observed a protective effect of HSP70 against apoptosis in cardiac cells (Figs. 3, 4). This is in accordance with a variety of other studies

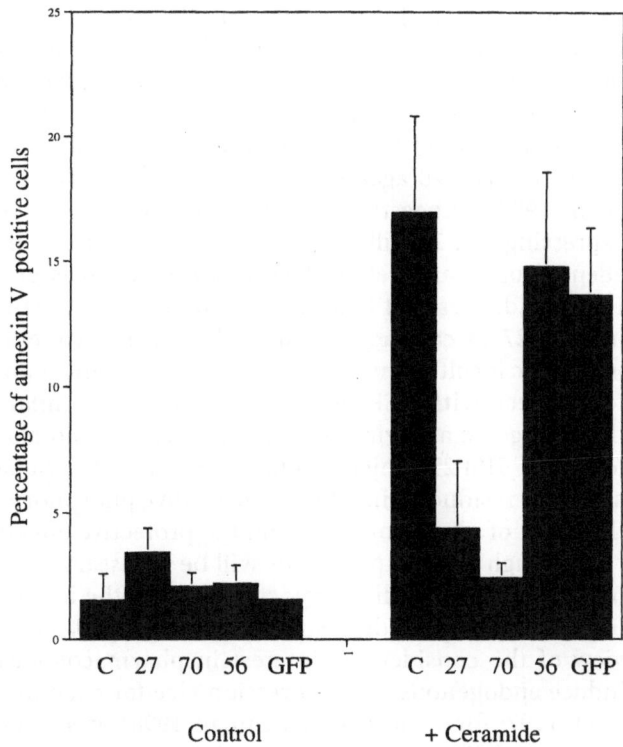

Fig. 4. Percentage of Annexin V-stained apoptotic cells in cardio-myocytes pre-infected with the indicated viruses or left untreated (*C*) and then 24 h after infection, either left untreated or treated for 6 h with 25 μM *ceramide*. The data represent the mean of two independent experiments whose standard error is indicated by the *bars*. Both HSP27 and HSP70-expressing viruses significantly reduce the number of apoptotic cells (*p*<0.05)

in non-neuronal cells which similarly demonstrate a protective effect of HSP70 against apoptotic stimuli (Dix et al. 1996; Li et al. 1996; Samali and Cotter 1996; Mosser et al. 1997). It is likely therefore that the protective effect of HSP70 against apoptosis is dependent upon the nature of the cell type and of the apoptotic stimuli with its protective effect in neuronal cells being less broad than in a variety of other cell types.

4
Conclusion

Although the great majority of studies on the protective effect of the heat shock proteins have concentrated on over-expression of HSP70, it is clear that HSP27 has a protective effect which is as potent as HSP70 in a variety of situations and indeed, has a clear protective effect in some situations where HSP70 is inactive.

Unlike HSP70, however, the studies with HSP27 have not so far been extended to the in vivo situation. It is therefore of considerable importance to prepare transgenic animals over-expressing HSP27 and to determine whether this produces a protective effect against cardiac or cerebral ischaemia. In particular, the relatively weak protective effect observed in transgenic mice over-expressing HSP70 against cerebral as opposed to cardiac ischaemia (Plumier et al. 1997), suggests that the effect of cerebral ischaemia in mice over-expressing HSP27 will be of considerable interest. Similarly, the availability of adenovirus (Martin et al. 1997) and HSV (Brar et al. 1999) vectors expressing HSP27 indicates that it should be possible to investigate the effect of delivering HSP27 to cardiac or neuronal cells in vivo, either prior to, or after an ischaemic insult to investigate the protective effect observed and compare this to that seen with HSP70 when delivered in a similar manner. Evidently, both the transgenic and viral vector approaches could also be extended to mutant forms of HSP27 which contain non-phosphorylatable alanine residues or aspartate residues mimicking constitutive phosphorylation, so as to determine the effect of phosphorylation on the protective effects of HSP27 in vivo.

Although these experiments will be necessary to confirm the role of HSP27 in neuronal and cardiac protection in vivo, the in vitro experiments described in this chapter provide clear evidence for a protective effect of this protein. In view of the considerable interest in pharmacological treatments which can induce endogenous HSP expression (see for example, Morris et al. 1996; Vigh et al. 1997; for review see Latchman 1998), it will be important to test such drugs for their ability to induce HSP27 rather than focusing on HSP70 alone. Indeed, it is likely that, at least in the brain, pharmacological treatments which can induce the endogenous HSP27 gene, or gene therapy treatments, which can deliver an exogenous HSP27 gene, may have a greater chance of producing benefit than procedures which induce or over-express HSP70.

Acknowledgements. I am most grateful to members of my laboratory and in particular, Dr B. Brar, Dr A. Stephanou and Dr M. Wagstaff for their contribution to our work described in this chapter and for helpful discussions. The work from my laboratory described in this chapter was supported by the British Heart Foundation and the Sir Jules Thorn Charitable Trust.

References

Amin V, Cumming DVE, Coffin RS, Latchman DS (1995) The degree of protection provided to neuronal cells by a pre-conditioning stress correlates with the amount of heat shock protein it induces and not with the similarity of the subsequent stress. Neurosci Lett 200:85–88

Amin V, Cumming DVE, Latchman DS (1996) Over-expression of heat shock protein 70 protects neuronal cells against both thermal and ischaemic stress but with different efficiencies. Neurosci Lett 206:45–48

Barbe M, Tytell M, Gower D, Welch W (1988) Hyperthermia protects against light damage in the rat retina. Science 241:1817–1820

Benjamin IJ, McMillan DR (1998) Stress (Heat Shock) proteins. Molecular chaperones in cardiovascular biology and disease. Circ Res 83:117–132

Bluhm WF, Martin JL, Mestril R, Dillmann WH (1998) Specific heat shock proteins protect microtubules during simulated ischemia in cardiac myocytes. Am J Physiol 275:H2243–H2249

Brar BS, Stephanou A, Wagstaff MJD, Coffin RS, Marber MS, Engelman G, Latchman DS (1999) Heat shock proteins delivered with a virus vector can protect cardiac cells against apoptosis as well as against thermal or ischaemic stress. J Mol Cell Cardiol 31:135–146

Carper SW, Rocheleau TA, Cimino D, Kristian Storm F (1997) Heat shock protein 27 stimulates recovery of RNA and protein synthesis following a heat shock. J Cell Biochem 66:153–164

ChoppM, Chen H, Ho K, Dereski MO, Brown E, Hetzel FW, Welch KM (1989) Transient hyperthermia protects against subsequent forebrain ischemic cell damage in the rat. Neurology 39:1396–1398

Cumming DVE, Heads RJ, Watson A, Latchman DS, Yellon DM (1996a) Differential protection of primary rat cardiocytes by transfection of specific heat stress proteins. J Mol Cell Cardiol 28:2343–2349

Cumming DVE, Heads RJ, Brand NJ, Yellon DM, Latchman DS (1996b) The ability of heat stress and metabolic preconditioning to protect primary rat cardiac myocytes. Basic Res Cardiol 9:79–85

Das DK, Engelman RM, Kimura Y (1993) Molecular adaptation of cellular defences following preconditioning of the heart by repeated ischaemia. Cardiovasc Res 27:578–584

Dix DJ, Allen JW, Collins BW, Mori C, Nakamura N, Poorman-Allen P, Goulding EH, Eddy EM (1996) Targeted gene disruption of hsp70-2 results in failed meiosis germ cell apoptosis, and male infertility. Proc Natl Acad Sci USA 93:3264–3268

Fink SL, Chang LK, Ho DY, Sapolsky RM (1997) Defective herpes simplex virus vectors expressing the rat brain stress-inducible heat shock protein 72 protect cultured neurons from severe heat shock. J Neurochem 68:961–969

Gray CC, Amrani M, Yacoub MH (1999) Heat stress proteins and myocardial protection: experimental model or potential clinical tool? Int J Biochem Cell Biol 31:559–573

Heads RJ, Latchman DS, Yellon DM (1995)Differential stress protein mRNA expression during early ischaemic preconditioning in the rabbit heart and its relationship to adenosine receptor function. J Mol Cell Cardiol 27:2133–2148

Heads RJ, Yellon DM, Latchman DS (1995) Differential cytoprotection against heat stress or hypoxia following expression of specific stress protein goes in myogenic cells. J Mol Cell Cardiol 27:1669–1678

Huot J, Roy G, Lambert H, Chrétien P, Landry J (1991) Increased survival after treatments with anticancer agents of Chinese hamster cells expressing the Human M_r 27,000 heat shock protein. Cancer Res 51:5245–5252

Huot J, Houle F, Spitz DR, Landry J (1996) HSP27 phosphorylation-mediated resistance against actin fragmentation and cell death induced by oxidative stress. Cancer Res 56:273–279

Hutter MM, Sievers RE, Barbosa V, Wolfe C (1994) Heat shock protein induction in rat hearts. A direct correlation between the amount of heat shock protein induced and the degree of myocardial protection. Circulation 89:353–360

Kitagawa K, Matsumoto M, Tagaya M, Hata R, Keda H, Ninobe M, Handa N, Fukunaga R, Kimura K, Mikshiba K, Kamada T (1990) "Ischemic tolerance" phenomenon found in the brain. Brain Res 528:21–24

Knauf U, Jakob U, Engel K, Buchner J, Gaestel M (1994) Stress- and mitogen-induced phosphorylation of the small heat shock protein Hsp25 by MAPKAP kinase 2 is not essential for chaperone properties and cellular thermoresistance. EMBO J 13:54–60

Latchman DS (1991) Heat shock proteins and human disease. J Royal College of Physicians and Surgeons 25:295–300

Latchman DS (1998) Heat shock proteins: protective effect and potential therapeutic use. Int J Mol Med 2:375–381

Latchman DS (1999) Herpes simplex virus vectors for gene therapy in Parkinson's disease and other diseases of the nervous system. J R Soc of Med 92:566–570

Latchman DS, Coffin RS (2000) Viral vectors and Parkinson's disease. Movement Disorders 15:9–17

Lavoie JN, Lambert H, Hickey E, Weber LA, Landry J (1995) Modulation of cellular thermoresistance and actin filament stability accompanies phosphorylation-induced changes in the oligomeric structure of heat shock protein 27. Mol Cell Biol 15:505–516

Li WX, Chen CH, Ling CC, Li GC (1996) Apoptosis in heat induced cell killing: the protective role of hsp70 and the sensitization effect of the c-myc gene. Radiat Res 145:324–330

Lowenstein DH, Chan PH, Miles MF (1991) The stress protein response in cultured neurons: characterization and evidence for a protective role in excitotoxicity. Neuron 7:1053–1060

Mailhos C, Howard MK, Latchman DS (1993) Heat shock protects neuronal cells from programmed cell death by apoptosis. Neuroscience 55:621–627

Mailhos C, Howard MK, Latchman DS (1994) Heat shock proteins hsp90 and hsp70 protect neuronal cells from thermal stress but not from programmed cell death. J Neurochem 63:1787–1795

Marber MS, Walker JM, Latchman DS, Yellon DM (1994) Myocardial protection following whole body heat stress in the rabbit is dependent on metabolic substrate and is related to the amount of the inducible 70 kb Dalton heat shock protein. J Clin Invest 93:1087–1094

Marber MS, Mestril R, Chi SH, Sayen MR (1995) Overexpression of the rat inducible 70 kDa heat shock protein in a transgenic mouse increases the resistance of the heart to ischemic injury. J Clin Invest 95:1446–1456

Martin JL, Mestril R, Hilal-Dandan R, Brunto LL, Dilmann WH (1997) Small heat shock proteins and protection against ischaemic injury in cardiac myocytes. Circulation 96:4343–4348

Martin JL, Hickey E, Weber LA, Dillmann WH, Mestril R (1999) Influence of phosphorylation and oligomerization on the protective role of the small heat shock protein 27 in rat adult cardiomyocytes. Gene Expression 7:349–355

Maulik N, Engelman RM, Wei Z, Liu X, Rousou JA, Flack JE, Deaton DW, Das DK (1995) Drug-induced heat-shock preconditioning improves postischemic ventricular recovery after cardiopulmonary bypass. Circulation 92 Suppl II:II-381–II-388

Mehlen P, Kretz-Remy C, Préville X, Arrigo A-P (1996a) Human hsp27, Drosophila hsp27 and human αB-crystallin expression-mediated increase in glutathione is essential for the protective activity of these proteins against TNFα-induced cell death. EMBO J 15:2695–2706

Mehlen P, Schulze-Osthoff K, Arrigo A (1996b) Small stress proteins as novel regulators of apoptosis. J Biol Chem 271:16510–16514

Meldrum DR, Meng X, Shames BD, Pomerantz B, Donnahoo KK, Banerjee A, Harken AH (1999) Liposomal delivery of heat-shock protein 72 into the heart prevents endotoxin-induced myocardial contractile dysfunction. Surgery 126:135–141

Mestril R, Chi SH, Sayen R, O'Reilly K, Dillmann WH (1994) Expression of inducible stress protein 70 in heart myogenic cells confers protection against simulated ischaemia-induced injury. J Clin Invest 93:759–767

Morris SD, Cumming DVE, Latchman DS, Yellon DS (1996) Specific induction of the 70 kDa heat stress proteins by the tyrosine kinase inhibitor herbimycin-A protects rat neonatal cardiomyocytes: a new pharmacological route to stress protein expression. J Clin Invest 97:706–712

Mosser DD, Caron AW, Bourget L, Denis-Larose C, Massie B (1997) Role of the human heat shock protein hsp70 in protection against stress induced apoptosis. Mol Cell Biol 17:5317–5327

Parsell DA, Lindquist S (1993) The functional of heat shock proteins in stress tolerance: degradation and reactivation of damaged proteins. Annu Rev Biochem 27:437–496

Plumier JCL, Ross BM, Currie RW, Angelidis CE, Kazlaris H, Kollias G, Pagoulatos GN (1995) Transgenic mice expressing the human heat shock protein 70 have improved postischemic myocardial recovery. J Clin Invest 95:1854–1860

Plumier JCL, Krueger AM, Currie RW, Kontoyiannis G, Kollias G, Pagoulatos GN (1997) Transgenic mice expressing the human inducible hsp70 have hippocampal neurons resistant to ischemic injury. Cell Stress Chaperones 2:162–167

Préville X, Schultz H, Knauf U, Gaestel M, Arrigo A-P (1998) Analysis of the role of Hsp25 phosphorylation reveals the importance of the oligomerization state of this small heat shock protein in its protective function against TNFα- and hydrogen peroxide-induced cell death. J Cell Biochem 69:436–452

Radford NB, Fina M, Benjamin IJ, Moreadith RW, Graves KH, Zhao P, Gavva S, Wiethoff A, Sherry AD, Malloy CR, Williams RS (1996) Cardioprotective effects of 70 kDa heat shock protein in transgenic mice. Proc Natl Acad Sci USA 93:2339–2342

Rogalla T, Ehrnsperger M, Preville X, Kotlyarov A, Lutsch G, Ducasse C, Paul C, Wieske M, Arrigo A-P, Buchner J, Gaestel M (1999) Regulation of Hsp27 oligomerization, chaperone function, and protective activity against oxidative stress/tumor necrosis factor α by phosphorylation. J Biol Chem 274:18947–18956

Rordorf G, Koroshetz WJ, Bonventre JV (1991) Heat shock protects cultured neurons from glutamate toxicity. Neuron 7:1043–1052

Samali A, Cotter TG (1996) Heat shock proteins increase resistance to apoptosis. Exp Cell Res 223:163–170

Suzuki K, Sawa Y, Kaneda Y, Ichikawa H, Shirakura R, Matsuda H (1997) In vivo gene transfection with heat shock protein 70 enhances myocardial tolerance to ischemia-reperfusion injury in rat. J Clin Invest 99:1645–1650

Uney JB, Kew CNN, Staley K, Tyers P, Sofroniew MV (1993) Transfection mediated expression of human hsp70i protects rat dorsal root ganglion neurones and glia from heat shock. FEBS Lett 334:313–317

Vigh L, Literati PN, Horvath I, Torok Z, Balogh G, Glatz A, Kovacs E, Boros I, Ferdinandy P, Farkas B, Jaszlits L, Jednakovits A, Koranyi L, Maresca B (1997) Bimoclomol: a non-toxic, hydroxylamine derivative with stress protein-inducing activity and cytoprotective effects. Nat Med 3:1150–1154

Wagstaff MJD, Collaco-Moraes Y, Smith J, de Belleroche J, Coffin RS, Latchman DS (1999) Protection of neuronal cells from apoptosis by hsp27 delivered with a herpes simplex virus-based vector. J Biol Chem 274:5061–5069

Wyatt S, Mailhos C, Latchman DS (1996) Trigeminal ganglion neurons are protected by the heat shock proteins hsp70 and hsp90 from thermal stress but not from programmed cell death following NGF withdrawal. Brain Res 39:52–56

Yellon DM, Latchman DS (1992) Stress proteins and myocardial protection. J Mol Cell Cardiol 24:113–124

Yenari MA, Fink SL, Sun GH, Chang LK, Patel MK, Kunis DM, Onley D, Ho DY, Sapolsky RM, Steinberg GK (1998) Gene therapy with HSP72 is neuroprotective in rat models of stroke and epilepsy. Ann Neurol 44:584–591

Subject Index